AF880402

Applied Environmental Science and Engineering for a Sustainable Future

Series Editors

Jega V. Jegatheesan, School of Engineering, RMIT University, Melbourne, VIC, Australia

Li Shu, LJS Environment, Edith Cowan University, Melbourne, Australia

Piet N. L. Lens, UNESCO-IHE Institute for Water Education, Delft, The Netherlands

Chart Chiemchaisri, Kasetsart University, Bangkok, Thailand

Applied Environmental Science and Engineering for a Sustainable Future (AESE) series covers a variety of environmental issues and how they could be solved through innovations in science and engineering. Our societies thrive on the advancements in science and technology which pave the way for better standard of living. The adverse effect of such improvements is the deterioration of the environment. Thus, better catchment management in order to sustainably manage all types of resources (including water, minerals and others) is of paramount importance. Water and wastewater treatment and reuse, solid and hazardous waste management, industrial waste minimisation, soil restoration and agriculture as well as a myriad of other topics need better understanding and application. This book series aims at fulfilling such a task in coming years.

Seyed Hossein Hoseinifar • Hien Van Doan
Editors

Novel Approaches Toward Sustainable Tilapia Aquaculture

Springer

Editors
Seyed Hossein Hoseinifar
Department of Fisheries, Faculty of Fisheries and Environmental Sciences
Gorgan University of Agricultural Sciences and Natural Resources
Gorgan, Iran

Hien Van Doan
Department of Animal and Aquatic Sciences, Faculty of Agriculture
Chiang Mai University
Chian Mai, Thailand

ISSN 2570-2165 ISSN 2570-2173 (electronic)
Applied Environmental Science and Engineering for a Sustainable Future
ISBN 978-3-031-38320-5 ISBN 978-3-031-38321-2 (eBook)
https://doi.org/10.1007/978-3-031-38321-2

© The Editor(s) (if applicable) and The Author(s), under exclusive license to Springer Nature Switzerland AG 2023
This work is subject to copyright. All rights are solely and exclusively licensed by the Publisher, whether the whole or part of the material is concerned, specifically the rights of translation, reprinting, reuse of illustrations, recitation, broadcasting, reproduction on microfilms or in any other physical way, and transmission or information storage and retrieval, electronic adaptation, computer software, or by similar or dissimilar methodology now known or hereafter developed.
The use of general descriptive names, registered names, trademarks, service marks, etc. in this publication does not imply, even in the absence of a specific statement, that such names are exempt from the relevant protective laws and regulations and therefore free for general use.
The publisher, the authors, and the editors are safe to assume that the advice and information in this book are believed to be true and accurate at the date of publication. Neither the publisher nor the authors or the editors give a warranty, expressed or implied, with respect to the material contained herein or for any errors or omissions that may have been made. The publisher remains neutral with regard to jurisdictional claims in published maps and institutional affiliations.

This Springer imprint is published by the registered company Springer Nature Switzerland AG
The registered company address is: Gewerbestrasse 11, 6330 Cham, Switzerland

Dedicated to Zahra, Tik, and our lovely kids
Sina
Nima
Gerda
Sheryn

Preface

First of all, we would like to thank the ten coauthors for donating their time and knowledge to this book. We hope that it will advance sustainable aquaculture, enhance public perception of tilapia culture, and contribute to the preservation of the ecosystem, upon which we all rely.

Sustainable agriculture has been broadly defined in the 1990 US Farm Bill as “a system that meets basic human food and fiber needs, is economically viable, improves environmental quality and the resource base on which agriculture depends over the long term, makes the most efficient use of nonrenewable resources and on-farm resources, and integrates biological cycles and controls where appropriate.” In the EU's current Blue Economy for a Sustainable Future, Green Deal, and F2F policies, aquaculture and sustainable aquaculture have emerged as crucial components. This method of food production has a lot of potential because it offers easily absorbed protein, has a small carbon footprint, and, with the appropriate strategy, favors short supply chains. The conviction behind this book, which was substantiated by study and practice, is that health management, feed additives, gut microbiota, probiotics, prebiotics, synbiotics, medicinal plants, alternative protein sources, and culture systems contribute to sustainable aquaculture. These novel approaches have played an important role in sustainable aquaculture development in the past few decades.

Our book was divided into nine chapters with approaches that contribute to sustainable Nile tilapia production. Chapter 1 covers the current status of world tilapia farming and highlights the advantages and challenges of commercial tilapia production including the economic aspects and the improvement of effluent quality to minimize the impact on the environment, which were covered in a different section. Chapter 2 discusses the health management of tilapia with a particular focus on various diseases and treatments. Chapter 3 emphasizes the effects of different feed additives on tilapia reproduction. A detailed reproductive physiology and breeding behavior of tilapia is discussed in this chapter with a special emphasis on the use of feed additives to control tilapia reproduction, sexual maturation, and sex determination. Chapter 4 covers the recent findings on the gut microbiota of tilapia with a particular focus on structure and modulation using environment-friendly approaches such as pro-, pre-, and synbiotics. Chapter 5 provides updated information regarding definitions, sources of bacterial probiotics, probiotic use in fish diets against pathogenic bacteria, mechanisms of action, beneficial aspects, and potential applications of probiotics in fish. It is anticipated that these will be of significant value for nutritionists, agricultural engineers, researchers, pharmacists, scientists, pharmaceutical industries, and veterinarians. Chapter 6 addresses the use of prebiotics and synbiotics as growth promotors and as alternatives to antibiotics in tilapia aquaculture, along with their possible mechanism of action. Moreover, the positive effects of pre- and synbiotics on tilapia health are also highlighted with emphases on growth, immune modulation, and alteration in gut microflora. Chapter 7 focuses on the use of different medicinal plants in tilapia aquaculture, their routes of administration, and their effects on fish physiology. Moreover, various modes of action of these medicinal plants are also discussed. Chapter 8 reviews the advances, novel ingredients, and new techniques that serve the mission for the future of aquafeed research on alternatives for fishmeal to achieve global aquaculture sustainability, with a focus on partial or full replacement of fishmeal especially in the feed of tilapia. Chapter 9 provides insights into the application of modern technology for sustainable tilapia production, which focuses on their effects on growth performance, immune response, and disease resistance.

Gorgan, Iran — Seyed Hossein Hoseinifar
Chian Mai, Thailand — Hien Van Doan

Contents

List of Figures

List of Tables

Chapter 1
Tilapia Fish for Future Sustainable Aquaculture

Ghasem Ashouri, Seyed Hossein Hoseinifar, Ehab El-Haroun, Roberta Imperatore, and Marina Paolucci

Abstract Lately, aquaculture has been recognized as the fast-growing industry in the food production sector, and it requires maintainable development to cover the world population's demand for aquatic and seafood products. Among the 400 farmed fish species, warm-water fish species such as tilapia need a little quantity of fishmeal in their diets compared to other species. Tilapia is classified as the second most widespread species whose production is increasing every year; Nile tilapia (*Oreochromis niloticus*) is easily adaptable to a large variety of environments, is capable of reproducing in cavities, has an excellent market position in Asia and Africa, is highly resistant to diseases, has good fillet quality, shows moderate feed conversion ratio and excellent growth rate on many natural and artificial diets. Nile tilapia (*O. niloticus*) is known in the market as "aquatic chicken" due to its high tolerance to adverse physical and environmental conditions and overcrowding, its capability to survive at low oxygen levels, and a wide range of salinity concentrations. Tilapia adapts easily to natural and artificial feeds, has good feed conversion value, grows moderately fast, has a final high yield potential, and is accepted by customers worldwide. In addition, tilapia can grow in different aquaculture systems, ranging from extensive, semi-intensive, and intensive; also it can be grown in monoculture or polyculture techniques. Since tilapia grows well in adverse environmental conditions, tolerates stress factors as handling, and is resilient to disease agents of pathogen infections and infectious diseases, it has become the most

Ghasem Ashouri, Seyed Hossein Hoseinifar, Ehab El-Haroun, Roberta Imperatore, and Marina Paolucci contributed equally with all other contributors.

G. Ashouri (✉) · R. Imperatore · M. Paolucci
Department of Sciences and Technologies (DST), University of Sannio, Benevento, Italy

S. H. Hoseinifar
Department of Fisheries, Faculty of Fisheries and Environmental Sciences, Gorgan University of Agricultural Sciences and Natural Resources, Gorgan, Iran

E. El-Haroun
Fish Nutrition Research Laboratory, Department of Animal Production, Faculty of Agriculture, Cairo University, Cairo, Egypt

© The Author(s), under exclusive license to Springer Nature Switzerland AG 2023

S. H. Hoseinifar, H. Van Doan (eds.), *Novel Approaches Toward Sustainable Tilapia Aquaculture*, Applied Environmental Science and Engineering for a Sustainable Future, https://doi.org/10.1007/978-3-031-38321-2_1

common species among farmers. Nile tilapia (*O. niloticus*) could be cultured in both fresh and saltwater; and in tropical, subtropical, and temperate climates. The authors in this chapter will cover the environmental and nutritional requirements of tilapia, defined as key factors in profit production. As mentioned above, tilapia has a wide tolerance for high stocking densities and environmental conditions. In addition to these advantages, tilapia grows very well in integrated culture systems with aquatic species such as carp and shrimps, as well as with crops like tomato and lettuce. However, the global extension of tilapia farming growing at a remarkably high rate is likely to pose environmental and socioeconomic threats. This chapter highlights the advantages and challenges of commercial tilapia production including the economic aspects, living behind the improvement of effluent quality to minimize the impact on the environment, which will be covered in a different section.

Keywords Tilapia · Farming · Sustainable · Production

1.1 Introduction

In the last few decades, aquaculture has become the fastest-growing sector in animal livestock production, securing a global food supply that reached 2018 ~115 million tons representing 263,400 million dollars (USD$) (FAO 2020). Feed formulation of warm-water fish species requires less fishmeal compared to other species. Tilapia production represents 40% of cultured fish (Prabu et al. 2019). Annual worldwide production of cultured tilapia was 3.4 million tons in 2011 and reached 6.2 million tons in 2019 (FAO 2020) (Fig. 1.1). Tilapia and catfish are considered to be native to the Middle East and Africa. Tilapia culture, though ancient (probably firstborn in Egypt simultaneously with Chinese polyculture), has experienced a recent commercial development. Today, tilapia has become one of the most attractive fish species in aquaculture due to several advantages such as i) massive adaptability to numerous environmental conditions, ii) easy reproduction in captivity, iii) resistance to environmental stress, diseases, and microbe pathogen infections, iv) good quality of flesh, v) feed on a low trophic level and excellent growth rate on a variety of diets (Welker and Lim 2011; Prabu et al. 2019). In 1980 tilapia was considered an ideal candidate for aquaculture in different regions of the world. Consequently, tilapia culture is currently growing commercially in at least 120 countries (Yue et al. 2016) all around the globe. Asian countries (e.g. China, Egypt, Indonesia, Philippines, and Thailand) are the major producers as well as consumers of tilapia (Chen et al. 2018). The most common cultured genus of tilapia is *Oreochromis*, and around 89% of these farmed fish are Nile tilapia (*Oreochromis niloticus*), due to their good growth performance in ponds (Ng and Hanim 2007).

Egypt and China are considered the main producers of Nile tilapia (*O. niloticus*) and represent one-third of the total global production (FAO 2020). The reasons for such a rapid expansion of the Nile tilapia (*O. niloticus*) culture could be attributed to

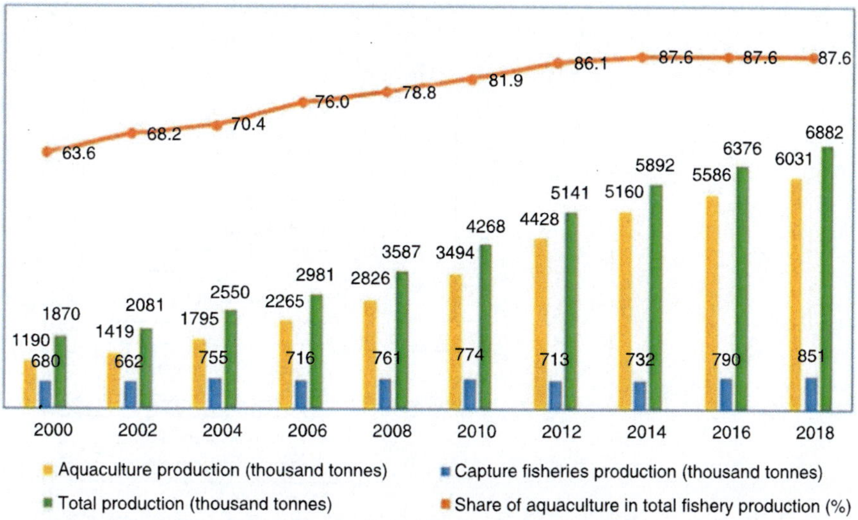

Fig. 1.1 Contribution of aquaculture to world tilapia production, 2000–2018. Data source: FAO (2020). FAO Global Fishery and Aquaculture Production Statistics (FishStat J; March 2020; www.fao.org/fishery/statistics/software/fishstatj/en)

technological advances associated with the intensification of cultural practices (Bhujel 2014b; Watanabe et al. 2002) including, i) development of novel strains and hybrids, ii) possibility to set up monosex male culture, iii) formulated diets, iv) use of a variety of production techniques such as semi-intensive and intensive systems (Ng and Romano 2013), and vi) marketing programmes aiming at enhancing the demand for tilapia on national and international markets (Wang and Lu 2016). Based on the increasing commercialization and continuing growth of the tilapia industry, the product is not only the second most important farmed fish globally (Fitzsimmons 2000), next to carp, but it is also described as the most important of all cultured fishes in the twenty-first century (Celik 2012).

The importance of tilapia among aqua farmers can be summarized as follows: (i) tilapia have fast growth and survival rate, (ii) reproduce easily in captivity, (iii) tolerate low water quality conditions and environmental variables such as temperature, salinity, low dissolved oxygen, etc., (iv) is easily adaptable to mono and polyculture techniques in intensive fish farming, (v) feed on low-quality diets and easily adapt to artificial diets, vi) show high profitability and low production costs, (vii) is highly resistant to stress and disease (lower risks for aqua farmers), and (viii) is highly accepted by consumers, with a good market request (Prabu et al. 2019; El-Sayed 2006b).

Although tilapia's culture is promising for aquaculture, in light of the many advantages above-mentioned, enhancing the production efficiency of tilapia has some challenges and research issues that are of the biggest concern to tilapia culturists (Yuan et al. 2017). For instance, these involve growth performance, unwanted reproduction (Gupta and Acosta 2004; Ng and Romano 2013; Chen

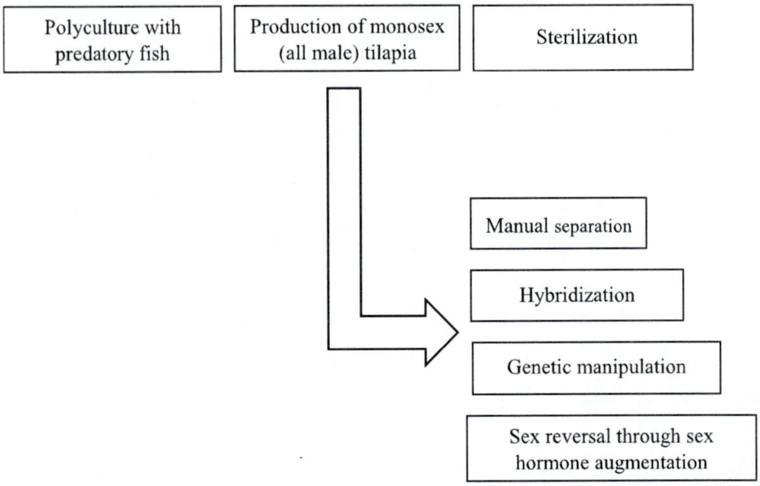

Fig. 1.2 Methods used for inhibiting reproduction or controlling the overpopulation of tilapias

et al. 2018), environmental tolerance (e.g. low temperatures and high salinity), disease resistance (Wang and Lu 2016; Li et al. 2016), quality of fillet yield (Yue et al. 2016), and increased production costs. Other issues of tilapia aquaculture are related to its negative effects on the environment and global biodiversity. Different protocols have been adopted to control and limit unwanted reproduction (Fig. 1.2). A brief overview of these methods is provided in one subsection of this chapter. However, none of these methods is considered 100% effective, and thus a combination of methods is suggested (Fuentes-Silva et al. 2013).

The poor aptitude of tilapia to tolerate low temperatures (<15°C) affects the geographical expansion of tilapia culture (El-Sayed 2006b; Lim and Webster 2006). The most cold-resistant species is blue tilapia, *Oreochromis aureus*, which is suitable for culture in regions with seasonal temperature changes and is usually used in the hybridization for the production of monosex. Moreover, most tilapia species are not tolerant to high salinity, although some (e.g., Mozambique tilapia, *Oreochromis mossambicus*; *Oreochromis spilurus*, redbelly tilapia, *Tilapia zillii*, and red tilapia hybrids) can grow in seawater. In comparison with other cultured fish (e.g. salmon), tilapia shows off-flavours and also minor levels of HUFAS, especially beneficial omega-3 fatty acids such as 20:5 n3 (EPA) and 22:6 n3 (DHA) which cause low market acceptance (Weaver et al. 2008). Finally, today, the disease resistance of tilapia has not received enough attention because rearing programmes have been focused only on growth efficiency and skin colour selection.

In brief, tilapia yield can be affected by several causes, so this chapter provides an overview of some aspects of tilapia culture. In particular, it focuses on some crucial factors for a successful production, such as management and nutritional requirements, evaluation of technological advances and different tilapia farming practices, environmental effects, and some constraints resulting from intensification practices.

1.2 Nutritional and Environmental Requirements

Tilapias possess highly desirable characteristics that make them good candidates for fish production under different production approaches as extensive, semi-intensive, and intensive, such as their ability to tolerate a wide range of environmental conditions (Chervinski 1982), high survival rate, and feed on a low trophic level which makes them attractive species to aquaculture investors. However, dietary requirements under different production techniques and the association with culture conditions are still not yet clear.

1.2.1 Nutritional Requirements

Fish feed generally accounts for as much as 60% or more of production costs in both semi-intensive and intensive aquaculture production systems (Montoya-Camacho et al. 2019). The nutritional requirements of tilapia have been comprehensively reviewed by Ng and Romano (2013), and Chavan et al. (2015). An obstacle to tilapia intensive culture is the rising cost and unpredictable reliability of fishmeal and fish oil global supplies (Ng and Romano 2013). Consequently, several alternative ingredients, in particular of plant origin, have been investigated, and some of them are currently used in tilapia feeds to reduce the fishmeal amount (Montoya-Camacho et al. 2019). Fish meal (FM) and fish oil (FO) are the main sources of, respectively, amino acids and fatty acids for many different species. The investigation of alternative ingredients of FM and FO requires a thorough understanding of the quantity and quality requirements of different nutrients of tilapia, besides the comprehension of the factors that may influence the nutrient utilization efficiency and consequently affect the production (Ng and Romano 2013).

1.2.1.1 Protein and Amino Acid Requirements of Tilapia

Proteins represent the most expensive components in aquafeeds (El-Sayed 1999). Several factors affect protein requirements such as protein source and amino acid (AA) profile, fish species, age, size, and life stage. Tilapia larvae, fry and fingerling require a high level of protein (30–40%) compared to tilapia broodstock (20–25%). Male tilapia requires a higher level of protein than females to reach optimal growth performance (Ng and Romano 2013; Abdel-Tawwab et al. 2010; Sweilum et al. 2005). In terms of protein, it is not just the quantity, but the quality and the essential amino acid (EAA) profile that will tremendously impact the total protein requirement. The ideal proteins to be introduced to the diet are represented by those whose amino acid composition is similar to the ratio required by the animal (Nguyen and Davis 2009). Furthermore, this will also decrease nitrogenous waste related to amino acids being underutilized or deaminated as an energy source (Abdel-Tawwab et al.

Table 1.1 The quantitative essential amino acid requirements of warm-water fish.[a] Table modified from Jauncey (2000)

	Common carp[b]	Channel cattfish[c]	Nile tilapia[d]	Mozambique tilapia[e]
Arginine	3.3	4.3	4.2	2.8
Histidine	2.1	1.5	1.7	1.1
Isoleucine	2.5	2.6	3.1	2.0
Leucine	3.3	3.5	3.4	3.4
Lysine	5.7	5.1	5.1	3.8
Methionine	2.1	2.3	2.7	1.0
Phenylalanine	3.4	5.0	3.8	2.5
Threonine	3.9	2.0	3.8	2.9
Tryptophan	0.8	0.5	1.0	0.4
Valine	3.6	3.0	2.5	2.2

[a]All values as % of dietary protein
[b]Experimentally determined data for common carp (*Cyprinus carpio*) from the review of Tacon (1987). Requirement estimated by the dose-response method
[c]Experimentally determined data for channel catfish (*Ictalurus punctatus*) from the review of Wilson (1991). Requirement estimated by the dose-response method
[d]Experimentally determined data for Nile tilapia (*Oreochronis niloticus*) from Santiago and Lovell (1988). Requirement estimated by the dose-response method
[e]Experimentally determined data for Mozambique tilapia (*O. Mossambicus*) from Jauncey (1983). Requirement estimated by the whole body and muscle amino acid composition method

2010). Prediction of EAA requirement of different fish species could be measured by either dose-response protocol or by measuring the amino acid profile of the whole body of fish. As known, warm-water species including the Nile tilapia (*O. niloticus*) require 10 essential amino acids (Table 1.1). EAA requirements could be covered using a mixture of plant and animal proteins (Montoya-Camacho et al. 2019), or by the inclusion of free amino acids (Nguyen and Davis 2009).

1.2.1.2 Lipid and Fatty Acid Requirements of Tilapia

Lipids and oils are considered to be the main source of digestible energy and essential fatty acids (EFAs) for the normal growth and development of fish. In addition, phospholipids play main functions in cell membrane structure and integrity, facilitate and control the absorption of fat-soluble vitamins, act as forerunners for sex hormones, and improve the texture and flavour of the diet.

Tilapia requirements of lipids rely on several factors including fish species, age, size, source of lipids, protein, and energy content (El-Sayed 2006b). For example, it was noticed that the level of protein decreased in the Nile tilapia (*O. niloticus*) diets from 33.1% to 25.6% by elevating lipid content from 5.2% to 9.1% and carbohydrates (CHO) from 31.7% to 36.7% (Li et al. 1991). The role of increasing lipid content could be described as a sparing protein effect, confirmed by Jauncey (2000) in hybrid tilapia (*O. niloticus* × *O. aureus*). Though increasing lipid levels up to 12% harm the growth of juvenile *O. aureus* × *O. niloticus* hybrids and augments the

Table 1.2 The essential fatty acid requirements of tilapia[a]

Species	Requirement	Reference
Tilapia zillii	1% 18:2n-6 or 1% 20:4n-6	Kanazawa et al. (1980)
Oreochromis niloticus	0.5–1% 18:2n-6 or 1% 20:4n-6	Teshima et al. (1982)
O. niloticus	0.5% 18:2n-6	Takeuchi et al. (1983)
O. aureus	18:2n-6 or 18:3n-3 ≤1%	Stickney and McGeachin (1983)

[a]Table adapted from Jauncey (2000)

accumulation of lipid in the carcass of the fish (Jauncey 2000), also it has a negative impact on the pelleting processing of the diets. However, an extruded feed where fat is added after the pelleting process solved the problem. In general, tilapia require about 10–15% dietary lipids (El-Sayed 2006b); however, the oil inclusion of commercial tilapia feed is typically about 4–5% (Orachunwong et al. 2001). The required EFAs cannot be synthesized by fish and must be provided by the diet (Jauncey 2000). Research on fatty acid requirements revealed that cold-water fish and marine fish require *w*-3 polyunsaturated fatty acids (n-3 PUFA), while freshwater and warm warm-water species require n-6 PUFA. Thus, warm-water species, including tilapia, utilize plant oil as a source of n-6 fatty acids more efficiently than FO and lipids as a source of n-3 fatty acids (El-Sayed 2006b). Several studies have indicated that tilapia requires n-6 EFA rather than n-3 EFA (Table 1.2).

The findings of previous research summarized that EFAs are considered a source of the fatty acid content in tilapia fillets and support the growth of fish. It has been suggested that diets for farmed tilapia should contain 0.5–1.0% of both n-3 and n-6 PUFA (Lim et al. 2011a; Ng 2005). Tilapia-fed diets containing high levels of n-3 PUFA have positive effects on the health of consumers, such as its positive impacts on the cardiovascular system (Lecerf 2009; Russo 2009), immune system (Ruxton et al. 2004) and inflammatory disorders (Calder 2006). In the last few decades, research conducted to find novel ingredients to substitute FO with vegetable oil in tilapia feed has been successfully carried out; however, high interest remains in using palm oil because of the low price compared to other vegetable oils and its easy availability on the market (Ng et al. 2001; Ng et al. 2006; Ng and Gibon 2010; Bahurmiz and Ng 2007). Fortunately, the use of palm oil in the diet does not reduce ($P \leq 0.05$) the performance of the Nile tilapia (Ng et al. 2001). However, the high inclusion of plant oil raises an important question to be addressed by scientists, that is, the plant oil's role in the fish diets and the impact on the fatty acid composition of the final product and, thus, the impact on human health (Huang et al. 1998; Young 2009; Bahurmiz and Ng 2007).

Recently, there has been an interest in conducting research using finishing diets rich in n-3 PUFA to investigate their effect on adjusting the final fatty acid profile of tilapia fillets to enhance their nutritional value before harvest (Ng and Chong 2004; Visentainer et al. 2005; Tonial et al. 2009; Trushenski et al. 2009; Dos Santos et al. 2011; Luo et al. 2012). In this context, Tonial et al. (2009) found that Nile tilapia (*O. niloticus*) fed diets containing soybean oil showed a decrease in the n-6/n-3 ratio in the fillet from 7.4 to 1.0 when reverted to a flaxseed-based diet which is rich in n-3

PUFA. In addition, Teoh et al. (2011) examined the FAs metabolism of both Genetically Improved Farmed Tilapia (GIFT) strain and red hybrid tilapia-fed purified diets with vegetable oil blends, and they found that FAs digestibility was not different among the tilapia strains.

1.2.1.3 Carbohydrates

Previous research carried out on tilapia requirement of carbohydrates (CHO) declared that tilapia does not have CHO specific requirements. Though, Wilson (1994) reported that warm-water fish species such as tilapia utilize CHO more efficiently than cold-water species. The main purpose of CHO inclusion in the diet is to act as an effective source of energy that can spare the protein as a source of energy to support fish growth. In addition, CHO act as a binder, facilitating the pelleting process, also CHO acts as a precursor of different metabolic components (NRC 1993). Fish species can utilize up to 35 to 40% digestible CHO (Anderson et al. 1984; El-Sayed and Garling Jr 1988). Several factors affect CHO digestion and assimilation, including the source of CHO since fish utilize complex carbohydrates (polysaccharides) more efficiently than mono and disaccharides (Shiau and Chuang 1995). It has been shown that increasing the dietary CHO/lipid ratio leads to increased glycolysis and lipogenesis but reduces gluconeogenesis and amino acid degradation in the liver of the Nile tilapia, *O. niloticus* (Shimeno et al. 1993). Moreover, it has been reported that CHO metabolism is influenced by their fibre content (Shiau and Yu 1999) and is affected by the dietary protein source (Shiau and Suen (1992). In this context, El-Sayed (1991) found that sugarcane bagasse could be included in *T. zillii* diets without a negative impact on both growth and feed digestibility, while the inclusion of sugarcane in Nile tilapia (*O. niloticus*) diets resulted in poor performance. Also, larger fish of the hybrid of *O. niloticus* × *O. aureus* utilized CHO better than smaller ones (Tung and Shiau 1993). Finally, previous research concluded that increasing feed frequency from 2 to 6 times/day enhanced growth, CHO utilization, and protein sparing effect (Tung and Shiau (1991); Shiau and Lei (1999); (Jauncey 2000).

1.2.1.4 Vitamin and Mineral Requirements of Tilapia

Micronutrients such as vitamins and minerals are essential cofactors in several metabolic mechanisms involved in different physiological processes in fish health and welfare. Different factors affect vitamin and mineral requirements, such as culture conditions and chemical dietary composition (Celik 2012). For example, in both extensive and semi-intensive fish production system, the inclusion of vitamins and minerals are not necessary since the fish consume natural food such as phytoplankton and zooplankton that contain enough amounts of vitamins and minerals that fulfil the fish requirements (El-Sayed 2006b) On the contrary, in intensive systems the presence and availability of natural food are limited or absent. Thus vitamins and minerals must be incorporated into the diets to support growth, health, and survival rate (Ng and Romano 2013). Fish feed manufacturers usually over-supplement feed with vitamins and minerals to counteract losses due to processing, storage, and leaching. It is well known that the nutritional requirements of vitamins and minerals depend on the life stage of tilapia. Tables 1.3 and 1.4 refer to the water-soluble and lipid-soluble vitamin requirements of tilapia fry and fingerlings stages, respectively. Shiau and Lung (1993) indicated that vitamin B12 is not required for tilapia hybrid (*O. niloticus* × *O. aureus*), likely due to the ability of gut bacteria to synthesize it (Shiau and Lung 1993; Shiau and Huang 2001; Barros et al. 2009). In terms of vitamin A requirement, Guo et al. (2010) reported that Nile tilapia (*O. niloticus*) does not require supplementation of vitamin A since cod liver oil, with its high content of vitamin A, is used as a source of lipids in the diet. In addition, Hu et al. (2006) reported that tilapia hybrids (*O. niloticus* × *O. aureus*) can synthesize vitamin A from β-carotene.

Mineral requirements of tilapia have been comprehensively reviewed by Ng and Romano (2013), and also Makwinja and Geremew (2020). Vital minerals are involved in many physiological processes, such as

- build skeletal structures
- osmoregulation (e.g., Na^+/K^+-ATPase)
- nerve and muscle contraction
- regulation of the pH of the blood and other body liquids
- metabolism-related enzyme activity (lipase, alkaline phosphatase) as cofactors
- key components of many enzymes, vitamins, hormones, and respiratory pigments

Nile tilapia (*O. niloticus*) usually require minerals from two major sources: water and feed. Brackish/marine environments are considered as main sources of minerals. However, since tilapias are mostly farmed in freshwater/low salinity waters, supplementing the diets with minerals is important to fulfil their needs to achieve optimal health and productivity. Previous research conducted to measure the mineral requirements of hybrid tilapia stated that the requirement of dietary NaCl or KCl ranged between 1.5 and 2–3 g kg^{-1} of diet, respectively (Shiau and Lu 2004); (Shiau and Hsieh 2001). Correct mineral requirement fulfilment is considered essential since the deficiency or excess of minerals leads to depressed growth performance.

Previous research confirmed the function of minerals as essential to support growth performance and the health status of different fish species. Robinson et al. (1987) reported that the inclusion of a 7 g Ca kg^{-1} purified diet was important to maintain the optimum growth of blue tilapia. Also, Shiau and Tseng (2007) found that hybrid tilapia-formulated diets should be supplemented with 2.7–3.3 g L^{-1} Ca and purified diets supplemented with 3.5–4.3 g Ca kg^{-1} to achieve optimum growth and feed efficiency. Research carried out on magnesium showed that 0.59 to 0.77 g and 0.50 to 0.65 g kg^{-1} diet were required for optimum performance of the Nile tilapia, *O. niloticus,* and blue tilapia, *O. aureus* (Dabrowska et al. 1989: Reigh et al. 1991), respectively. Moreover, trials conducted on phosphorus showed that a 5 g P kg^{-1} diet of phosphorus is required for the best growth and bone mineralization of *O. aureus* (Robinson et al. 1987).

Table 1.3 Dietary water-soluble vitamin requirements of tilapia species. Units are mg kg^{-1} of diet unless otherwise stated

Vitamin	Species and size (g)	Recommended level	Deficiency signs	Reference
Choline	*O. niloticus* (3.0)	3000		Kasper et al. (2000)
	O. niloticus × *O. aureus* (0.6)	1000	poor growth, reduced survival, reduced blood triglyceride and phospholipids concentrations	Shiau and Lo (2000)
Thiamin (B1)	*O. mossambicus*	2.5	anorexia, light coloration, nervous disorder, poor growth and poor feed efficiency, low hematicrit and red blood cell count, and increased serum pyruvate	Lim et al. (1991)
	O. niloticus	4.0		Lim et al. (2000)
	O. niloticus (4–5)	3.5		Lim et al. (2011b)
Riboflavin (B2)	*O. aureus* (0.7)	6.0	lethargy, anorexia, poor growth, high mortality, loss of colour, fin erosion, short body dwarfism, and cataracts	Soliman and Wilson (1992b)
	O. mossambicus × *O. niloticus*	5.0		Lim et al. (1993)
Niacin (B3)	*O. niloticus* × *O. aureus* (2.2)	26 (fed glucose) 121 (fed dextrin)	haemorrhage, deformed snout, gill and skin oedema, fin and mouth lesions	Shiau and Suen (1992)
Pantothenic acid (B5)	*O. aureus* (0.7)	10	poor growth, haemorrhage, sluggishness, high mortality, anaemia, hyperplasia of gill lamellae, and increased mortality	Soliman and Wilson (1992a)
	O. niloticus (13.7)	10		Ayyat et al. (2011)
Pyridoxine (B6)	*O. mossambicus* × *O. niloticus*	3		Lim et al. (1995)
	O. niloticus × *O. aureus* (0.7)	1.7–9.5 (28% CP) 15.0–16.5 (36% CP)	poor growth, high mortality, abnormal neurological signs, caudal fin erosion, mouth lesion, and convulsions	Shiau and Hsieh (1997)
Biotin (B7)	*O. niloticus* × *O. aureus* (1)	0.06	poor growth reduced hepatic pyruvate carboxylase and acetyl CoA carboxylase activities	Shiau and Chin (1999)
Folic acid (B9)	*O. niloticus* ()	0.5	anorexia, reduced growth, lethargy, increased mortality, low haematocrit	Lim and Klesius (2001)
	O. niloticus × *O. aureus* (0.4)	0.8		Shiau and Huang (2001)

	O. niloticus (14.5)	0.5–1.0		Barros et al. (2009)
Vitamin B12	*O. niloticus* × *O. aureus* (1.0)	Not required		Shiau and Lung (1993)
Vitamin C[a]	*O. niloticus* × *O. aureus* (1.5)	80 (Ascorbic acid) (41–48) C2S (37–42) C2MP-Mg	lordosis, scoliosis, poor growth, feed efficiency, poor wound healing, haemorrhage, anaemia, exophthalmia and gill, and operculum deformity	Shiau and Hsu (1999)
Inositol	*O. niloticus* (5.8)	Not required		Peres et al. (2004)
	O. niloticus × *O. aureus* (0.5)	400	poor growth, higher muscle/liver lipid	Shiau and Su (2005)

O. Oreochromis sp., *CP* crude protein
[a]Since ascorbic acid is unstable, storage, more stable forms of ascorbic acid such as L-Ascorbyl-2-sulphate (C2S) and L-ascorbyl-2-monophosphate-magnesium (C2MP-Mg) should be used

Table 1.4 Dietary lipid-soluble vitamin requirements of tilapia species. Units are mg kg^{-1} of diet unless otherwise stated

Vitamin	Species and size (g)	Recommended level	Deficiency signs	Reference
Vitamin A	*O. niloticus* (11.4)	5000 IU kg^{-1}	poor growth, abnormal swimming, restlessness, exophthalmia, high mortality, haemorrhage, and reduce the mucus secretion	Saleh et al. (1995)
	O. niloticus × *O. aureus* (1.6)	6000–7000 IU kg^{-1}		Hu et al. (2006)
	O. niloticus (7.7)	Not required		Guo et al. (2010)
β-carotene	*O. niloticus* × *O. aureus* (0.5)	29.0–44.0		Hu et al. (2006)
Vitamin D	*O. niloticus* × *O. aureus* (0.8)	375 IU kg^{-1}	poor growth, low haemoglobin, reduced hepatosomatic index, and alkaline phosphatase	Shiau and Hwang (1993)
	O. aureus (2.3)	Not required		O'Connell and Gatlin (1994)
Vitamin E	*O. niloticus* (7.5)	20–100		Eleraky et al. (1995)
	O. niloticus × *O. aureus*	63–206 (12% CL)	poor growth, anorexia, haemorrlage, impaired erythropoiesis, muscle degeneration, ceroid deposition in liver and spleen and lack of skin colour	Shiau and Shiau (2001)
	O. niloticus × *O. aureus* (0.7)	42–44 (5% CL)		Huang et al. (2003)
Vitamin K	*O. niloticus* × *O. aureus*	5.2	poor growth and low plasma prothrombin	Lee (2003)

O. Oreochromis sp., *CL* crud lipid

Chromium (Cr), zinc (Zn), copper (Cu), selenium (Se), and Iron (Fe) are considered trace elements, and research on these elements showed their importance in improving the growth and health status of tilapia. Shiau and Shy (1998) found that the inclusion of 140 mg Cr kg^{-1} improved the growth of hybrid tilapia due to its role as a cofactor in the activity of insulin and enhancing carbohydrate utilization. Furthermore, (Eid and Ghonim 1994) and do Carmo Sá et al. (2004) stated that the inclusion of 0.03 g Zn kg^{-1} for fry and 0.08 Zn kg^{-1} for fingerling diet is required for optimum growth. More research conducted by Watanabe et al. (1988) found that Nile tilapia (*O. niloticus*) required 2 to 3 mg Cu kg^{-1}. Shiau and Su (2003) found that hybrid tilapia (*O. niloticus* × *O. aureus*) required 150 to 160 mg kg^{-1} Fe.

1.2.1.5 Nutrition and Immunity

The production of the Nile tilapia (*O. niloticus*) under a semi-intensive and intensive production system exposes the fish to pathogen infections and disease outbreaks. The main strategy to maintain fish health in aquaculture is the provision of a balanced diet supplemented with immune stimulants that help boost the immune system and keep it under control of disease outbreaks. Functional feed additives such as prebiotics, bioactive compounds derived from medicinal plants, and probiotics could avoid the use of antibiotics and chemotherapy and aid in limiting disease outbreaks by controlling fish mortality in intensive aquatic farms (Merrifield et al. 2010; Hoseinifar et al. 2016: Dawood and Koshio 2016). Functional feed additives modify the gut microbiome, increase the activity of beneficial bacteria, increase the secretion of digestive enzymes, and decrease harmful bacteria. In addition, these feed additives upregulate gene expression related to immunity and inflammatory cytokines such as *IL-1*, *IL-8*, and *Lyz*, oxidative enzymes such as catalyse, superoxide dismutase, and glutathione, and growth genes such as *GH* and *IGF-1*. Such compounds act as health factors capable of modulating the immune responses in tilapia (Table 1.5 included as supplementary data), and other cultivated fish species.

Furthermore, functional feed additives have a positive impact on the immune system defence system by (i) stimulating the production of plasma proteins (globulin and albumin), which play a vital role in the synthesis of antibodies (immunoglobulins), (ii) enhancing the activity of lysozyme, (iii) increasing the production of defence cells such as leukocytes and lymphocytes that produce antibodies), (iv) stimulating the production of macrophages, which are responsible for the phagocytosis, and (v) modulating the composition of the gut flora and improving gut health via the increase in villi length, width goblet cells, improving mucus secretion, and reducing gut inflammation.

Probiotics, generally defined as live microorganisms, are provided via the diet or rearing water (when supplied in an adequate amount). They possess different beneficial characteristics leading to the exclusion of pathogenic bacteria and the modulation of the immune system of the host, by improving the microbial balance of the host (Merrifield et al. 2010). The most common probiotics used in aqua feeds are

Table 1.5 Tilapia studies to evaluate the effects of functional feed additives on immunity*

Feed additives	Tilapia species	Immune responses	Immune-related genes	Stress-regulating genes	Anti-oxidative capacity	Haematological and biochemical parameters	Disease challenge	Disease resistance and survival rate	Reference
Probiotics									
L. rhamnosus	Nile tilapia (*O. niloticus*)	LZM↑, PA↑, ACH50↑	–	–	–	–	*E. tarda*	↑	Pirarat et al. (2006)
Commercial probiotics[a]		LZM↑, BA↑, NBT→	–	–	–	Neutrophil↑, Hb↑, Plasma protein↑	*E. tarda*	↑	Taoka et al. (2006)
S. cerevisiae		BA↑	–	–	–	Neutrophil↑, RBCs↑, Hb↑, Ht↑, Albumin↑, Globulin↑	*A. hydrophila*	↑	Abdel-Tawwab et al. (2008)
B. subtilis + *L. acidophilus*		NBT↑, LZM↑, BA↑	–	–	–	Ht↑	Mixed species pathogen[b]	↑	Aly et al. (2008)
E. faecium		LZM↑, C3↑, RBA↑	–	–	MPO↑	Serum protein→, Albumin→, Globulin→	–	–	Wang et al. (2008)
P. acidilactici		LZM↑, RBA→	–	–	–	WBCs↑, RBCs→, Serum protein→, Albumin→, Globulin→	–	–	Ferguson et al. (2010)
B. coagulans		LZM→, RBA↑	–	–	SOD↑, CAT↑, MPO↑	Serum protein↑, Albumin→, Globulin↑	–	–	Zhou et al. (2010a)
Lactococcus lactis		LZM↑, RBA↑	–	–	SOD↑, MPO↑	–	*A. hydrophila*	↑	Zhou et al. (2010b)

L. rhamnosus		LZM→, PA↑, ACH50↑, RBA→, BA↓	*TNF-α*↑, *IL-1*↑	–	–	–	–	–	Pirarat et al. (2011)
P. acidilactici		LZM→, IEL↑, GC↑	*TNFα*↑	–	–	WBCs↑, Neutrophils↑, Monocytes↑,	–	–	Standen et al. (2013)
L. acidophilus		BA↑	*IL-1β*	–	–	–	*A. hydrophila*	↑	Villamil et al. (2014)
B. licheniformis		LZM↑, C3↑	–	–	SOD↑	–	*S. iniae*	↑	Han et al. (2015)
B. subtilis + *S. cerevisiae*		RBA↑	–	–	–	WBCs↑, Lymphocyte↑, Monocyte↑, RBCs↑, Ht↑, Hb→,	*A. hydrophila* *S. iniae*	↑	Iwashita et al. (2015)
L. plantarum		Total Ig↑, PA↑, LZM↑	*IL-4*↑, *IL-12*↑, *IFN-γ*↑,	–	–	WBCs↑, RBCs↑, Hb↑, Serum protein↑, Albumin↑, Globulin↑	*A. hydrophila*	↑	Hamdan et al. (2016)
AquaStar® Growout[c]		IEL↑, GC↑	*TLR2*↑, *TNF-α*↑, *IL-1β*↑, *IL-10*↑, *TGF-β*↑	*Caspase-3*↑, *PCNA*↑, *HSP70*↑	–	–	–	–	Standen et al. (2016)
AquaStar® Growout[c]	Nile tilapia (*O. niloticus*)	ACH50↑, RBA→, LZM→	–	–	CAT↑, GPx↑, GR↑, GSH↑	WBCs→, RBCs→, HT→	–	–	Ramos et al. (2017)
L. plantarum/ *B. velezensis*		Serum ACH50↑, RBA↑, PA↑; Serum and mucus LZM↑,	–	–	Serum and mucus Perox-idase activity↑	–	*S. agalactiae*	↑	Van Doan et al. (2018c)

(continued)

Table 1.5 (continued)

Feed additives	Tilapia species	Immune responses	Immune-related genes	Stress-regulating genes	Anti-oxidative capacity	Haematological and biochemical parameters	Disease challenge	Disease resistance and survival rate	Reference
		Peroxidase activity↑,							
L. rhamnosus / *Lactococcus lactis*		–	*TNF-α*↑, *IFN-γ*↑, *Lyzc*↑, *IL-1β*↑	*HSP70*↑	–	–	*S. agalactiae*	↑	Xia et al. (2018)
B. subtilis/ *S. cerevisiae*		LZM↑	–	–	–	WBCs↑, RBCs↑, Hb↑, Serum protein↑, Albumin↑, Globulin↑	–	–	Opiyo et al. (2019)
R. stabekisii		RBA↑, PA↑; serum LZM↑,	*TNF-α*↑, *IL-1β*↑, *TGF-β*↑	*HSP70*↑	SOD↑	–	*A. hydrophila*	↑	Tan et al. (2019)
Bacillus sp.		LZM↑, IgM↑, GC↑	–	–	SOD↑, CAT↑	ALP↑	*A. hydrophila*	↑	Kuebutornye et al. (2020)
B. subtilis / *B. cereus*		–	*Lyzc*↑	–	–	–	*S. agalactiae*	↑	Xia et al. (2020)
S. cerevisiae	Hybrid tilapia (*O. niloticus* ♀ × *O. aureus* ♂)	LZM↑, PA↑, RBA↑, C3↑, C4↑	–	–	–	–	–	–	He et al. (2009)
B. subtilis		–	*TNF-α*↑, *IL-1β*↑, *TGF-β*↑	*HSP70*↓	–	–	–	–	He et al. (2013)
L. brevis/ *L. acidophilus*		–		*HSP70*↑	–	–	*A. hydrophila*	↑	Liu et al. (2013)

			TNF-α↑, *IL-1β*↑, *TGF-β*↑						
L. plantarum			*TNF-α*↑, *IL-1β*↑	*HSP70*↑			*A. hydrophila*	↑	Ren et al. (2013)
Lycogen™[d]	Red tilapia (*O. mossambicus* × *O. niloticus*)	LZM↑, ACH50↑	–	–	–	–	–	–	Chiu and Liu (2014)
Bacillus spp.		–	*IFN-γ*↑, *IL-8*↑, *IRF-3*↑, *MX*↑, *RSAD-2*↑	–	–	–	TiLV[e]	↑	Waiyamitra et al. (2020)
Prebiotic									
β-glucan	Nile tilapia (*O. niloticus*)	LZM↑	–	–	–	–	*S. iniae*	↑	Whittington et al. (2005)
Sangrovit®[f]		LZM→	–	–	–	WBCs↑, RBCs→, Hb→, ALT→	–	–	Rawling et al. (2009)
Inulin		LZM↑, NBT↑	–	–	–	Ht↑	*A. hydrophila*	↑	Ibrahem et al. (2010)
GroBiotic®-A[g]	Nile tilapia (*O. niloticus*)	LZM↑, NBT↑	–	–	SOD↓, CAT↓	–	*A. hydrophila*	↑	Zheng et al. (2011)
GroBiotic®-A[g]		LZM→, ACH50→, total Ig→	–	–	–	Serum total Protein→	*S. iniae*	→	Vechklang et al. (2012)
β-glucan		Serum LZM→, ACH50→, RBA↑, BA→	–	–	–	WBCs→, RBCs→, Hb→, Ht→	*S. iniae*	→	Welker et al. (2012)
Inulin or JA[h]		LZM↑, ACH50↑, total Ig↑, GC↑	–	–	–	Albumin↑, Total Protein↑	–	–	Tiengtam et al. (2015)

(continued)

Table 1.5 (continued)

Feed additives	Tilapia species	Immune responses	Immune-related genes	Stress-regulating genes	Anti-oxidative capacity	Haematological and biochemical parameters	Disease challenge	Disease resistance and survival rate	Reference
MOS		LZM→, RBA→	–	–	–	Serum total Protein→	–	–	Yuji-Sado et al. (2015)
FOS		LZM↑, serum IgM↑, NO↑	–	–	SOD↑CAT↑, GPx↑, MDA↑	–	–	–	Abd El-Gawad et al. (2016)
LMWSA		LZM↑, Serum ACH50↑, RBA↑, PA↑	–	–	–	–	*S. agalactiae*	↑	Van Doan et al. (2016b)
XOS		Serum and mucus LZM↑, Serum ACH50↑, RBA↑, PA↑	–	–	Serum and mucus Perox-idase activity↑	–	*S. agalactiae*	↑	Van Doan et al. (2018b)
COS	Hybrid tilapia (*O. niloticus* ♀ × *O. aureus* ♂)	–	*TNF-α*↓, *TGF-β*↑	*HSP70*↓	–	–	*A. hydrophila*	↑	Qin et al. (2014)
GOS	Red hybrid tilapia red tilapia (*O. niloticus* × *O. mossambicus*)	–	–	–	–	WBCs↑, RBCs↑, Ht→	*S. iniae*	↑	Plongbunjong et al. (2011)
Synbiotic									
S. cerevisiae + β-glucan	Nile tilapia (*O. niloticus*)	LZM→, total Ig→, ACH50→	–	–	–	Serum total protein→	–	–	Shelby et al. (2009)
L. plantarum + LMWSA		Serum LZM↑, ACH50↑, RBA↑, PA↑	–	–	–	–	*S. agalactiae*	↑	Van Doan et al. (2016a)

B. subtilis + Malic acid		–	–	–	–	WBCs↑, RBCs↑, Ht↑, Hb↑, total Protein↑, Albumin↑, globulin↑, ALT↓, AST↓	–	↑	Hassaan et al. (2018)
L. plantarum + CCM[i]	Nile tilapia (*O. niloticus*)	Mucus LZM↑, Serum LZM↑, ACH50↑, RBA↑, PA↑	–	–	Serum and mucus Perox-idase activity↑	–	–	–	Van Doan et al. (2017a)
Kefir[j] + LMWSA		Serum LZM↑, ACH50↑, RBA↑, PA↑	–	–	–	–	*S. agalactiae*	↑	Van Doan et al. (2017b)
ASP[k] + β-glucan		NBT↑, LZM↑, IgM, BA↑, PA↑	–	–	SOD↑, CAT↑, GPx↑, MDA↓	WBCs↑, RBCs↑, Ht↑, Hb↑, total Protein↑	–	–	Dawood et al. (2020)
L. plantarum + XOS		Mucus LZM↑, Serum LZM↑, ACH50↑, RBA↑, PA↑	–	–	Serum and mucus Perox-idase activity↑	–	*S. agalactiae*	↑	Van Doan et al. (2020b)
Lactobacillus spp. + FOS	Hybrid tilapia (*O. niloticus* ♀ × *O. aureus* ♂)	–	–	–	–	–	*A. hydrophila*	↑	Liu et al. (2017)
Plant compounds									
Garlic (*Allium sativum*)	Nile tilapia (*O. niloticus*)	–	–	–	–	WBCs↑, RBCs↑, Ht↑, Hb↑, ALT↓, AST↓	*A. hydrophila*	↑	Shalaby et al. (2006)
		Serum LZM↑, RBA↑, PA↑	–	–	–	–	–	–	Yin et al. (2006)

(continued)

Table 1.5 (continued)

Feed additives	Tilapia species	Immune responses	Immune-related genes	Stress-regulating genes	Anti-oxidative capacity	Haematological and biochemical parameters	Disease challenge	Disease resistance and survival rate	Reference
Chinese herbs (*Astragalus radix*)									
Garlic (*Allium sativum*)		NBT↑	–	–	–	–	*P. fluorescens*	↑	Diab et al. (2008)
Ginseng Herb (Ginsana® G115)		–	–	–	–	WBCs↑, RBCs↑, Ht↑, Hb↑	–	–	Goda (2008)
Psidium guajava		–	–	–	–	–	*A. hydrophila*	↑	Pachanawan et al. (2008)
Garlic (*Allium sativum*)		–	–	–	SOD↑, CAT↑, GPx↑, MDA↓	Serum total Protein↑, ALT↓, AST↓	–	–	Metwally (2009)
Echnicacea and Garlic[1]		–	–	–	–	WBCs↑, Ht↑, Neutrophil↑, Lymphocyte↑	*A. hydrophila*	↑	Aly and Mohamed (2010)
Rosmarinus officinalis	Nile tilapia (*O. niloticus*)	–	–	–	–	–	*S. iniae*	↑	Zilberg et al. (2010)
Ginger (*Zingiber officinale*)		LZM↑, IgM↑	–	–	–	–	*A. hydrophila*	↑	El-Sayed et al. (2014)
Aloe vera		LZM→	–	–	–	WBCs↑, RBCs↑, Ht↑, Hb↑, Neutrophil↑, monocytes↑	*S. iniae*	↑	Gabriel et al. (2015)

Moringa oleifera		–	–	–	–	ALT↓, AST↓	*A. hydrophila*	↑	Gbadamosi et al. (2016)
Ginger (*Z. officinale*)		–	–	–	SOD↑, CAT↑, MDA↓	WBCs↑, RBCs↑, Ht↑, Hb↑, Neutrophil↑, monocytes↑	*A. hydrophila*	↑	Şahan et al. (2016)
Orange pectin		Mucus LZM↑, Serum LZM↑, ACH50, PA↑↑, RBA→	–	–	Serum and mucus Peroxidase activity↑	–	*S. agalactiae*	↑	Van Doan et al. (2018a)
Withania sominefera		NBT↑	–	–	Liver SOD↑, CAT↑, GST↑, GSH↑, GPx↑, MDA↓, serum TAC↑	–	*A. hydrophila*	↑	Zahran et al. (2018)
Ocimum basilicum		Serum LZM↑	–	–	–	Ht→, Hb↑, Plasma total Protein↑, ALT↓,	–	–	de Souza et al. (2019)
Thai ginseng (*Boesenbergia rotunda*)		Mucus LZM↑, Serum LZM↑, ACH50↑, RBA↑, PA↑	–	–	Serum and mucus Peroxidase activity↑	–	*S. agalactiae*	↑	Van Doan et al. (2019)
Berberine powder		Mucus LZM↑, Serum LZM↑, ACH50↑, RBA↑, PA↑	–	–	Serum and mucus Peroxidase activity↑	–	*S. agalactiae*	↑	Van Doan et al. (2020a)
Organic acids									
Alginic acid	Nile tilapia (*O. niloticus*)	–	–	–	–	–	–	–	Merrifield et al. (2011)
calcium lactate		–	–	–	–		–	–	

(continued)

Table 1.5 (continued)

Feed additives	Tilapia species	Immune responses	Immune-related genes	Stress-regulating genes	Anti-oxidative capacity	Haematological and biochemical parameters	Disease challenge	Disease resistance and survival rate	Reference
						Ht↑, Hb↑, Plasma total Protein↑, ALT↓, AST↓			Hassaan et al. (2014)
L-malic acid		LZM↑	–	–	SOD↑, MDA↑	–	–	–	Chen et al. (2016)
OAB[m]	Red hybrid tilapia red tilapia (*O. niloticus* × *O. mossambicus*)	–	–	–	–	–	*S. agalactiae*	↑	Ng et al. (2009)

Arrows indicate an increase (↑), decrease (↓), or no change (→) in the response

Probiotic genera abbreviations: *B.*: *Bacillus*, *E.*: *Entercoccus*, *L.*: *Lactobacillus*, *P.*: *Pediococcus*, *R.*: *Rummeliibacillus*

Pathogens genera abbreviations: *A.*: *Aeromonas*, *E.*: *Edwardsiella*, *P.*: *Pseudomonas*, *S.*: *Streptococcus*

Prebiotic abbreviations: *COS:* Chito-oligosaccharides, *FOS:* Fructooligosaccharide, *GOS:* Galactooligosaccharide, *MOS:* Mannan oligosaccharide, *XOS:* Xylooligosaccharides, *LMWSA:* Low molecular weight sodium alginate

*Parameters investigated abbreviations: *ACH50:* alternative complement haemolytic 50 activity, *BA:* Bactericidal activity, *C3 & C4:* Complement component 3 & 4, *Ig:* Immunoglobulins, *LZM:* Lysozyme activity, *PA:* Phagocytic activity, *RBA:* Respiratory burst activity, *NBT:* Nitroblue tetrazolium, *NO:* Nitric oxide activity, *RBCs:* Red blood cells, *WBCs:* Leucocytes, *Ht:* Haematocrit, *Hb:* Haemoglobin, *ALP:* Alkaline phosphatase, *ALT:* Alanine aminotransferase activity, *AST:* Aspartate aminotransferase activity, *SOD:* Superoxide dismutase, *CAT:* Catalase, *GPx:* Glutathione peroxidase, *GR:* Glutathione reductase, *GSH:* Glutathione, *GST:* Glutathione S-transfrase, *TSA:* Total antioxidant capacity, *MPO:* Myeloperoxidase, *MDA:* Malondialdehyde activity, *IEL:* Intraepithelial leucocyte levels in the intestine, *GC:* Gablet cells, *IL:* Interleukin, *TNFα:* Tumour necrosis factor-α, *TLR:* Toll-like receptors, *IFN-γ:* Interferon gama, *TGF-β:* Transforming growth factor beta, *IRF-3:* Interferon regulatory factor, *Mx:* Transcription of *mx*, *RSAD-2:* Radical S-Adenosyl Methionine Domain Containing 2 gene (*VIPERIN* gen), *PCNA:* Proliferating cell nuclear antigen, *HSP70:* Heat shock 70 kDa proteins, *Lyzc:* C-type lysozyme

[a]The commercial probiotics contained *Bacillus subtilis*, *Lactobacillus acidophilus*, *Clostridium butyricum*, and *Saccharomyces cerevisiae*.

[b]The mixed pathogens include *A. hydrophila*, *P. fluorescens*, and *S. iniae*

[c]A commercial probiotic product contains *Bacillus* spp., *Pediococcus* spp., *Enterococcus* spp., and *Lactobacillus* spp.

[d]A commercial carotenoid product from the extract of probiotic Rhodobacter sphaeroides mutant strain WLAPD911 (Lycogen™) contains neurosporene, β-carotene, spheroidenone and methoxyneurosporene rather than lycopene.
[e]Tilapia lake virus (*Tilapia tilapinevirus*).
[f]A commercial product containing the isoquinoline alkaloid sanguinarine.
[g]GroBiotic®-A is a mixture of partially autolyzed brewer's yeast Saccharomyces cerevisiae, dairy components, and fermentation products such as β-glucan and oligosaccharides.
[h]Jerusalem artichoke (*Helianthus tuberosus*).
[i]Mushroom (*Cordyceps militaris*) substrate
[j]Kefir is a complex community of yeasts (*Kluyveromyces, Saccharomyces,* and *Torula*), lactobacilli (*Lactobacillus*, *Lactococcus*, *Leuconostoc*, and *Streptococcus* spp.), and acetic acid bacteria (*Acetobacter*).
[k]*Aspergillus oryzae.*
[l]*Echinacea purpurea* and *Allium sativum.*
[m]OAB, organic acids blend (Sunzen Corporation, Malaysia); PDF, potassium diformate (FORMIs, ADDCON, Norway).

Bacillus spp., lactic acid bacteria (LAB), certain Gram-negative spp., and yeast. They are incorporated into fish diets and lead to high levels of colonization and moderate gut microbial ecology populations (Merrifield et al. 2010).

A prebiotic is defined as a feed additive derived from vegetables and fruit that enhances the fish's performance and/or modifies the community of gastrointestinal beneficial bacteria, which consequent improvement of the host's well-being and health (Gibson and Roberfroid 1995). The criteria used to identify and recognize compounds as prebiotics are: (i) compounds that are neither hydrolysed nor assimilated by the gut; (ii) these compounds should be fermented by the gastrointestinal microbe community, (iii) they should be beneficial for bacteria to the colon through the enhancement of their growth and/or metabolism activation; finally, (iv) they should be able to transfer the colonic flora into healthier compounds to the host (Gibson and Roberfroid 1995; Lauzon et al. 2014).

Ringø et al. (2014) reported that the most common prebiotics used in aquafeeds are fructooligosaccharides (FOS), mannanoligosaccharides (MOS), galactooligosaccharides (GOS), and many commercial products that could be fermented by the gut flora and recognized as non-digestible compounds. Bioactive compounds and oils derived from medicinal plants have a positive impact on the host, including enhancement of performance, immune system response, modification of the gut flora, reduction of free radicals from the metabolic process, and enhancement of the host health and welfare against pathogen microbes (Alemayehu et al. 2018; Jeney et al. 2015).

Organic acids are also employed as a functional feed additive to enhance gut health and performance in fish (Lim et al. 2015). Organic acids have three different mechanisms of action in the gastrointestinal tract of fish: (i) the pH-reducing action of organic acids in the gastrointestinal tract leads to enhanced activity of the digestive enzymes, (ii) the reduction of pH inhibits the growth of pathogens bacteria in the gastrointestinal tract, and (iii) the organic acids integrated into aquafeeds decrease the potential risk of microbial contamination including pathogenic bacteria and moulds or fungi during feed storage (Lim et al. 2015).

1.2.2 Environmental Requirements

The tilapia genus is recognized as the most widely cultured group of species since they are raised in different regions of the world and are highly adaptable to a wide range of environmental conditions (Table 1.6).

Tilapia is tolerant to a wide range of rearing conditions such as salinity, ammonia, pH, dissolved oxygen (DO), and temperature. Among tilapia species, the least adaptable species is Nile tilapia (*O. niloticus*), while the most tolerant species to saltwater are Mozambique tilapia (*T. mossambicus*), blue tilapia (*O. aureus*), and red belly tilapia (*T. zilli*) (El-Sayed 2006b, c).

El-Sayed (2006c) reported that the best salinity level to attain optimum growth in different tilapia species is up to 19 ‰. The optimal growing temperatures range

Table 1.6 Limits and optima of water quality parameters for tilapia

Parameter	Range	Optimum for growth	Reference
Salinity (‰)	Up to 36	<19	El-Sayed (2006c)
DO (mg L^{-1})	Down to 0.1	>3	Ross (2000)
Temperature (°C)	8–42	22–29	Mjoun et al. (2010)
pH	3.7–11	7–9	Ross (2000)
Ammonia (mg L^{-1})	Up to 7.1	<0.05	El-Sherif and El-Feky (2008)

between 22 and 29 °C, while the best temperature for spawning is higher than 22 °C. (El-Sayed 2006c) stated that tilapia can survive below 10° C and Nile tilapia (*O. niloticus*) is the least tolerant to cold water while *O. aureus* is the most tolerant to cold water. (Ross 2000) found that tilapia can tolerate a wide range of dissolved oxygen (DO); however, the optimum level of DO to achieve better growth performance is higher than 3 mg $L^{-1,}$ and the best DO for optimum feed utilization efficiency is 7 mg L^{-1} (Bergheim 2007). (Shelton and Popma 2006) found that Tilapia requires a pH of ~7 or slightly higher than 7 to achieve the best growth. In general, tilapia tolerate a wide range of pH between 3.7 and 11, but the optimum growth rate is achieved between pH 7.5 to 9.5 (Ross 2000).

In terms of ammonia, a concentration higher than 2 mg L^{-1} is considered toxic (Shelton and Popma 2006). While ammonia at a concentration of 1.2 mg L^{-1} causes low appetite and reduces growth (Celik 2012; El-Sherif and El-Feky 2008). (El-Sherif and El-Feky 2008) reported that the optimum ammonia concentration for Nile tilapia (*O. niloticus*) is estimated to be below 0.05 mg L^{-1}.

1.3 Comparative Assessment of Tilapia Culture Systems

The selection of the culture technique is principally affected by the purpose of the aquaculturists or determined by the geographical conditions which involve site selection, substructure, environmental and physical conditions (especially climate), socioeconomic aspects, technical knowledge, and marketing potential. Different types of tilapia production are well known all over the world as monoculture or polyculture in different rearing units such as cages, ponds, and tanks. Also new production systems are practised in different regions of the world as aquaponics, hydroponics, and biofloc. Tilapia production is divided into three categories: (i) extensive: where the stocking density is low and organic and inorganic fertilizers used to promote the natural food, (ii) Semi-intensive: where the source of food combined between fertilization to promote natural food includes phytoplankton and zooplankton plus supplemented diets; (iii) intensive: based on high stocking density, high water exchange rate, and balanced complete feeds are provided. The approximate annual yields of polyculture systems including tilapia with carps reach or exceed 5 tonnes/ha (Prabu et al. 2019).

The major challenges of tilapia production are deprived growth, pathogen infections, and disease outbreaks. However, there have been several solutions that could help to expand tilapia production outside their tropical and sub-tropical regions. Some of these include compiled intensive-extensive systems, closed-cycle that aids in controlling the environmental variables, and intensive system indoor using RAS system with advanced water treatment methods such as biofloc. The common production approach of Nile tilapia (*O. niloticus*) in Asia and Latin America in lakes and reservoirs in different countries is represented by the intensive system in floating cages. The success of intensive technique using cage/tank systems relies on several factors such as (i) cage/tank size and shape, (ii) stocking density and iii) water exchange/water flow rate. Cage culture offers several advantages as follows: (i) minimizes fertilization, (ii) allows the recovery of eggs by spawning fish, (iii) allows the fish to grow in a real environment. The pros of using cage production techniques could be summarized in the following points as stated by (Bhujel 2014a):

- Use of water bodies that cannot be drained or seined and would otherwise not be suitable for aquaculture.
- Flexible and convenient for management.
- No accumulation of ammonia, nitrite, nitrate, and other waste products that are quickly flushed out with water flow from the cage.
- High feeding rates are feasible, and a higher fish growth rate could be achieved.
- Predators (e.g., birds, otters, etc.) can be easily controlled.
- Ease and low cost of harvesting.
- Easy monitoring of fish feeding and health status.
- Low capital investment compared to other production techniques.

Nevertheless, some cons of cage culture involve: (i) poaching risk, (ii) failure of ability to prevent poor water quality conditions because of pollution by industries or factories, iii) difficulties in treating disease and parasites, (iv) need to frequently scrub the cages, (v) risk of fish escape from the cage, (vi) inability to provide natural foods and dependence on manufactured fish diets (complete in nutritional composition), and finally (vii) there may be criticism raised by environmentalists (Bhujel 2014a).

Thereafter, tanks and raceways can be another option to replace pond or cage culture if the conditions are not suitable for the cage technique. (Liao and Chen 2001) stated that in Asia, the intensive tilapia production system in tanks is commonly practised in Taiwan, Malaysia, and Philippines. Taiwan is considered a pioneer for tilapia-intensive culture in concrete tanks (small to medium-level operation), with a production of over 55,000 tons/year. In comparison to the pond production system, tanks and raceways are easy to monitor and handle the stocks and possess a high degree of control of the environmental conditions, specifically water quality parameters. On the contrary, the disadvantages of using tank and raceway culture are given by the high cost of construction, operation, and production and the requirement of proficient attention due to the higher risk of major fish mortality caused by disease outbreaks.

1.3.1 Biofloc Technology

Wastewater with high ammonia, nitrite, and nitrate concentrations damages the fish culture system and pollutes other natural water bodies, so there is a need to look for alternative culture techniques to decrease the environmental damages caused by aquaculture (Gutierrez-Wing and Malone 2006). "Biofloc" Technology (BFT) is considered a promising alternative technique to avoid the harmful environmental impact caused by aquaculture. Biofloc technology through aeration and the addition of extra carbon to the aquaculture system increases the nitrifying bacterial colonies that maintain water quality and, together with phytoplankton and zooplankton are considered additional foods in the aquaculture farming system (Gutierrez-Wing and Malone 2006). The basic concept and function of biofloc are to stimulate the growth of heterotrophic microbial bacteria that convert a toxic source of nitrogen "NH_3" to a safe source of nitrogen, in addition to converting nitrogenous waste to a microbial protein that could be used as another source of protein to the fish (De Schryver et al. 2008). To date, BFT technology has achieved worthwhile attention due to its higher production efficiency, protein recycling from food leftover through nitrifying bacteria, water quality improvement, and a novel approach to infectious disease prevention (Ekasari et al. 2015a). In addition, Ekasari et al. (2015a, b) reported that BFT technique employment could boost the reproductive performance of Nile tilapia (*O. niloticus*) broodstock by enhancing fecundity and larval survival rate and also improving the immune system against infectious diseases. Therefore, all the above advantages of biofloc technology attract the attention of scientists to conduct their research in BFT systems to guarantee and secure the factors within the recommended levels to achieve aquaculture sustainability and food security.

1.4 Genetic Improvement

Research for genetic improvement, particularly quantitative genetic approaches, can have a tremendous impact on aquaculture and can be responsible for increased production efficiency and improved productivity in aquatic animals. Over the past decades, the importance of tilapia in global aquaculture has increased, as well as the intensity and diversity of research for genetic improvement (Gupta and Acosta 2004). In addition, aquatic animals give the opportunity to scientists to implement different approaches for improving fish genetics, including hybridization, selective breeding, sex control, and crossbreeding. Recently research in fish genetics proved that using and implementing new approaches help discover new strains with high growth rate, feed efficiency, survival rate, tolerance to a wide range of environmental conditions, and disease resistance. Moreover, the adoption and implementation of new genetic approaches could help discover new strains of fish and shrimp that could grow in either freshwater or/and brackish water systems (Nguyen et al. 2010; Ninh et al. 2014). Nowadays, most tilapia genetic research has focused on hybridization and monosex male fry production. The most relevant findings of this research indicate that tilapia males grow faster than females, and tilapia shows early maturation, which leads to consecutive spawning during the growing period and thus inhibits and limits growth. Thereafter, female tilapia shows difficulty growing uniformly, so male fry is preferred (Chen et al. 2018; Martínez et al. 2014). This shows the importance of searching for novel techniques and modification and improvement of the existing ones such as manual sexing, interspecific hybridization, androgenesis, triploid, transgenesis, hormonal sex reversal, and YY male technology to produce monosex tilapia for successful and intensive tilapia production (Beardmore et al. 2001; Ponzoni et al. 2011). The current approaches to producing 'all male' tilapia have limitations that make them expensive, unmaintainable, or not acceptable. For example, manual sexing is labour-consuming, and attention to maintaining broodstock for interspecific hybridization is needed. Although the most widespread approach to producing monosex offspring is tilapia masculinization with hormone therapies or adrenal malfunction, hormone therapies have posed concerns because they may influence consumer acceptance and marketability of the fish, and hormone residues may have irreversible effects on biodiversity and water quality. In the case of the YY approach, the production of YY males requires at least three generations of breeding. Moreover, the employment of YY technology depends on the contribution of an outstanding laboratory with advanced facilities for the creation of YY males (Baroiller et al. 1995; Abucay et al. 1999; Baroiller and D'Cotta 2001; Tessema et al. 2006). Therefore, there is a need to look for alternative techniques to create 'all male' tilapia. Some research indicates an effect of temperature on genotype interaction, such as increasing the male: female ratio in response to thermal treatment. For example, thermal treatment of about 38 °C for 10 days after post-hatching succeeded in producing above 80% of males (Ponzoni et al. 2011). It should be mentioned that of all the genetic techniques, just selective breeding presents the chance of permanent genetic achievements because the achievements can be transferred from generation to generation. Finally, a combination of classic selective breeding using marker-assisted selection (MAS) and polygenic selection could considerably promote the male-to-female ratio as a response to thermosensitivity.

1.5 Environmental Impacts

Currently, tilapias have been introduced as exotic species in most countries worldwide, with successful growth and reproduction in new habitats. The traditional tilapia culture in semi-intensive, small-scale systems with minimum negative effects on the environment is now being replaced with intensive, large-scale farming systems. Since the use of manufactured inputs such as artificial feed, chemotherapeutic agents, and hormones will become inevitable in intensive culture systems. The worldwide expansion of tilapia rearing at an extremely high rate is very presumably to cause environmental and socioeconomic risks. In the last few decades, Nile tilapia (*O. niloticus*) production expanded to increase the seafood supply and fulfil the global demand for animal protein and food security. Nevertheless, although tilapia produces substantial economic growth, its fast expansion has caused numerous environmental threats like the destruction of wild habitats, the interaction between alien and endemic species, disturbance of wildlife, use of artificial culture inputs (e.g. chemotherapeutic agents, antibiotics, hormones, and fuels), and eutrophication because of the aquaculture wastewater (El-Sayed 2006a). Thus, the use of advanced and efficient management approaches is necessary. In this regard, some innovative methods have been recommended to boost responsible aquaculture activities including the amalgamation of aquaculture practices with livestock farming and agriculture (e.g. aquaponic, hydroponic, etc.), and also using the BFT and RAS systems that could help to control or manage infectious diseases outbreak and discharges of aquaculture farms (Wang and Lu 2016; Forio and Goethals 2020). These approaches can be the basis for effective long-term solutions for eco-friendly and green aquaculture in the future. Therefore, novel approaches are required if sustainable and green aquaculture is to be meaningfully understood and implemented (Montoya-Camacho et al. 2019). Since eutrophication, a process that is caused by the excessive input of nutrients (e.g., phosphorous and nitrogen), is largely recognized as a serious threat to the environment (Nakano et al. 2016), in the past few decades, researchers have been investigating techniques to reduce aquaculture waste outputs, mainly phosphorus and nitrogen, from aquaculture operations to obtain satisfaction of environmentalists (Azim and Little 2008; Pinho et al. 2017; Boyd 2019). The well-known techniques to achieve eco-friendly aquaculture are the BFT technique and the integrated multi-trophic approach. The biofloc approach has been accomplishing acceptance as an efficient alternative water quality management system (Emerenciano et al. 2013; Dauda 2020). This technique presents the elimination of nutrients from water with the production of microbial communities, which can be consumed by the culture species in situ as natural foods (De Schryver et al. 2008). The other approach is the integrated multi-trophic aquaculture which is defined as a unique self-cleaning approach for aquaculture ponds since the waste from one species (including uneaten feed, faeces, and metabolic excretion) is the source of feed to support the growth of other species (Sampantamit et al. 2020).

Another factor that could affect the tilapia aquaculture community environment is the introduction of alien species destroying the ecosystem compositions and posing risks the global biodiversity (Brown et al. 2018; Anton et al. 2019). Although transgenic tilapia provide several advantages for tilapia farming, the rate of genetic alteration in transgenic tilapia is such that their phenotypic and behavioural attributes cannot be easily predicted (Mair 2002). Furthermore, transgenic tilapias are a new tool that introduces new strains into the community of wild tilapia and may have negative effects on the environment and other native species. The negative impact of transgenic is the replacement of native populations with novel strains that could become a part of the gene pool and also change the hierarchy of the natural populations (Dunham 1999).

Unfortunately, despite the negative effects that the extension of tilapia rearing may have on the environment, most introductions have not been preceded by any environmental impact evaluation. Instead, in most cases, the evaluation was performed after the introductions took place. In such a case, modulating and managing the impacts of introduced tilapia in their new environments will be very challenging, or even unfeasible. Therefore, cautious and thorough assessment, as also proper management plans and programmes must be adopted to be carried out before any introductions or transfer of tilapia.

1.6 Some Constraints and Suggestions for Solutions to Tilapia Farming

Although tilapia farming holds great promise, there are some constraints to its development. Some constraints and their possible solutions are reported as follows:

1. Training resource allocation related to tilapia farming. In general, aquaculture needs education about new technologies and farm management, because education can play a substantial role in enhancing the skills and experiences of farmers and also resolving the restrictions and challenges facing aquaculture (Olaoye et al. 2013). Previous research expressed the importance of the aquaculturists' education level to select reasonable technologies and manage the facilities efficiently (Ogunmefun and Achike (2017); Uddin et al. (2021). Therefore, resource allocation for training courses and workshops for tilapia farmers has to be considered.
2. Insufficient supply of tilapia fry. The lack of larvae production to respond to the growing world demand is one of the major bottleneck restrictions to the development of the tilapia-intensive culture (El-Sayed 2002). One of the most important obstacles to high-quality tilapia fry production is the poor reproductive performance of bloodstock due to asynchronous spawning cycles and low fecundity rate.

   Bhujel (2000) stated that the monitoring and management of the environmental and nutritional status of brood stocks can improve their efficiency. Moreover, the selective breeding of superior brood stocks and strains with preferable size and age for breeding objectives could remarkably improve larvae production. Among the environmental factors, high salinity and low temperature might be helpful for the control and synchronization of broodstock reproduction when fry demand is low (Bhujel 2000).
3. Environmental tolerance. Environmental tolerance is the major factor in controlling the success of tilapia production. Although some tilapia like Mozambique tilapia can grow in seawater, most tilapia species are categorized as freshwater fish and not tolerant to high salinity (El-Sayed 2006b; Shelton and Popma 2006). One of the promising techniques to improve salinity tolerance is the crossbreeding between Nile tilapia (*O. niloticus*) and Mozambique tilapia

(*O. mossambicus*). Furthermore, the limited ability of tilapias to tolerate low temperatures (<15 °C) (El-Sayed 2006b; Shelton and Popma 2006) restricts the expansion of tilapia culture in a different geographic zone. Using warm water such as cooling water of some industries, thermal effluents, and/or warm springs and also maintaining tilapia in a greenhouse or indoor ponds because of their non-resistance to cold water can help them to overwinter in subtropical scopes of tilapia culture.

4. Early maturation. Tilapia's first maturity occurs at an early age (2–3 months old) and a short length (10–12 cm length). The most preferable and cost-effective technique is to create 'all male' tilapia because males grow faster than females and also have a more standard size (see Sect. 1.4).
5. Genetic deterioration. In some conditions, there is evidence of genetic deterioration. Genetic deterioration of introduced stocks is widely attributed to poor broodstock management resulting in inbreeding and introgression of less favourable genes. With the rapid advance of next-generation sequencing techniques (Metzker 2010), marker-assisted selection and genomic selection will significantly accelerate the genetic improvement of tilapias (Sonesson 2011; Yue 2014). In general, the desirable characteristics of improved tilapia have been focused on higher production efficiency, better appearance, tolerance to certain environmental conditions, and, especially, control of unwanted breeding.

   Remarkable interests in improved growth rate and performance of tilapia under farm conditions have been shown from breeding programmes for selection and sex control. The achievements of the implementation of such breeding strategies have been introduced to aquaculture through technically and economically sustainable programmes (Mair 2002). For example, farming the Genetically Improved Farmed Tilapia (GIFT) strain rather than the not-improved local strain of the Nile tilapia (*O. niloticus*) can enhance the growth rate and production efficiency of tilapia. Since feed accounts for over 50% of the cost of production, the higher feed conversion ratio of the GIFT strain would decrease production costs. The GIFT strain has a remarkably higher growth performance, better feed conversion ratio, and higher production efficiency than the local strain (Ridha and Cruz 2002).

   The other important aspect is the genetically improved strains dissemination to achieve the targeted beneficiaries effectively, including monitoring of the impact and adoption of improved breeds. On the other hand, genetic improvement programmes will require the development of production stocks that are acceptable to each environment. Therefore, success in genetic improvement programmes will require long-term support and collaboration between partners from the government, university, and industry.
6. Disease resistance. Since tilapia, cultural practices have been intensified, and the densities of tilapia have increased in different systems and culture has expanded into the colder climatic zones, where suitable environmental factors are more difficult to maintain, infectious diseases have emerged (Watanabe et al. 2002). Developing approaches for fish health stability through genetic improvement,

water quality management, stress reduction, and the use of preventive immunostimulants are required to control infectious diseases.

7. Lack of access to freshwater. Tilapias are freshwater fish; however, access to freshwater resources is one of the most critical environmental issues in developed and industrialized countries (Hankins 2002). Therefore, it is suggested to use water supplies that are not suitable for human consumption or agriculture such as brackish water or seawater. Moreover, the use of recirculating systems is another fit approach for water quality and quantity management.
8. Negative impacts on the environment and global biodiversity. It is clear from the previous section (see Sect. 1.5) that mismanaged transfers and/or introductions of tilapia can lead to destructive environmental impacts. If tilapias are established in their new environment, it would be approximately infeasible to control and reduce their catastrophic impacts. It is hence necessary that strict regulations be established to control the introduction of tilapias in a new habitat and also precise monitoring programmes and certification of tilapia farms are mandatory to protect the environment and aquatic biodiversity (Bush et al. 2013).

   Yue et al. (2016) expressed that the recirculating aquaculture systems and cage culture can mitigate the adverse effects of tilapia culture on the environment and aquatic biodiversity. Therefore, they would probably be developed in the production and technological advances of tilapia culture. These systems facilitate aquaculture and hence will be the key parts of next-generation aquaculture.
9. Flesh quality problems of tilapias. Tilapia, flesh quality issues, are as follows: (i) odour and flavour, which are attributed to culture conditions, (ii) high percentage of bone if the harvest has occurred in small-size fish, and (iii) farmed tilapia species contain low levels of omega-3 fatty acids compared with other fish, especially salmon (Weaver et al. 2008). Likely, selective breeding (Gjedrem and Baranski 2010) and supplementing tilapia feeds with marine microalgae containing a high level of omega-3 (Tadesse et al. 2003) can increase the essential fatty acid content of tilapias. Moreover, finishing diets and also the GIFT breeds (reviewed by Eknath and Hulata 2009) and transgenic tilapia can moderate this problem. Although, before the commercialization of transgenic fish, food safety issues should be addressed.
10. Failed marketing of products. The lack of attention given to marketing and business has also been recognized as one of the restrictions to the achievement of commercial tilapia production. The assessment of the tilapia market is rarely undertaken by aquaculturists due to time and expense and difficulties in attracting the cooperation of wholesalers and retailers, which should be considered (Watanabe et al. 1997).

1.7 Conclusions and Recommendations

This chapter may be considered as a short preface to tilapia rearing needs. Tilapia culture has gained significance increasingly in the world. Tilapia has a lot of positive characteristics that make it proper for culture. Amongst these are its general resilience, high tolerance to unfavourable environmental conditions and adaptability to high stocking densities, its potential ability to tolerate low levels of dissolved oxygen and a wide range of salinity concentrations, and its infectious diseases resistance. Tilapia can utilize and grow in a wide variety of natural and artificial feeds, has a high survival rate, acceptable feed conversion ratio, fast growth rate, and high yield potential, and is accepted by a wide range of farmers and consumers. Moreover, tilapia can grow well in different aquaculture systems, ranging from extensive systems with simple substructures to more intensive systems with complex infrastructure. With the increasing demand for tilapia products, tilapia farming will continue to be a source of different business benefits, since it is a cheap and easy source of affordable and inexpensive animal protein and provides several job opportunities to the community in developing countries. Finally, suitably designed fish farms, precisely selective breeding of tilapia strains, selection of a proper tilapia production system by the aquaculturists, government support on supply seed, feed, and instruments, training, extension services, and advice to the aquaculturists regarding tilapia culture, and the development of an organized marketing agenda would increase the commercial profitability and sustainability of tilapia production in many countries around the world.

Therefore, it can be unavoidable to conduct research studies on resolving the issues met in tilapia culture because tilapia culture will guarantee the socioeconomic advantages and food security of developing countries. For example, although tilapia feed on a wide range of natural and artificial feeds, specific dietary requirements are yet lacking, and the interactions among nutrients and with cultured conditions and tilapia health and welfare are not completely known. Nowadays, research and interest in dietary feed additives such as immunostimulants and growth stimulants, especially phytobioactive compounds to improve fish health and growth performance are likely to continue, which will fill existing research gaps. Moreover, more research work and resource and management development are required to improve breeds that are more cold-tolerant, salt-tolerant, and disease resistant.

References

Abd El-Gawad E, Abd El-Latif A, Shourbela R (2016) Enhancement of antioxidant activity, non-specific immunity and growth performance of Nile tilapia, *Oreochromis niloticus* by dietary fructooligosaccharide. J Aquac Res Dev 7(5):427–433. https://doi.org/10.4172/2155-9546.1000427

Abdel-Tawwab M, Abdel-Rahman AM, Ismael NE (2008) Evaluation of commercial live bakers' yeast, *Saccharomyces cerevisiae* as a growth and immunity promoter for Fry Nile tilapia,

Oreochromis niloticus (L.) challenged in situ with *Aeromonas hydrophila*. Aquaculture 280(1-4):185–189. https://doi.org/10.1016/j.aquaculture.2008.03.055

Abdel-Tawwab M, Ahmad MH, Khattab YA, Shalaby AM (2010) Effect of dietary protein level, initial body weight, and their interaction on the growth, feed utilization, and physiological alterations of Nile tilapia, *Oreochromis niloticus* (L.). Aquaculture 298(3-4):267–274. https://doi.org/10.1016/j.aquaculture.2009.10.027

Abucay JS, Mair GC, Skibinski DO, Beardmore JA (1999) Environmental sex determination: the effect of temperature and salinity on sex ratio in *Oreochromis niloticus* L. Aquaculture 173(1-4): 219–234. https://doi.org/10.1016/S0044-8486(98)00489-X

Alemayehu TA, Geremew A, Getahun A (2018) The role of functional feed additives in tilapia nutrition. Fish Aquac J 9(2):1–6. https://doi.org/10.4172/2150-3508.1000249

Aly S, Mohamed M (2010) *Echinacea purpurea* and *Allium sativum* as immunostimulants in fish culture using Nile tilapia (*Oreochromis niloticus*). J Anim Physiol Anim Nutr 94(5):e31–e39. https://doi.org/10.1111/j.1439-0396.2009.00971.x

Aly SM, Ahmed YA-G, Ghareeb AA-A, Mohamed MF (2008) Studies on *Bacillus subtilis* and *Lactobacillus acidophilus*, as potential probiotics, on the immune response and resistance of Tilapia nilotica (*Oreochromis niloticus*) to challenge infections. Fish Shellfish Immunol 25(1-2):128–136. https://doi.org/10.1016/j.fsi.2008.03.013

Anderson J, Jackson A, Matty A, Capper B (1984) Effects of dietary carbohydrate and fibre on the tilapia *Oreochromis niloticus* (Linn.). Aquaculture 37(4):303–314. https://doi.org/10.1016/0044-8486(84)90296-5

Anton A, Geraldi NR, Lovelock CE, Apostolaki ET, Bennett S, Cebrian J, Krause-Jensen D, Marbà N, Martinetto P, Pandolfi JM, Santana-Garcon J, Duarte CM (2019) Global ecological impacts of marine exotic species. Nat Ecol Evol 3:787–800. https://doi.org/10.1038/s41559-019-0851-0

Ayyat M, El-Marakby H, Sharaf SM (2011) Effect of dietary protein level, stocking density, and dietary pantothenic acid supplementation rate on performance and blood components of Nile tilapia, *Oreochromis niloticus*. 23(2):122–135. https://doi.org/10.1080/10454438.2011.581572

Azim ME, Little DCJA (2008) The biofloc technology (BFT) in indoor tanks: water quality, biofloc composition, and growth and welfare of Nile tilapia (*Oreochromis niloticus*). Aquaculture 283(1-4):29–35. https://doi.org/10.1016/j.aquaculture.2008.06.036

Bahurmiz OM, Ng W-K (2007) Effects of dietary palm oil source on growth, tissue fatty acid composition and nutrient digestibility of red hybrid tilapia, *Oreochromis* sp., raised from stocking to marketable size. Aquaculture 262(2-4):382–392

Baroiller JF, Chourrout D, Fostier A, Jalabert B (1995) Temperature and sex chromosomes govern sex ratios of the mouthbrooding cichlid fish *Oreochromis niloticus*. J Exp Zool 273(3):216–223. https://doi.org/10.1002/jez.1402730306

Baroiller J-F, D'Cotta H (2001) Environment and sex determination in farmed fish. Comp Biochem Physiol Part C Toxicol Pharmacol 130(4):399–409. https://doi.org/10.1016/S1532-0456(01)00267-8

Barros MM, Ranzani-Paiva MJT, Pezzato LE, Falcon DR, Guimarães IG (2009) Haematological response and growth performance of Nile tilapia (*Oreochromis niloticus* L.) fed diets containing folic acid. Aquac Res 40(8):895–903. https://doi.org/10.1111/j.1365-2109.2009.02175.x

Beardmore J, Mair G, Lewis R (2001) Monosex male production in finfish as exemplified by tilapia: applications, problems, and prospects. Aquaculture 197(1–4):283–301. https://doi.org/10.1016/S0044-8486(01)00590-7

Bergheim A (2007) Water quality criteria I recirculation system for tilapia. IRIS-International Research Institute of Stavanger 4068, Norway

Bhujel RC (2000) A review of strategies for the management of Nile tilapia (*Oreochromis niloticus*) broodfish in seed production systems, especially hapa-based systems. Aquaculture 181(1-2): 37–59. https://doi.org/10.1016/S0044-8486(99)00217-3

Bhujel RC (2014a) Grow-out in cage. In: Bhujel RC (ed) A manual for tilapia business management. CABI, Wallingford, Oxfordshire, UK, pp 102–113. https://doi.org/10.1079/9781780641362.0000_5

Bhujel RC (2014b) Introduction. In: Bhujel RC (ed) A manual for tilapia business management. CABI, Wallingford, Oxfordshire, UK, pp 1–11. https://doi.org/10.1079/9781780641362.0000_1

Boyd CE (2019) Water quality: an introduction. doi:https://doi.org/10.1007/978-3-030-2335-8

Brown PM, Roy DB, Harrower C, Dean H, Rorke S, Roy H (2018) Spread of a model invasive alien species, the harlequin ladybird *Harmonia axyridis* in Britain and Ireland. Sci Data 5(1):180239. https://doi.org/10.1038/sdata.2018.239

Bush SR, Belton B, Hall D, Vandergeest P, Murray FJ, Ponte S, Oosterveer P, Islam MS, Mol AP, Hatanaka M, Kruijssen F, Ha T, Little D, Kusmawati R (2013) Certify sustainable aquaculture? Science 341(6150):1067–1068. https://doi.org/10.1126/science.1237314

Calder PC (2006) n−3 Polyunsaturated fatty acids, inflammation, and inflammatory diseases. Am J Clin Nutr 83(6):1505S–1519S. https://doi.org/10.1093/ajcn/83.6.1505S

do Carmo e Sá MVC, Pezzato LE, MMBF L, de Magalhães Padilha P (2004) Optimum zinc supplementation level in Nile tilapia *Oreochromis niloticus* juveniles diets. Aquaculture 238(1-4):385–401. https://doi.org/10.1016/j.aquaculture.2004.06.011

Celik E (2012) Tilapia culture review, MS thesis. Norwegian University of Life Sciences, Ås

Chavan B, Yakupitiyage A, Kamble MT, Medhe SV (2015) Tilapia as food fish: enhancement of Ω-3 polyunsaturated fatty acids in tilapia (*Oreochromis spp.*). Int J Agric Sci 7:671–677

Chen J, Fan Z, Tan D, Jiang D, Wang D (2018) A review of genetic advances related to sex control and manipulation in tilapia. J World Aquac Soc 49(2):277–291. https://doi.org/10.1111/jwas.12479

Chen YJ, Luo L, Zhang GZ, Li Z, Bai FJ, Shi YQ, Yang HS (2016) Effect of dietary L-malic acid supplementation on growth, feed utilization and digestive function of juvenile GIFT tilapia *Oreochromis niloticus* (Linnaeus, 1758). J Appl Ichthyol 32(6):1118–1123. https://doi.org/10.1111/jai.13119

Chervinski J (1982) Environmental physiology of tilapias. In: Pullin RSV, Low-McConnell RH (eds) The biology and culture of Tilapia. Proceedings ofthe 7th ICLARM Conference, International Center for Livin Aquatic Resources Management, Manila, Philippines, pp 119–128

Chiu K-H, Liu W-S (2014) Dietary administration of the extract of *Rhodobacter sphaeroides* WL-APD911 enhances the growth performance and innate immune responses of seawater red tilapia (*Oreochromis mossambicus* × *Oreochromis niloticus*). Aquaculture 418:32–38. https://doi.org/10.1016/j.aquaculture.2013.10.007

Dabrowska H, Meyer-Burgdorff K, Günther K-D (1989) Interaction between dietary protein and magnesium level in tilapia (*Oreochromis niloticus*). Aquaculture 76(3-4):277–291. https://doi.org/10.1016/0044-8486(89)90081-1

Dauda AB (2020) Biofloc technology: a review on the microbial interactions, operational parameters and implications to disease and health management of cultured aquatic animals. Rev Aquac 12(2):1193–1210. https://doi.org/10.1111/raq.12379

Dawood MA, Eweedah NM, Moustafa EM, Shahin MG (2020) Synbiotic effects of *Aspergillus oryzae* and β-glucan on growth and oxidative and immune responses of nile Tilapia, *Oreochromis niloticus*. Probiotics Antimicrob 12(1):172–183. https://doi.org/10.1007/s12602-018-9513-9

Dawood MA, Koshio S (2016) Recent advances in the role of probiotics and prebiotics in carp aquaculture: a review. Aquaculture 454:243–251. https://doi.org/10.1016/j.aquaculture.2015.12.033

De Schryver P, Crab R, Defoirdt T, Boon N, Verstraete W (2008) The basics of bio-flocs technology: the added value for aquaculture. Aquaculture 277(3-4):125–137. https://doi.org/10.1016/j.aquaculture.2008.02.019

Diab AS, Aly SM, John G, Abde-Hadi Y, Mohammed MF (2008) Effect of garlic, black seed and Biogen as immunostimulants on the growth and survival of Nile tilapia, *Oreochromis niloticus*

(Teleostei: Cichlidae), and their response to artificial infection with *Pseudomonas fluorescens*. Afr J Aquat Sci 33(1):63–68. https://doi.org/10.2989/AJAS.2007.33.1.7.391

Dos Santos L, Furuya W, Da Silva L, Matsushita M, de Castro ST (2011) Dietary conjugated linoleic acid (CLA) for finishing Nile tilapia. Aquac Nutr 17(2):e70–e81. https://doi.org/10.1111/j.1365-2095.2009.00735.x

Dunham RA (1999) Utilization of transgenic fish in developing countries: potential benefits and risks. J World Aquac Soc 30(1):1–11. https://doi.org/10.1111/j.1749-7345.1999.tb00312.x

Eid AE, Ghonim SI (1994) Dietary zinc requirement of fingerling *Oreochromis niloticus*. Aquaculture 119(2-3):259–264. https://doi.org/10.1016/0044-8486(94)90180-5

Ekasari J, Rivandi DR, Firdausi AP, Surawidjaja EH, Zairin M Jr, Bossier P, De Schryver P (2015a) Biofloc technology positively affects Nile tilapia (*Oreochromis niloticus*) larvae performance. Aquaculture 441:72–77. https://doi.org/10.1016/j.aquaculture.2015.02.019

Ekasari J, Zairin M Jr, Putri DU, Sari NP, Surawidjaja EH, Bossier P (2015b) Biofloc-based reproductive performance of Nile tilapia *Oreochromis niloticus* L. broodstock. Aquac Res 46(2):509–512. https://doi.org/10.1111/are.12185

Eknath AE, Hulata G (2009) Use and exchange of genetic resources of Nile tilapia (*Oreochromis niloticus*). Rev Aquac 1(3-4):197–213. https://doi.org/10.1111/j.1753-5131.2009.01017.x

Eleraky W, Saleh G, Gropp J (1995) A short note on the vitamin E requirement of Tilapia nilotica (*Oreochromis niloticus*)—effects on health and growth. J Appl Ichthyol 11(3-4):375–377. https://doi.org/10.1111/j.1439-0426.1995.tb00044.x

El-Sayed AFM (2002) Effects of stocking density and feeding levels on growth and feed efficiency of Nile tilapia (*Oreochromis niloticus* L.) fry. Aquac Res 33(8):621–626. https://doi.org/10.1046/j.1365-2109.2002.00700.x

El-Sayed A-FM (1991) Evaluation of sugarcane bagasse as a feed ingredient for the tilapias, *Oreochromis niloticus* and *Tilapia Zilli*. Asian Fish Sci 4(1991):53–60

El-Sayed A-FM (1999) Alternative dietary protein sources for farmed tilapia, *Oreochromis spp*. Aquaculture 179(1-4):149–168

El-Sayed A-FM (2006a) Environmenal impacts. In: El-Sayed A-FM (ed) Tilapia culture. CABI, Oxfordshire, UK, pp 207–215. https://doi.org/10.1079/9780851990149.0000_12

El-Sayed A-FM (2006b) Tilapia culture. CABI, Oxfordshire, UK, p 293. https://doi.org/10.1079/9780851990149.0000

El-Sayed A-FM (2006c. Avances en Nutrición Acuícola VIII) Tilapia culture in salt water: environmental requirements, nutritional implications and economic potentials. In: Suárez LEC, Marie DR, Salazar MT et al (eds) VIII Simposium Internacional de Nutrición Acuícola, 15 - 17 Novembre. Universidad Autónoma de Nuevo León, Monterrey, Nuevo León, México, pp 95–106

El-Sayed A-FM, Garling DL Jr (1988) Carbohydrate-to-lipid ratios in diets for *Tilapia zillii* fingerlings. Aquaculture 73(1-4):157–163. https://doi.org/10.1016/0044-8486(88)90050-6

El-Sayed HM, Hakim Y, El-Sayed BM (2014) Dietary effect of ginger (*Zingiber officinale Roscoe*) on growth performance, immune response of Nile tilapia (*Oreochromis niloticus*) and disease resistance against Aeromonas hydrophila. Abbassa Int J Aqua 7:35–52

El-Sherif M, El-Feky AM Effect of ammonia on Nile tilapia (*O. niloticus*) performance and some hematological and histological measures. In: Proceeding of the 8th International Symposium on Tilapia in Aquaculture. 12-14 October, Cairo, Egypt, 2008. Citeseer, pp 513–530

Emerenciano M, Gaxiola G, Cuzon G (2013) Biofloc technology (BFT): a review for aquaculture application and animal food industry. In: Matovic MD (ed) Biomass now-cultivation utilization. IntechOpen, Queen's University, Belfast, Canada, pp 301–328. https://doi.org/10.5772/53902

FAO (2020) Sustainability in action, vol 200. State of World Fisheries and Aquaculture, Rome. https://doi.org/10.4060/ca9229en

Ferguson R, Merrifield DL, Harper GM, Rawling MD, Mustafa S, Picchietti S, Balcàzar JL, Davies SJ (2010) The effect of *Pediococcus acidilactici* on the gut microbiota and immune status of on-growing red tilapia (*Oreochromis niloticus*). J Appl Microbiol 109(3):851–862. https://doi.org/10.1111/j.1365-2672.2010.04713.x

Fitzsimmons K (2000) Tilapia: the most important aquaculture species of the 21st century. In: Proceedings from the fifth International Symposium on tilapia Aquaculture, Rio de Janeiro. ISTA Rio de Janeiro, pp 3-8

Forio MAE, Goethals PL (2020) An integrated approach of multi-community monitoring and assessment of aquatic ecosystems to support sustainable development. Sustainability 12(14): 5603. https://doi.org/10.3390/su12145603

Fuentes-Silva C, Soto-Zarazúa GM, Torres-Pacheco I, Flores-Rangel A (2013) Male tilapia production techniques: a mini-review. Afr J Biotechnol 12(36):5496–5502. https://doi.org/10.5897/AJB11.4119

Gabriel NN, Qiang J, He J, Ma XY, Kpundeh MD, Xu P (2015) Dietary *Aloe vera* supplementation on growth performance, some haemato-biochemical parameters and disease resistance against *Streptococcus iniae* in tilapia (GIFT). Fish Shellfish Immunol 44(2):504–514. https://doi.org/10.1016/j.fsi.2015.03.002

Gbadamosi OK, Fasakin AE, Adebayo OT (2016) Hepatoprotective and stress-reducing effects of dietary *Moringa oleifera* extract against *Aeromonas hydrophila* infections and transportation-induced stress in Nile tilapia, *Oreochromis niloticus* (Linnaeus 1757) fingerlings. Int J Environ Agric Res 2(7):121–128

Gibson GR, Roberfroid MB (1995) Dietary modulation of the human colonic microbiota: introducing the concept of prebiotics. J Nutr 125(6):1401–1412. https://doi.org/10.1093/jn/125.6.1401

Gjedrem T, Baranski M (2010) Selective breeding in aquaculture: an introduction. doi:https://doi.org/10.1007/978-90-481-2773-3

Goda AMS (2008) Effect of dietary Ginseng herb (Ginsana® G115) supplementation on growth, feed utilization, and hematological indices of Nile Tilapia, *Oreochromis niloticus* (L.), fingerlings. J World Aquac Soc 39(2):205–214. https://doi.org/10.1111/j.1749-7345.2008.00153.x

Guo R, Lim C, Xia H, Aksor MY, Li M (2010) Effect of various dietary vitamin A levels on growth performance and immune response of tilapia (*Oreochromis niloticus*). Front Agric China 4(4): 507–512. https://doi.org/10.1007/s11703-010-1048-0

Gupta MV, Acosta BO (2004) A review of global tilapia farming practices. Aquac Asia Magazine 10(1):7–12

Gutierrez-Wing MT, Malone RF (2006) Biological filters in aquaculture: trends and research directions for freshwater and marine applications. Aquac Eng 34(3):163–171. https://doi.org/10.1016/j.aquaeng.2005.08.003

Hamdan A, El-Sayed A, Mahmoud M (2016) Effects of a novel marine probiotic, *Lactobacillus plantarum* AH 78, on growth performance and immune response of Nile tilapia (*Oreochromis niloticus*). J Appl Microbiol 120(4):1061–1073. https://doi.org/10.1111/jam.13081

Han B, Long W-q, He J-y, Liu Y-j, Si Y-q, Tian L-x (2015) Effects of dietary *Bacillus licheniformis* on growth performance, immunological parameters, intestinal morphology and resistance of juvenile Nile tilapia (*Oreochromis niloticus*) to challenge infections. Fish Shellfish Immunol 46(2):225–231. https://doi.org/10.1016/j.fsi.2015.06.018

Hankins JA (2002) Perspective on the role of government, industry, and research in advancing the environmental compatibility and sustainability of aquaculture. In: Libey GS, Timmons MB, Flick GJ, Rakestraw TT (eds) Proceedings of the 3rd. International Conference on Recirculating Aquaculture, 20-23 July, Roanoke, Virginia. Virginia Polytechnic and State University, pp 80–87

Hassaan M, Soltan M, Jarmołowicz S, Abdo H (2018) Combined effects of dietary malic acid and *Bacillus subtilis* on growth, gut microbiota and blood parameters of N ile tilapia (*Oreochromis niloticus*). Aquac Nutr 24(1):83–93. https://doi.org/10.1111/anu.12536

Hassaan M, Wafa M, Soltan M, Goda A, Mogheth N (2014) Effect of dietary organic salts on growth, nutrient digestibility, mineral absorption and some biochemical indices of Nile tilapia; *Oreochromis niloticus* L. fingerlings. World Appl Sci J 29(1):47–55. https://doi.org/10.5829/idosi.wasj.2014.29.01.81237

He S, Zhang Y, Xu L, Yang Y, Marubashi T, Zhou Z, Yao B (2013) Effects of dietary *Bacillus subtilis* C-3102 on the production, intestinal cytokine expression and autochthonous bacteria of hybrid tilapia *Oreochromis niloticus*♀ × *Oreochromis aureus*♂. Aquaculture 412:125–130. https://doi.org/10.1016/j.aquaculture.2013.06.028

He S, Zhou Z, Liu Y, Shi P, Yao B, Ringø E, Yoon I (2009) Effects of dietary *Saccharomyces cerevisiae* fermentation product (DVAQUA®) on growth performance, intestinal autochthonous bacterial community and non-specific immunity of hybrid tilapia (*Oreochromis niloticus*♀ × *O. aureus*♂) cultured in cages. Aquaculture 294(1-2):99–107. https://doi.org/10.1016/j.aquaculture.2009.04.043

Hoseinifar SH, Ringø E, Shenavar Masouleh A, Esteban MÁ (2016) Probiotic, prebiotic and synbiotic supplements in sturgeon aquaculture: a review. Rev Aquac 8(1):89–102. https://doi.org/10.1111/raq.12082

Hu C-J, Chen S-M, Pan C-H, Huang C-H (2006) Effects of dietary vitamin A or β-carotene concentrations on growth of juvenile hybrid tilapia, *Oreochromis niloticus* × *O. aureus*. Aquaculture 253(1-4):602–607. https://doi.org/10.1016/j.aquaculture.2005.09.003

Huang C-H, Chang R-J, Huang S-L, Chen W (2003) Dietary vitamin E supplementation affects tissue lipid peroxidation of hybrid tilapia, *Oreochromis niloticus* ×*O. aureus*. Comp Biochem Physiol Part B Biochem Mol Biol 134(2):265–270. https://doi.org/10.1016/s1096-4959(02)00256-7

Huang C-H, Huang M-C, Hou P-C (1998) Effect of dietary lipids on fatty acid composition and lipid peroxidation in sarcoplasmic reticulum of hybrid tilapia, *Oreochromis niloticus* × *O. aureus*. Comparative Biochemistry Physiology Part B: Biochemistry. Mol Biol 120(2): 331–336. https://doi.org/10.1016/S0305-0491(98)10022-6

Ibrahem MD, Fathi M, Mesalhy S, Abd El-Aty A (2010) Effect of dietary supplementation of inulin and vitamin C on the growth, hematology, innate immunity, and resistance of Nile tilapia (*Oreochromis niloticus*). Fish Shellfish Immunol 29(2):241–246. https://doi.org/10.1016/j.fsi.2010.03.004

Iwashita MKP, Nakandakare IB, Terhune JS, Wood T, Ranzani-Paiva MJT (2015) Dietary supplementation with *Bacillus subtilis*, *Saccharomyces cerevisiae* and *Aspergillus oryzae* enhance immunity and disease resistance against *Aeromonas hydrophila* and *Streptococcus iniae* infection in juvenile tilapia *Oreochromis niloticus*. Fish Shellfish Immunol 43(1):60–66. https://doi.org/10.1016/j.fsi.2014.12.008

Jauncey K (1983) The quantitative essential amino acid requirements of *Oreochromis* (*Sarotherodon*) *mossambicus*. In: International Symposium on Tilapia in Aquaculture, 1st Proceedings, 8-13 May, Nazareth, Israel, 8-13 May. Tel Aviv University, pp 328–337

Jauncey K (2000) Nutritional requirments. In: Beveridge MCM, McAndrew BJ (eds) Tilapias: biology and exploitation. Fish and Fisheries Series, vol 25. Springer, Dordreht, Netherlands, pp 327–375. https://doi.org/10.1007/978-94-011-4008-9_9

Jeney G, De Wet L, Jeney Z, Yin G (2015) Plant extracts. In: Lee CS, Lim C, Gatlin DM III, Webster CD (eds) Dietary nutrients, additives, and fish health. Wiley-Blackwell, Oxford, UK, pp 321–332. https://doi.org/10.1002/9781119005568.ch16

Kanazawa A, Teshima S, Sakamoto M, Awal A (1980) Requirements of *Tilapia zillii* for essential fatty acids. Bull Japan Soc Sci Fish 46(11):1353–1356. https://doi.org/10.2331/suisan.46.1353

Kasper CS, White MR, Brown PB (2000) Choline is required by tilapia when methionine is not in excess. J Nutr 130(2):238–242. https://doi.org/10.1093/jn/130.2.238

Kuebutornye FK, Wang Z, Lu Y, Abarike ED, Sakyi ME, Li Y, Xie CX, Hlordzi V (2020) Effects of three host-associated *Bacillus* species on mucosal immunity and gut health of Nile tilapia, *Oreochromis niloticus* and its resistance against *Aeromonas hydrophila* infection. Fish Shellfish Immunol 97:83–95. https://doi.org/10.1016/j.fsi.2019.12.046

Lauzon HL, Dimitroglou A, Merrifield DL, Ringø E, Davies SJ (2014) Probiotics and prebiotics: concepts, definitions and history. In: Merrifield D, Ringø E (eds) Aquaculture nutrition: gut health, probiotics prebiotics. Wiley-Blackwell, Oxford, UK, pp 169–184. https://doi.org/10.1002/9781118897263.ch7

Lecerf J-M (2009) Fatty acids and cardiovascular disease. Nutr Rev 67(5):273–283. https://doi.org/10.1111/j.1753-4887.2009.00194.x

Lee J (2003) Vitamin K requirements of juvenile hybrid tilapia (*Oreochromis niloticus* × *O. aureus*) and grouper (*Epinephelus malabaricus*). Master Thesis, National Taiwan Ocean University, Keelung, Taiwan

Li K, Liu L, Clausen JH, Lu M, Dalsgaard A (2016) Management measures to control diseases reported by tilapia (*Oreochromis* spp.) and whiteleg shrimp (*Litopenaeus vannamei*) farmers in Guangdong, China. Aquaculture 457:91–99. https://doi.org/10.1016/J.AQUACULTURE.2016.02.008

Li Z, Ye W, He X (1991) The nutritional value of commercial feed ingredients for Nile tilapia (*Oreochromis niloticus*) in China. In: Fish nutrition research in Asia. Proceedings of the fourth Asian Fish Nutrition Workshop, Manila, Philippines. Asian Fish Soc Spec Publ, Asian Fish Soc, pp 101–106

Liao D, Chen C-W (2001) Socio-economics and technical efficiency of tilapia production in Taiwan. 25-30 November. In: 6th Asian Fisheries Forum, p. 147, National Sun Yat-Sen University, Kaohsiung, Taiwan. Asian Fish Soc, p 365

Lim C, Barros MM, Klesius PH, Shoemaker CA (2000) Thiamin requirement of Nile tilapia, *Oreochromis niloticus*. Page 201 in book of abstracts. Paper presented at the Aquaculture America 2000, 2-5 February, New Orleans, Louisiana,

Lim C, Klesius P (2001) Influence of dietary levels of folic acid on growth response and resistance of Nile tilapia, *Oreochromis niloticus* to *Streptococcus iniae*. Page 150 in book of abstracts. In: 6th Asian Fisheries Forum. 25-30 November, kaohsiung, Taiwan. Asian Fish Soc

Lim C, Leamaster B, Brock J (1991) Thiamin requirement of red hybrid tilapia grown in seawater. Page 39 in programs and abstracts. In: 22nd Annual Conference & Exposition, 16-20 June, San Juan, Puert Rico. Baton Rouge, Louisiana: World Aquac Soc

Lim C, Leamaster B, Brock JA (1993) Riboflavin requirement of fingerling red hybrid tilapia grown in seawater. J World Aquac Soc 24(4):451–458. https://doi.org/10.1111/j.1749-7345.1993.tb00573.x

Lim C, LeaMaster BR, Brock JA (1995) Pyridoxine requirement of fingerling red hybrid tilapia grown in seawater. J Appl Aquac 5(2):49–60. https://doi.org/10.1300/J028v05n02_05

Lim C, Lückstädt C, Webster CD, Kesius P (2015) Organic acids and their salts. In: Lee CS, Lim C, Gatlin DM III, Webster CD (eds) Dietary nutrients, additives, and fish health. Willey-Blackwell, Oxford, UK, pp 305–319. https://doi.org/10.1002/9781119005568.ch15

Lim C, Yildirim-Aksoy M, Barros MM, Klesius P (2011b) Thiamin requirement of Nile tilapia, *Oreochromis niloticus*. J World Aquac Soc 42(6):824–833. https://doi.org/10.1111/j.1749-7345.2011.00531.x

Lim C, Yildirim-Aksoy M, Klesius P (2011a) Lipid and fatty acid requirements of tilapias. N Am J Aquac 73(2):188–193. https://doi.org/10.1080/15222055.2011.579032

Lim CE, Webster CD (2006) Nutrient requirements. In: Lim CE, Webster CD (eds) Tilapia: biology, culture, and nutrition. CRC Press, New York, US, pp 469–501. https://doi.org/10.1300/5513_12

Liu W, Ren P, He S, Xu L, Yang Y, Gu Z, Zhou Z (2013) Comparison of adhesive gut bacteria composition, immunity, and disease resistance in juvenile hybrid tilapia fed two different *Lactobacillus* strains. Fis Shellfish Immunol 35(1):54–62. https://doi.org/10.1016/j.fsi.2013.04.010

Liu W, Wang W, Ran C, He S, Yang Y, Zhou Z (2017) Effects of dietary scFOS and lactobacilli on survival, growth, and disease resistance of hybrid tilapia. 470:50–55. https://doi.org/10.1016/j.aquaculture.2016.12.013

Luo Z, Tan XY, Liu CX, Li XD, Liu XJ, Xi WQ (2012) Effect of dietary conjugated linoleic acid levels on growth performance, muscle fatty acid profile, hepatic intermediary metabolism and antioxidant responses in genetically improved farmed tilapia strain of Nile tilapia, *Oreochromis niloticus*. Aquac Res 43(9):1392–1403. https://doi.org/10.1111/j.1365-2109.2011.02942.x

Mair GC (2002) Tilapia genetics and breeding in Asia. In: Guerrero RI, Guerrero-del Castillo M (eds) Tilapia Farming in the 21st Century, Proceedings of the International Forum on Tilapia Farming in the 21st Century, 25-27 February, Los Baños, Laguna Philippines. Philippine Fish Assoc Inc., pp 100–123

Makwinja R, Geremew A (2020) Roles and requirements of trace elements in tilapia nutrition. Egypt J Aquac Res 46(3):281–287. https://doi.org/10.1016/j.ejar.2020.05.001

Martínez P, Viñas AM, Sánchez L, Díaz N, Ribas L, Piferrer F (2014) Genetic architecture of sex determination in fish: applications to sex ratio control in aquaculture. Front Genet 5:340. https://doi.org/10.3389/fgene.2014.00340

Merrifield DL, Dimitroglou A, Foey A, Davies SJ, Baker RT, Bøgwald J, Castex M, Ringø E (2010) The current status and future focus of probiotic and prebiotic applications for salmonids. Aquaculture 302(1-2):1–18. https://doi.org/10.1016/j.aquaculture.2010.02.007

Merrifield DL, Harper GM, Mustafa S, Carnevali O, Picchietti S, Davies SJ (2011) Effect of dietary alginic acid on juvenile tilapia (*Oreochromis niloticus*) intestinal microbial balance, intestinal histology and growth performance. 344(1):135–146. https://doi.org/10.1007/s00441-010-1125-y

Metwally M (2009) Effects of garlic (*Allium sativum*) on some antioxidant activities in tilapia nilotica (*Oreochromis niloticus*). World J Fish Marine Sci 1(1):56–64

Metzker ML (2010) Sequencing technologies—the next generation. Nat Rev Genet 11(1):31–46. https://doi.org/10.1038/nrg2626

Mjoun K, Rosentrater K, Brown ML (2010) Tilapia: environmental biology and nutritional requirements. South Dakota Cooperetive extension service. Fact Sheets . Paper 164. http://openprairie.sdstate.edu/extension_fact/164

Montoya-Camacho N, Marquez-Ríos E, Castillo-Yáñez FJ, Cárdenas López JL, López-Elías JA, Ruíz-Cruz S, Jiménez-Ruíz EI, Rivas-Vega ME, Ocaño-Higuera VM (2019) Advances in the use of alternative protein sources for tilapia feeding. Rev Aquac 11(3):515–526. https://doi.org/10.1111/raq.12243

Nakano S-I, Yahara T, Nakashizuka T (2016) Aquatic biodiversity conservation and ecosystem services. doi:https://doi.org/10.1007/978-981-10-0780-4

Ng W (2005) Lipid nutrition of farmed tilapia. Glob Aquacult Advocate, October: 60-61

Ng W-K, Chong C-Y An overview of lipid nutrition with emphasis on alternative lipid sources in tilapia feeds. In: Proceedings of the Sixth International Symposium on Tilapia in Aquaculture, Manila, Philippines, Sept. 12–16, 2004. Citeseer, pp 241-248

Ng W-K, Gibon V (2010) Palm oil and saturated fatty acid-rich vegetable oils. In: Turchini GM, Ng W-K, Tocher DR (eds) Fish oil replacement alternative lipid sources in aquaculture feeds

Ng WK, Hanim R (2007) Performance of genetically improved Nile tilapia compared with red hybrid tilapia fed diets containing two protein levels. Aquac Res 38(9):965–972. https://doi.org/10.1111/j.1365-2109.2007.01758.x

Ng WK, Koh CB, Din Z (2006) Palm oil-laden spent bleaching clay as a substitute for marine fish oil in the diets of Nile tilapia, *Oreochromis niloticus*. Aquac Nutr 12(6):459–468. https://doi.org/10.1111/j.1365-2095.2006.00449.x

Ng WK, Koh CB, Sudesh K, Siti-Zahrah A (2009) Effects of dietary organic acids on growth, nutrient digestibility and gut microflora of red hybrid tilapia, *Oreochromis* sp., and subsequent survival during a challenge test with *Streptococcus agalactiae*. Aquac Res 40(13):1490–1500. https://doi.org/10.1111/j.1365-2109.2009.02249.x

Ng W-K, Lim P-K, Sidek H (2001) The influence of a dietary lipid source on growth, muscle fatty acid composition and erythrocyte osmotic fragility of hybrid tilapia. Fish Physiol Biochem 25(4):301–310. https://doi.org/10.1023/A:1023271901111

Ng WK, Romano N (2013) A review of the nutrition and feeding management of farmed tilapia throughout the culture cycle. Rev Aquac 5(4):220–254. https://doi.org/10.1111/raq.12014

Nguyen NH, Ponzoni RW, Abu-Bakar KR, Hamzah A, Khaw HL, Yee HY (2010) Correlated response in fillet weight and yield to selection for increased harvest weight in genetically improved farmed tilapia (GIFT strain), *Oreochromis niloticus*. 305(1-4):1–5. https://doi.org/10.1016/j.aquaculture.2010.04.007

Nguyen TN, Davis AD (2009) Methionine requirement in practical diets of juvenile Nile tilapia, *Oreochromis niloticus*. J World Aquac Soc 40(3):410–416. https://doi.org/10.1111/j.1749-7345.2009.00261.x

Ninh NH, Thoa NP, Knibb W, Nguyen NH (2014) Selection for enhanced growth performance of Nile tilapia (*Oreochromis niloticus*) in brackish water (15–20 ppt) in Vietnam. Aquaculture 428:1–6. https://doi.org/10.1016/j.aquaculture.2014.02.024

NRC (1993) Nutrient requirements of fish. National Academies Press, Washington D.C., p 114

O'Connell JP, Gatlin DM (1994) Effects of dietary calcium and vitamin D3 on weight gain and mineral composition of the blue tilapia (*Oreochromis aureus*) in low-calcium water. Aquaculture 125(1-2):107–117. https://doi.org/10.1016/0044-8486(94)90287-9

Ogunmefun S, Achike A (2017) Socioeconomic characteristics and constraints of pond fish farmers in Lagos State, Nigeria. Agric Sci Res J 7(10):304–317

Olaoye O, Ashley-Dejo S, Fakoya E, Ikeweinwe N, Alegbeleye W, Ashaolu F, Adelaja O (2013) Assessment of socio-economic analysis of fish farming in Oyo State, Nigeria. GJSFRAVet 13(9):45–55

Opiyo MA, Jumbe J, Ngugi CC, Charo-Karisa H (2019) Dietary administration of probiotics modulates non-specific immunity and gut microbiota of Nile tilapia (*Oreochromis niloticus*) cultured in low input ponds. Int J Vet Sci Med 7(1):1–9. https://doi.org/10.1080/23144599.2019.1624299

Orachunwong C, Thammasart S, Lohawatanakul C 2001 Recent developments in tilapia feeds. In: Recent developments in tilapia feeds. In: Subasinghe S & Singh T (eds) Tilapia: production, marketing and technical developments. Proceedings of the Tilapia 2001 International Technical and Trade Conference on Tilapia, Infofish, Kuala Lumpur, Malaysia. pp. 113–122

Pachanawan A, Phumkhachorn P, Rattanachaikunsopon P (2008) Potential of *Psidium guajava* supplemented fish diets in controlling *Aeromonas hydrophila* infection in tilapia (*Oreochromis niloticus*). J Biosci Bioeng 106(5):419–424. https://doi.org/10.1263/jbb.106.419

Peres H, Lim C, Klesius PH (2004) Growth, chemical composition and resistance to *Streptococcus iniae* challenge of juvenile Nile tilapia (*Oreochromis niloticus*) fed graded levels of dietary inositol. Aquaculture 235(1-4):423–432. https://doi.org/10.1016/j.aquaculture.2003.09.021

Pinho SM, Molinari D, de Mello GL, Fitzsimmons KM, Emerenciano MGC (2017) Effluent from a biofloc technology (BFT) tilapia culture on the aquaponics production of different lettuce varieties. Ecol Eng 103:146–153. https://doi.org/10.1016/j.ecoleng.2017.03.009

Pirarat N, Kobayashi T, Katagiri T, Maita M, Endo M (2006) Protective effects and mechanisms of a probiotic bacterium *Lactobacillus rhamnosus* against experimental *Edwardsiella tarda* infection in tilapia (*Oreochromis niloticus*). Vet Immunol Immunopathol 113(3-4):339–347. https://doi.org/10.1016/j.vetimm.2006.06.003

Pirarat N, Pinpimai K, Endo M, Katagiri T, Ponpornpisit A, Chansue N, Maita M (2011) Modulation of intestinal morphology and immunity in nile tilapia (*Oreochromis niloticus*) by *Lactobacillus rhamnosus* GG. Res Vet Sci 91(3):e92–e97. https://doi.org/10.1016/j.rvsc.2011.02.014

Plongbunjong V, Phromkuntong W, Suanyuk N, Viriyapongsutee B, Wichienchot S (2011) Effects of prebiotics on growth performance and pathogenic inhibition in sex-reversed red tilapia (*Oreochromis niloticus* × *Oreochromis mossambicus*). Thai J Agric Sci 44(5):162–167

Ponzoni RW, Nguyen NH, Khaw HL, Hamzah A, Bakar KRA, Yee HY (2011) Genetic improvement of Nile tilapia (*Oreochromis niloticus*) with special reference to the work conducted by the WorldFish Center with the GIFT strain. Rev Aquac 3(1):27–41. https://doi.org/10.1111/j.1753-5131.2010.01041.x

Prabu E, Rajagopalsamy CBT, Ahilan B, Jeevagan IJMA, Renuhadevi M (2019) Tilapia–an excellent candidate species for world aquaculture: a review. Ann Res Rev Biol 33(3):1–14. https://doi.org/10.9734/ARRB/2019/v31i330052

Qin C, Zhang Y, Liu W, Xu L, Yang Y, Zhou Z (2014) Effects of chito-oligosaccharides supplementation on growth performance, intestinal cytokine expression, autochthonous gut bacteria and disease resistance in hybrid tilapia *Oreochromis niloticus*♀ × *Oreochromis aureus*♂. Fish Shellfish Immunol 40(1):267–274. https://doi.org/10.1016/j.fsi.2014.07.010

Ramos M, Batista S, Pires M, Silva A, Pereira L, Saavedra M, Ozório R, Rema P (2017) Dietary probiotic supplementation improves growth and the intestinal morphology of Nile tilapia, *Oreochromis niloticus*. Anim Sci 11(8):1259–1269. https://doi.org/10.1017/S1751731116002792

Rawling MD, Merrifield DL, Davies SJ (2009) Preliminary assessment of dietary supplementation of Sangrovit® on red tilapia (*Oreochromis niloticus*) growth performance and health. Aquaculture 294(1-2):118–122. https://doi.org/10.1016/j.aquaculture.2009.05.005

Reigh RC, Robinson EH, Brown PB (1991) Effects of dietary magnesium on growth and tissue magnesium content of blue tilapia *Oreochromis aureus*. J World Aquac Soc 22(3):192–200. https://doi.org/10.1111/j.1749-7345.1991.tb00734.x

Ren P, Xu L, Yang Y, He S, Liu W, Ringø E, Zhou Z (2013) *Lactobacillus planarum* subsp. *plantarum* JCM 1149 vs. *Aeromonas hydrophila* NJ-1 in the anterior intestine and posterior intestine of hybrid tilapia *Oreochromis niloticus*♀ × *Oreochromis aureus*♂: An *ex vivo* study. Fish Shellfish Immunol 35(1):146–153. https://doi.org/10.1016/j.fsi.2013.04.023

Ridha MT, Cruz E (2002) Evaluation of growth, feed conversion, salinity tolerance and survival of an improved strain of the Nile tilapia (*Oreochromis niloticus*). Kuwait Inst Sci Res, Kuwait

Ringø E, Dimitroglou A, Hoseinifar SH, Davies SJ (2014) Prebiotics in finfish: an update. In: Merrifield DL, Ringø E (eds) Aquaculture nutrition: gut health, probiotics and prebiotics. Wiley-Blackwell, Oxford, UK, pp 360–400. https://doi.org/10.1002/9781118897263.ch14

Robinson EH, LaBomascus D, Brown PB, Linton TL (1987) Dietary calcium and phosphorus requirements of *Oreochromis aureus* reared in calcium-free water. Aquaculture 64(4):267–276. https://doi.org/10.1016/0044-8486(87)90189-X

Ross L (2000) Environmental physiology and energetics. In: Beveridge MCM, McAndrew BJ (eds) Tilapias: biology and exploitation. Fish Fish, vol 25. Springer, Dordrecht, Netherlands., pp 89–128. https://doi.org/10.1007/978-94-011-4008-9_4

Russo GL (2009) Dietary n−6 and n−3 polyunsaturated fatty acids: from biochemistry to clinical implications in cardiovascular prevention. Biochem Pharmacol 77(6):937–946. https://doi.org/10.1016/j.bcp.2008.10.020

Ruxton C, Reed SC, Simpson M, Millington K (2004) The health benefits of omega-3 polyunsaturated fatty acids: a review of the evidence. J Hum Nutr Diet 17(5):449–459. https://doi.org/10.1111/j.1365-277X.2004.00552.x

Şahan A, Özütok S, Kurutaş EB (2016) Determination of some hematological parameters and antioxidant capacity in Nile tilapia (*Oreochromis niloticus* Linnaeus, 1758) fed ginger (*Zingiber officinale Roscoe*) to *Aeromonas hydrophila*. Turkish J Fish Aquat Sci 16(1):197–204. https://doi.org/10.4194/1303-2712-v16_1_20

Saleh G, Eleraky W, Gropp J (1995) A short note on the effects of vitamin A hypervitaminosis and hypovitaminosis on health and growth of Tilapia nilotica (*Oreochromis niloticus*). J Appl Ichthyol 11(3-4):382–385. https://doi.org/10.1111/j.1439-0426.1995.tb00046.x

Sampantamit T, Ho L, Lachat C, Sutummawong N, Sorgeloos P, Goethals P (2020) Aquaculture production and its environmental sustainability in Thailand: challenges and potential solutions. Sustainability 12(5):2010. https://doi.org/10.3390/su12052010

Santiago CB, Lovell RT (1988) Amino acid requirements for growth of Nile tilapia. J Nutr 118(12): 1540–1546. https://doi.org/10.1093/jn/118.12.1540

Shalaby A, Khattab Y, Abdel Rahman A (2006) Effects of Garlic (*Allium sativum*) and chloramphenicol on growth performance, physiological parameters and survival of Nile tilapia (*Oreochromis niloticus*). J Venom Anim Toxins Incl Trop Dis 12(2):172–201. https://doi.org/10.1590/S1678-91992006000200003

Shelby RA, Lim C, Yildirim-Aksoy M, Welker TL, Klesius PH (2009) Effects of yeast oligosaccharide diet supplements on growth and disease resistance in juvenile Nile tilapia, *Oreochromis niloticus*. J Appl Aquac 21(1):61–71. https://doi.org/10.1080/10454430802694728

Shelton WL, Popma TJ (2006) Biology. In: Lim CE, Webster CD (eds) Tilapia, biology, culture and nutrition. CRC, New York, US, pp 1–49. https://doi.org/10.1300/5513_01

Shiau S, Shiau L (2001) Re-evaluation of the vitamin E requirements of juvenile tilapia (*Oreochromis niloticus* × *O. aureus*). Anim Sci 72(3):529–534. https://doi.org/10.1017/S135772980005205X

Shiau S-Y, Chin Y-H (1999) Estimation of the dietary biotin requirement of juvenile hybrid tilapia, *Oreochromis niloticus* × *O. aureus*. Aquaculture 170(1):71–78. https://doi.org/10.1016/S0044-8486(98)00391-3

Shiau S-Y, Chuang J-C (1995) Utilization of disaccharides by juvenile tilapia, *Oreochromis niloticus* x *O. aureus*. Aquaculture 133(3-4):249–256. https://doi.org/10.1016/0044-8486(95)00018-W

Shiau S-Y, Hsieh H-L (1997) Vitamin B6 requirements of tilapia *Oreochromis niloticus* x *O. aureus* fed two dietary protein concentrations. Fish Sci 63(6):1002–1007. https://doi.org/10.2331/fishsci.63.1002

Shiau S-Y, Hsieh J-F (2001) Quantifying the dietary potassium requirement of juvenile hybrid tilapia (*Oreochromis niloticus* × *O. aureus*). Br J Nutr 85(2):213–218. https://doi.org/10.1079/BJN2000245

Shiau S-Y, Hsu T-S (1999) Quantification of vitamin C requirement for juvenile hybrid tilapia, *Oreochromis niloticus* × *Oreochromis aureus*, with L-ascorbyl-2-monophosphate-Na and L-ascorbyl-2-monophosphate-Mg. Aquaculture 175(3-4):317–326. https://doi.org/10.1016/S0044-8486(99)00103-9

Shiau S-Y, Huang S-Y (2001) Dietary folic acid requirement for maximum growth of juvenile tilapia *Oreochromis niloticus* × *O. aureus*. Fish Sci 67(4):655–659. https://doi.org/10.1046/j.1444-2906.2001.00302.x

Shiau S-Y, Hwang J-Y (1993) Vitamin D requirements of juvenile hybrid Tilapia *Oreochromis niloticus* × *O. aureus*. Nippon Suisan Gakkaishi 59(3):553–558. https://doi.org/10.2331/suisan.59.553

Shiau S-Y, Lei M-S (1999) Feeding strategy does affect carbohydrate utilization by hybrid tilapia *Oreochromis niloticus* × *O. aureus*. Fish Sci 65(4):553–557. https://doi.org/10.2331/fishsci.65.553

Shiau S-Y, Lo P-S (2000) Dietary choline requirements of juvenile hybrid tilapia, *Oreochromis niloticus* × *O. aureus*. J Nutr 130(1):100–103. https://doi.org/10.1093/jn/130.1.100

Shiau S-Y, Lu L-S (2004) Dietary sodium requirement determined for juvenile hybrid tilapia (*Oreochromis niloticus* × *O. aureus*) reared in fresh water and seawater. BJN 91(4):585–590. https://doi.org/10.1079/BJN20041091

Shiau S-Y, Lung C-Q (1993) No dietary vitamin B12 required for juvenile tilapia *Oreochromis niloticus* × *O. aureus*. Comp Biochem Physiol Part A Physiol 105(1):147–150. https://doi.org/10.1016/0300-9629(93)90187-9

Shiau S-Y, Shy S-M (1998) Dietary chromic oxide inclusion level required to maximize glucose utilization in hybrid tilapia, *Oreochromis niloticus* × *O. aureus*. Aquaculture 161(1-4):357–364. https://doi.org/10.1016/S0044-8486(97)00283-4

Shiau S-Y, Su L-W (2003) Ferric citrate is half as effective as ferrous sulfate in meeting the iron requirement of juvenile tilapia, *Oreochromis niloticus* × *O. aureus*. J N 133(2):483–488. https://doi.org/10.1093/jn/133.2.483

Shiau S-Y, Su S-L (2005) Juvenile tilapia (*Oreochromis niloticus* × *Oreochromis aureus*) requires dietary myo-inositol for maximal growth. Aquaculture 243(1-4):273–277. https://doi.org/10.1016/j.aquaculture.2004.10.002

Shiau S-Y, Suen G-S (1992) Estimation of the niacin requirements for tilapia fed diets containing glucose or dextrin. J Nutr 122(10):2030–2036. https://doi.org/10.1093/jn/122.10.2030

Shiau SY, Tseng HC (2007) Dietary calcium requirements of juvenile tilapia, *Oreochromis niloticus* × *O. aureus*, reared in fresh water. Aquac Nutr 13(4):298–303. https://doi.org/10.1111/j.1365-2095.2007.00481.x

Shiau S-Y, Yu Y-P (1999) Dietary supplementation of chitin and chitosan depresses growth in tilapia, *Oreochromis niloticus* × *O. aureus*. Aquaculture 179(1-4):439–446. https://doi.org/10.1016/S0044-8486(99)00177-5

Shimeno S, Ming D, Takeda M (1993) Metabolic response to dietary carbohydrate to lipid ratios in *Oreochromis niloticus*. Bull Japan Soc Sci Fish 59:827–833

Soliman AK, Wilson RP (1992a) Water-soluble vitamin requirements of tilapia. 1 Pantothenic acid requirement of blue tilapia, *Oreochromis aureus*. Aquaculture 104(1-2):121–126. https://doi.org/10.1016/0044-8486(92)90142-8

Soliman AK, Wilson RP (1992b) Water-soluble vitamin requirements of tilapia. 2. Riboflavin requirement of blue tilapia, *Oreochromis aureus*. Aquaculture 104(3-4):309–314. https://doi.org/10.1016/0044-8486(92)90212-4

Sonesson A (2011) Genomic selection for aquaculture: principles and procedures. In: Liu ZJ (ed) Next generation sequencing and whole genome selection in aquaculture. Blackwell Publishing, Des Moines, IA, pp 151–164

de Souza EM, de Souza RC, Melo JF, da Costa MM, de Souza AM, Copatti CE (2019) Evaluation of the effects of *Ocimum basilicum* essential oil in Nile tilapia diet: growth, biochemical, intestinal enzymes, haematology, lysozyme and antimicrobial challenges. Aquaculture 504:7–12. https://doi.org/10.1016/j.aquaculture.2019.01.052

Standen B, Peggs D, Rawling M, Foey A, Davies S, Santos G, Merrifield D (2016) Dietary administration of a commercial mixed-species probiotic improves growth performance and modulates the intestinal immunity of tilapia, *Oreochromis niloticus*. Fish Shellfish Immunol 49:427–435. https://doi.org/10.1016/j.fsi.2015.11.037

Standen B, Rawling M, Davies S, Castex M, Foey A, Gioacchini G, Carnevali O, Merrifield D (2013) Probiotic *Pediococcus acidilactici* modulates both localised intestinal-and peripheral-immunity in tilapia (*Oreochromis niloticus*). Fish Shellfish Immunol 35(4):1097–1104. https://doi.org/10.1016/j.fsi.2013.07.018

Stickney R, McGeachin R Responses of *Tilapia aurea* to semipurified diets of differing fatty acid composition. In: Proceedings of the International Symposium on Tilapia in Aquaculture, Nazareth, Israel, 1983. Tel Aviv University Press, pp 346-355

Sweilum MA, Abdella MM, Salah El-Din SA (2005) Effect of dietary protein-energy levels and fish initial sizes on growth rate, development and production of Nile tilapia, *Oreochromis niloticus* L. Aquac Res 36(14):1414–1421. https://doi.org/10.1111/j.1365-2109.2005.01362.x

Tacon AG (1987) The nutrition and feeding of farmed fish and shrimp a training manual: Part 1. The essential nutrients. FAO, Brasilia, Brazil

Tadesse Z, Boberg M, Sonesten L, Ahlgren G (2003) Effects of algal diets and temperature on the growth and fatty acid content of the cichlid fish *Oreochromis niloticus* L.–A laboratory study. Aquat Ecol 37(2):169–182. https://doi.org/10.1023/A:1023942711822

Takeuchi T, Satoh S, Watanabe T (1983) Dietary lipids suitable for the practical feed of *Tilapia nilotica*. Bull Japan Soc Sci Fish 49:1361–1365

Tan HY, Chen S-W, Hu S-Y (2019) Improvements in the growth performance, immunity, disease resistance, and gut microbiota by the probiotic *Rummeliibacillus stabekisii* in Nile tilapia (*Oreochromis niloticus*). Fish Shellfish Immunol 92:265–275. https://doi.org/10.1016/j.fsi.2019.06.027

Taoka Y, Maeda H, Jo J-Y, Kim S-M, Park S-I, Yoshikawa T, Sakata T (2006) Use of live and dead probiotic cells in tilapia *Oreochromis niloticus*. Fish Sci 72(4):755–766. https://doi.org/10.1111/j.1444-2906.2006.01215.x

Teoh C-Y, Turchini GM, Ng W-K (2011) Genetically improved farmed Nile tilapia and red hybrid tilapia showed differences in fatty acid metabolism when fed diets with added fish oil or a vegetable oil blend. Aquaculture 312(1-4):126–136. https://doi.org/10.1016/j.aquaculture.2010.12.018

Teshima SI, Kanazawa A, Sakamoto M (1982) Essential fatty acids of *Tilapia nilotica*. Mem FacFish Kagoshima Univ 31:201–204

Tessema M, Müller-Belecke A, Hörstgen-Schwark G (2006) Effect of rearing temperatures on the sex ratios of *Oreochromis niloticus* populations. Aquaculture 258(1-4):270–277. https://doi.org/10.1016/j.aquaculture.2006.04.041

Tiengtam N, Khempaka S, Paengkoum P, Boonanuntanasarn S (2015) Effects of inulin and Jerusalem artichoke (*Helianthus tuberosus*) as prebiotic ingredients in the diet of juvenile Nile tilapia (*Oreochromis niloticus*). Anim Feed Sci Tech 207:120–129. https://doi.org/10.1016/j.anifeedsci.2015.05.008

Tonial I, Stevanato F, Matsushita M, De Souza N, Furuya W, Visentainer J (2009) Optimization of flaxseed oil feeding time length in adult Nile tilapia (*Oreochromis niloticus*) as a function of muscle omega-3 fatty acids composition. Aquac Nutr 15(6):564–568. https://doi.org/10.1111/j.1365-2095.2008.00623.x

Trushenski JT, Boesenberg J, Kohler CC (2009) Influence of grow-out feed fatty acid composition on finishing success in Nile tilapia. N Am J Aquac 71(3):242–251. https://doi.org/10.1577/A08-051.1

Tung P-H, Shiau S-Y (1991) Effects of meal frequency on growth performance of hybrid tilapia, *Oreochromis niloticus* × *O. aureus*, fed different carbohydrate diets. Aquaculture 92:343–350. https://doi.org/10.1016/0044-8486(91)90039-A

Tung P-H, Shiau S-Y (1993) Carbohydrate utilization versus body size in tilapia *Oreochromis niloticus* × *O. aureus*. Comp Biochem Physiol Part A Physiol 104(3):585–588. https://doi.org/10.1016/0300-9629(93)90468-J

Uddin MN, Kabir KH, Roy D, Hasan MT, Sarker MA, Dunn ES (2021) Understanding the constraints and its related factors in tilapia (*Oreochromis* sp.) fish culture at farm level: A case from Bangladesh. Aquaculture 530:735927. https://doi.org/10.1016/j.aquaculture.2020.735927

Van Doan H, Hoseinifar SH, Chitmanat C, Jaturasitha S, Paolucci M, Ashouri G, Dawood MA, Esteban MÁ (2019) The effects of Thai ginseng, *Boesenbergia rotunda* powder on mucosal and serum immunity, disease resistance, and growth performance of Nile tilapia (*Oreochromis niloticus*) fingerlings. Aquaculture 513:734388. https://doi.org/10.1016/j.aquaculture.2019.734388

Van Doan H, Hoseinifar SH, Dawood MA, Chitmanat C, Tayyamath K (2017a) Effects of Cordyceps militaris spent mushroom substrate and *Lactobacillus plantarum* on mucosal, serum immunology and growth performance of Nile tilapia (*Oreochromis niloticus*). Fish Shellfish Immunol 70:87–94. https://doi.org/10.1016/j.fsi.2017.09.002

Van Doan H, Hoseinifar SH, Elumalai P, Tongsiri S, Chitmanat C, Jaturasitha S, Doolgindachbaporn S (2018a) Effects of orange peels derived pectin on innate immune response, disease resistance and growth performance of Nile tilapia (*Oreochromis niloticus*) cultured under indoor biofloc system. Fish Shellfish Immunol 80:56–62. https://doi.org/10.1016/j.fsi.2018.05.049

Van Doan H, Hoseinifar SH, Faggio C, Chitmanat C, Mai NT, Jaturasitha S, Ringø E (2018b) Effects of corncob derived xylooligosaccharide on innate immune response, disease resistance, and growth performance in Nile tilapia (*Oreochromis niloticus*) fingerlings. Aquaculture 495:786–793. https://doi.org/10.1016/j.aquaculture.2018.06.068

Van Doan H, Hoseinifar SH, Jaturasitha S, Dawood MA, Harikrishnan R (2020a) The effects of berberine powder supplementation on growth performance, skin mucus immune response, serum immunity, and disease resistance of Nile tilapia (*Oreochromis niloticus*) fingerlings. Aquaculture 520:734927. https://doi.org/10.1016/j.aquaculture.2020.734927

Van Doan H, Hoseinifar SH, Tapingkae W, Seel-Audom M, Jaturasitha S, Dawood MA, Wongmaneeprateep S, Thu TTN, Esteban MÁ (2020b) Boosted growth performance, mucosal and serum immunity, and disease resistance Nile tilapia (*Oreochromis niloticus*) fingerlings using corncob-derived xylooligosaccharide and *Lactobacillus plantarum* CR1T5. Probiotics Antimicrob Proteins 12(2):400–411. https://doi.org/10.1007/s12602-019-09554-5

Van Doan H, Hoseinifar SH, Khanongnuch C, Kanpiengjai A, Unban K, Srichaiyo S (2018c) Host-associated probiotics boosted mucosal and serum immunity, disease resistance and growth performance of Nile tilapia (*Oreochromis niloticus*). Aquaculture 491:94–100. https://doi.org/10.1016/j.aquaculture.2018.03.019

Van Doan H, Hoseinifar SH, Tapingkae W, Khamtavee P (2017b) The effects of dietary kefir and low molecular weight sodium alginate on serum immune parameters, resistance against *Streptococcus agalactiae* and growth performance in Nile tilapia (*Oreochromis niloticus*). Fish Shellfish Immunol 62:139–146. https://doi.org/10.1016/j.fsi.2017.01.014

Van Doan H, Hoseinifar SH, Tapingkae W, Tongsiri S, Khamtavee P (2016a) Combined administration of low molecular weight sodium alginate boosted immunomodulatory, disease resistance and growth enhancing effects of *Lactobacillus plantarum* in Nile tilapia (*Oreochromis niloticus*). Fish Shellfish Immunol 58:678–685. https://doi.org/10.1016/j.fsi.2016.10.013

Van Doan H, Tapingkae W, Moonmanee T, Seepai A (2016b) Effects of low molecular weight sodium alginate on growth performance, immunity, and disease resistance of tilapia, *Oreochromis niloticus*. Fish Shellfish Immunol 55:186–194. https://doi.org/10.1016/j.fsi.2016.05.034

Vechklang K, Lim C, Boonanuntanasarn S, Welker T, Ponchunchuwong S, Klesius PH, Wanapu C (2012) Growth performance and resistance to *Streptococcus iniae* of juvenile Nile tilapia (*Oreochromis niloticus*) fed diets supplemented with GroBiotic®-A and Brewtech dried brewers yeast. J Appl Aquac 24(3):183–198. https://doi.org/10.1080/10454438.2012.678786

Villamil L, Reyes C, Martínez-Silva M (2014) In vivo and in vitro assessment of *Lactobacillus acidophilus* as probiotic for tilapia (*Oreochromis niloticus*, Perciformes: Cichlidae) culture improvement. Aquac Res 45(7):1116–1125. https://doi.org/10.1111/are.12051

Visentainer JV, de Souza NE, Makoto M, Hayashi C, Franco MRB (2005) Influence of diets enriched with flaxseed oil on the α-linolenic, eicosapentaenoic and docosahexaenoic fatty acid in Nile tilapia (*Oreochromis niloticus*). Food Chem 90(4):557–560. https://doi.org/10.1016/j.foodchem.2004.05.016

Waiyamitra P, Zoral MA, Saengtienchai A, Luengnaruemitchai A, Decamp O, Gorgoglione B, Surachetpong W (2020) Probiotics modulate tilapia resistance and immune response against tilapia lake virus infection. Pathogens 9(11):919. https://doi.org/10.3390/pathogens9110919

Wang M, Lu M (2016) Tilapia polyculture: a global review. Aquac Res 47(8):2363–2374. https://doi.org/10.1111/are.12708

Wang Y-B, Tian Z-Q, Yao J-T, Li W-f (2008) Effect of probiotics, *Enteroccus faecium*, on tilapia (*Oreochromis niloticus*) growth performance and immune response. Aquaculture 277(3-4):203–207. https://doi.org/10.1016/j.aquaculture.2008.03.007

Watanabe T, Satoh S, Takeuchi T (1988) Availability of minerals in fish meal to fish. Asian Fish Sci 1(2):75–195

Watanabe WO, Losordo TM, Fitzsimmons K, Hanley F (2002) Tilapia production systems in the Americas: technological advances, trends, and challenges. Rev Fish Sci 10(3-4):465–498. https://doi.org/10.1080/20026491051758

Watanabe WO, Olla BL, Wicklund RI, Head WD (1997) Saltwater culture of the Florida red tilapia and other saline-tolerant tilapias: a review. Tilapia Aquac Am 1:54–141

Weaver KL, Ivester P, Chilton JA, Wilson MD, Pandey P, Chilton F (2008) The content of favorable and unfavorable polyunsaturated fatty acids found in commonly eaten fish. J Am Diet Assoc 108(7):1178–1185. https://doi.org/10.1016/j.jada.2008.04.023

Welker TL, Lim C (2011) Use of probiotics in diets of tilapia, Aquac Res Dev. S1::014. https://doi.org/10.4172/2155-9546.S1-014

Welker TL, Lim C, Yildirim-Aksoy M, Klesius PH (2012) Use of diet crossover to determine the effects of β-glucan supplementation on immunity and growth of Nile Tilapia, *Oreochromis niloticus*. J World Aquac Soc 43(3):335–348. https://doi.org/10.1111/j.1749-7345.2012.00569.x

Whittington R, Lim C, Klesius PH (2005) Effect of dietary β-glucan levels on the growth response and efficacy of *Streptococcus iniae* vaccine in Nile tilapia, *Oreochromis niloticus*. Aquaculture 248(1-4):217–225. https://doi.org/10.1016/j.aquaculture.2005.04.013

Wilson R (1994) Utilization of dietary carbohydrate by fish. Aquaculture 124(1-4):67–80. https://doi.org/10.1016/0044-8486(94)90363-8

Wilson RP (1991) Handbook of nutrient requirements of finfish. CRC Press, Florida, p 204

Xia Y, Lu M, Chen G, Cao J, Gao F, Wang M, Liu Z, Zhang D, Zhu H, Yi M (2018) Effects of dietary *Lactobacillus rhamnosus* JCM1136 and *Lactococcus lactis* subsp. *lactis* JCM5805 on the growth, intestinal microbiota, morphology, immune response and disease resistance of juvenile Nile tilapia, *Oreochromis niloticus*. Fish Shellfish Immunol 76:368–379. https://doi.org/10.1016/j.fsi.2018.03.020

Xia Y, Wang M, Gao F, Lu M, Chen G (2020) Effects of dietary probiotic supplementation on the growth, gut health and disease resistance of juvenile Nile tilapia (*Oreochromis niloticus*). Anim Nutr 6(1):69–79. https://doi.org/10.1016/j.aninu.2019.07.002

Yin G, Jeney G, Racz T, Xu P, Jun X, Jeney Z (2006) Effect of two Chinese herbs (*Astragalus radi* x and *Scutellaria radix*) on non-specific immune response of tilapia, *Oreochromis niloticus*. Aquaculture 253(1-4):39–47. https://doi.org/10.1016/j.aquaculture.2005.06.038

Young K (2009) Omega-6 (n-6) and omega-3 (n-3) fatty acids in tilapia and human health: a review. Int J Food Sci Nutr 60(sup5):203–211. https://doi.org/10.1080/09637480903140503

Yuan Y, Yuan Y, Dai Y, Gong Y (2017) Economic profitability of tilapia farming in China. Aquac Int 25(3):1253–1264. https://doi.org/10.1007/s10499-017-0111-8

Yue GH (2014) Recent advances of genome mapping and marker-assisted selection in aquaculture. Fish Fish 15(3):376–396. https://doi.org/10.1111/faf.12020

Yue GH, Lin HR, Li JL (2016) Tilapia is the fish for next-generation aquaculture. Int J Marine Sci Ocean Technol 3(1):11–13. https://doi.org/10.19070/2577-4395-160003

Yuji-Sado R, Raulino-Domanski F, de Freitas PF, Baioco-Sales F (2015) Growth, immune status and intestinal morphology of Nile tilapia fed dietary prebiotics (mannan oligosaccharides-MOS). Lat Am J Aquat Res 43(5):944–952. https://doi.org/10.3856/vol43-issue5-fulltext-14

Zahran E, Abd El-Gawad EA, Risha E (2018) Dietary Withania sominefera root confers protective and immunotherapeutic effects against *Aeromonas hydrophila* infection in Nile tilapia (*Oreochromis niloticus*). Fish Shellfish Immunol 80:641–650. https://doi.org/10.1016/j.fsi.2018.06.009

Zheng Z, Wang K, Gatlin DM III, Ye J (2011) Evaluation of the ability of GroBiotic®-A to enhance growth, muscle composition, immune responses, and resistance against *Aeromonas hydrophila* in Nile tilapia, *Oreochromis niloticus*. J World Aquac Soc 42(4):549–557. https://doi.org/10.1111/j.1749-7345.2011.00497.x

Zhou X, Tian Z, Wang Y, Li W (2010a) Effect of treatment with probiotics as water additives on tilapia (*Oreochromis niloticus*) growth performance and immune response. Fish physiol Biochem 36(3):501–509. https://doi.org/10.1007/s10695-009-9320-z

Zhou X, Wang Y, Yao J, Li W (2010b) Inhibition ability of probiotic, *Lactococcus lactis*, against *A. hydrophila* and study of its immunostimulatory effect in tilapia (*Oreochromis niloticus*). Int J Eng Sci Technol 2(7):73–80. https://doi.org/10.4314/IJEST.V2I7.63743

Zilberg D, Tal A, Froyman N, Abutbul S, Dudai N, Golan-Goldhirsh A (2010) Dried leaves of *Rosmarinus officinalis* as a treatment for streptococcosis in tilapia. J Fish Dis 33(4):361–369. https://doi.org/10.1111/j.1365-2761.2009.01129.x

Chapter 2
Health Management for Sustainable Tilapia Farming

Farid Firouzbakhsh, Seyed Hossein Hoseinifar, and Mehwish Faheem

Abstract Aquatic animals, especially fish, tend to react more quickly to changes in the environment than terrestrial farm species. In an aquatic environment, there is a profound and inverse relationship between environmental quality and the disease status of fish. Health maintenance is a concept in which animals are reared under conditions that optimize the growth rate, feed conversion efficiency, reproduction, and survival while minimizing problems related to infectious, nutritional, and environmental diseases, all within an economic situation. Health management is, therefore, controlling the environmental fluctuations to reduce the magnitude of change and produce a more economical, healthier, and better-quality product.

Keywords Viral Diseases · Farming · Health · Parasites · Treatments

2.1 Principles of Health Maintenance

Health maintenance involves a series of principles that apply to most farm-raised animals. However, fish tend to react more quickly to environmental change than terrestrial farm animals. Because of their homeothermic nature, most terrestrial farm animals respond comparatively slowly to poor environmental conditions. In contrast, fish—being poikilothermic—respond quickly and often fatally to handling, temperature change, excessive or inadequate dissolved gases in the water, metabolites, or chemical additives, and so forth, to which they are unable to adapt. These elements

F. Firouzbakhsh (✉)
Department of Fisheries, Faculty of Animal Science and Fisheries, Sari Agricultural Sciences and Natural Resources University, Sari, Iran

S. H. Hoseinifar
Department of Fisheries, Faculty of Fisheries and Environmental Sciences, Gorgan University of Agricultural Sciences and Natural Resources, Gorgan, Iran

M. Faheem
Department of Zoology, GC University, Lahore, Pakistan

© The Author(s), under exclusive license to Springer Nature Switzerland AG 2023
S. H. Hoseinifar, H. Van Doan (eds.), *Novel Approaches Toward Sustainable Tilapia Aquaculture*, Applied Environmental Science and Engineering for a Sustainable Future, https://doi.org/10.1007/978-3-031-38321-2_2

also increase fish susceptibility to infectious agents and compromise their immune response (Plumb and Hanson 2011).

The objective of health maintenance is to help control environmental fluctuations through management practices, thus reducing the magnitude of change and producing a more economical, healthier, and better-quality product. The final goals of health management are disease prevention, reduction of infectious disease incidence, and reduction of disease severity when it occurs. Successful health maintenance and disease prevention or control do not depend on any single procedure but are the culmination of the application of integrated concepts and exercising management options (Plumb and Hanson 2011).

Health maintenance is a concept in which animals are reared under conditions that optimize the growth rate, feed conversion efficiency, reproduction, and survival while minimizing problems related to infectious, nutritional, and environmental diseases, all within an economic situation. As culture becomes more intensive, need for intervention increases accordingly, and principles of health maintenance become of greater importance.

In an aquatic environment, there is a profound and inverse relationship between environmental quality and the disease status of fish. As environmental conditions deteriorate, the severity of infectious diseases increases; therefore, sound health maintenance practices can play a major role in maintaining a suitable environment where healthy fish can be grown.

Health maintenance does not simply target infectious diseases but emphasizes proper utilization of physical facilities; use of genetically improved fish and certified "specific pathogen-free" (SPF) stocks whenever available and/or feasible; environmental control; prophylactic therapy; feed quality and quantity, pond, cage, raceway, tank, or recirculating system management; control of vegetation; aeration and use of other water quality maintenance practices; and a management commitment to provide an optimum habitat in terms of water quality for fish being cultured.

Stress generally is the reaction of an animal to a physical, physiological, or chemical insult. Experience has shown that a wide variety of viral, bacterial, parasitic, and other fish diseases will cause mortality if cultured fish are held in unfavourable environmental conditions (Wedemeyer 1996).

Multiple stressful parameters have a compounding synergistic effect on the fish, and even though the fish may be able to handle the stressors individually, the combined effect can be lethal (Boyd 1990).

A theory of host/pathogen/environment relationship was applied to fish with regard to the development of infectious diseases by Snieszko (1973). This theory is based on the premise that to have an infectious disease, a host and pathogen are required but an unfavourable environmental condition often acts as a trigger for the disease to improve. The interaction of these factors is expressed in the equation: $H(A + S^2) = D$,

where:

H = Species or strain of host (natural resistance).
A = Etiological agent.

S = Environmental stressors.
D = Disease.

Environmental stressors are squared because as fish approach adaptation limits, stressors increase accumulatively rather than additively. Also, when more than one stressor is involved (oxygen, ammonia, CO_2, temperature, etc.), detrimental factors act synergistically (Plumb and Hanson 2011). Intensive fish culture causes unique but manageable environmental problems for the aquaculturist (Boyd 1990; Wedemeyer 1996). All fish require adequate water maintained at a suitable temperature and oxygen concentration level for proper growth and reproduction.

2.2 Diseases of Tilapia

Tilapias are the most extensively cultured fish in the world, especially in China, Indonesia, and Egypt (FAO 2017) and they are recognized as one of the most important species in tropical freshwater aquaculture. Tilapia are hardy fish with good resistance to pathogens as long as they are kept under good water quality conditions, in the proper temperature range, and using proper husbandry practices. As culture systems intensify (cages, raceways, or recirculating systems), stress due to improper handling, exposure to poor water quality, high fish density requiring high feeding rates, or low water temperatures can exacerbate the impact of some pathogens on tilapia (Plumb and Hanson 2011).

2.2.1 Bacterial Diseases

The outbreak of bacterial diseases is the main limiting factor that threatens tilapia production, specifically when cultured under intensive conditions (Abdel-Latif et al. 2020). Fish diseases often occur following an initial stressor. By some accounts, bacterial diseases may be the number one threat to the future of the tilapia aquaculture industry. The most serious bacterial diseases in cultured tilapia are arranged in Table 2.1.

2.2.1.1 Streptococcosis

The first reports of *Streptococcus* spp. infections in fish date back to the mid-1950s (Hoshina et al. 1958), but the first case involving tilapia was a decade later (Wu 1970). Streptococcosis may be sub-acute but is often chronic in tilapia. Although multiple species of *Streptococcus* have at various times been reported from fish, *S. iniae* and *S. agalactiae* are the two that most frequently cause serious disease in tilapia. *Streptococcus* spp. infections have been reported in tilapia from

Table 2.1 The most bacterial diseases of tilapia

Disease	Agent	Clinical sings
Streptococcosis	*Streptococcus iniae, Enterococcus* sp.	Exophthalmia, haemorrhages in eyes, pale liver, spiral swimming
Francisellosis	*Francisella* spp.	Ulcer, ascites, exophthalmia, loss of appetite
Motile Aeromonas septicaemia (MAS)	*Aeromonas hydrophila* and related species	Swim lethargically, gasping
Edwardsiellosis	*Edwardsiella tarda*	Ascites, pale liver, swollen spleen and kidney, swollen abdomen
Columnaris	*Flavobacterium columnare*	Frayed fin, grey patches on the skin, pale gills, and necrotic
Phothbacterium infection	*Photobacterium damselae*	Dark skin, swollen kidney and spleen, ascites
Vibriosis	*Vibrio* spp.	Similar to MAS
Pseudomonas infections	*Pseudomonas* spp.	Similar to MAS

North, South, and Central America, Australia, Asia, South Africa, and other African countries, Great Britain, Norway, and Middle Eastern countries. Tilapia-associated streptococci are essentially worldwide in distribution (Klesius et al. 2008).

Clinical signs of streptococcosis in tilapia are not always specific, but in most species of fish, eye disease and meningoencephalitis are common. Affected fish generally are lethargic, and exhibit spiral swimming, abdominal distention, and exophthalmia with haemorrhage and diffuse haemorrhage in the skin and base of fins. Internal findings in affected fish include exudate in the abdominal cavity, pale livers, and a greatly enlarged, black spleen.

Streptococcus spp. infections occur in both fresh- and saltwater-grown tilapia. *S. iniae* is more common in freshwater systems, and *S. agalactiae is* more common in brackish water. Tilapia in water with 15–30 ppt salinity at 25 and 30 °C are more susceptible to *Streptococcus* isolated from fish in saltwater than when in freshwater at the same temperature (Chang and Plumb 1996). Mortality of tilapia infected with *Streptococcus* varies from low to high, depending on other circumstances. Under culture conditions, mortality is as high as 75% in naturally infected tilapia. In experimental infections with *S. agalactiae* in tilapia, mortality can reach as high as 90%, but mortality in natural infections is generally lower (Evans et al. 2002). Tilapia susceptibility to *Streptococcus* is usually associated with environmental stress, skin injury, scale loss, and other factors associated with intensive aquaculture (Chang and Plumb 1996).

2.2.1.2 Francisellosis

The bacterium belonged to the genus *Francisella* (Hsieh et al. 2006) and is an intracellular bacterium from tilapia and a limited number of other warm-water fish species were initially referred to as a piscirickettsia like organism (PLO) or

Francisella-like organism (FLO) and was believed to be non-culturable on cell-free media. The organism from hybrid striped bass and tilapia was identified by isolation of bacterial DNA from the tissue of diseased fish and amplification of the 16S rRNA gene and comparing the sequences to a comparative database (Mauel et al. 2007). The bacterium was later cultured on agar media and subjected to additional molecular studies where it was determined that warm-water strains of *Francisella* from tilapia were distinct from cold-water isolates from cod (Soto et al. 2009). Members of the genus *Francisella* affect a wide range of animals including humans (*F. tularenses*) and francisellosis has become a serious health problem for cultured tilapia in many geographical regions. The species of *Francisella* associated with disease in warm-water fish, *F. asiatica,* has been found in Taiwan, the United States, and Latin America (Mauel et al. 2007; Soto et al. 2009), where it affects both freshwater and saltwater fish species. The pathogen has been implicated in the disease of several species of tilapia (*O. mossambicus, O. niloticus,* and *Sarotherodon melanotheron*) (Mauel et al. 2003). *F. noatunensis* affects cold-water fish species such as the Atlantic cod. These reports indicate a wide geographical range and fish species susceptibility to *Francisella* spp.

Francisellosis may be either an acute, sub-acute or chronic disease, depending on culture conditions and water temperatures. Affected fish are dark, swim lethargically, and have a loss of appetite; they display skin haemorrhages and ulcers; exophthalmia, and ascites; the spleen and kidneys are enlarged and contain distinct white nodules of varying sizes (Mauel et al. 2007; Soto et al. 2009). The gills exhibit primary and secondary lamellar fusion because of epithelial hyperplasia. Also, multiple white granulomas occur in the gills, spleen, kidney, and testes but seldom in the liver; occasionally black granulomas are seen internally. Special stains such as Giemsa revealed small, pleomorphic coccobacilli inside and outside the cells. It should be pointed out that some fish may show no clinical signs (Mauel et al. 2003).

Francisellosis affects all ages and sizes of tilapia, from small fingerlings to adults. Mortality of tilapia infected with *Francisella* spp. ranged from 5% to 80%, with an average of 50% in Nile tilapia in Latin America (Mauel et al. 2007). The disease has the propensity to occur during cooler months of the year, and infections are stimulated by cold stress on farms in which it is endemic (Mauel et al. 2003). In temperature studies, tilapia maintained between 21.5 and 26.5 °C, initial mortalities occurred on day 15, and mortality doubled almost daily thereafter (Mauel et al. 2003). Tilapia maintained between 26.5 and 29.5 °C and showed no sign of disease or mortality.

2.2.1.3 Motile Aeromonas Septicaemia

In tilapia, motile Aeromonas septicaemia (MAS) is associated with several different species of bacteria, the most common of which is *Aeromonas hydrophila.* Although MAS is not uncommon in cultured tilapia, there are few published "case reports" of its occurrence, but when it occurs, it is often a secondary problem. Tilapia with MAS lose their equilibrium, swim lethargically, gasp at the surface, and generally display the same clinical signs as other fish species.

2.2.1.4 Pseudomonas

Pseudomonas fluorescens occasionally produces clinical signs and pathology similar to *A. hydrophila* and can cause significant mortality in tilapia. The infection occurred during winter and spring when water temperatures were 15–20 °C resulting in mortalities of 0.2–0.3% per day.

2.2.1.5 Vibriosis

Vibriosis in tilapia is often mild to chronic, and clinical signs do not differ significantly from those for MAS. Mortality of infected tilapia is usually chronic, with relatively low daily losses, but cumulative mortality can be significant. Epizootiology of vibriosis in tilapia is similar to that of MAS in the respect that both diseases are usually secondary infections (Sakata 1988). In saltwater, *V. anguillarum* or *V. vulnificus* are involved, while in freshwater, *V. mimicus*or *V. cholerae* is found, and *V. parahaemolyticus* can occur in either environment. Sakata (1988) reported that Nile tilapia suffered 10–20% mortality due to a vibriosis infection following transfer from freshwater to saltwater pens at 18–20 °C. Decreasing water temperatures, coupled with high salinity, are considered compounding stressors on tilapia populations.

2.2.1.6 Columnaris

When columnaris occur in tilapia, pale areas form on the body and frayed fins are the most frequently observed clinical sign of disease, and infected fish will swim lethargically or float at the surface. Nile tilapia are more susceptible to columnaris infections and have higher mortality when the pH is either very acidic or alkaline (Marzouk and Bakeer 1991). So physical injury and/or environmental stress can precipitate these infections.

2.2.1.7 Edwardsiellosis

Edwardsiella tarda affects tilapia cultured under high density and other stressful conditions in either freshwater or marine environments. The sign of Tilapia infected with *E. tarda is* swim lethargy, enlarged abdominal area, and swollen, opaque, and haemorrhaged eyes. Internally, focal areas of necrosis are seen; the liver is often pale and mottled, the spleen is dark red and swollen, the kidney is swollen and soft, and the intestine is inflamed and usually void of food.

2.2.1.8 Photobacterium

Photobacterium damselae subsp. *Damselae* was shown to cause disease in freshwater wild and cultured Nile tilapia in Egypt (Khalil and Aly 2008). Infected tilapia swam lethargically and had dark skin. Internally they had swollen kidney, spleen, and liver with bloody fluid in the abdominal cavity. Also, pinhead nodules were present on the liver, spleen, and kidney. The mortality of experimentally infected fish was 20–40%.

2.2.2 *Viral Diseases*

Compared with the research history of bacterial and parasitic diseases of aquatic animals, the study of fish viral diseases is relatively new. Infectious pancreatic necrosis virus (IPNV) of salmonids was the first proven viral fish disease. The infectious pancreatic necrosis virus was described by Wood et al. (1955). Several viruses have been isolated and identified in association with mortality cases of tilapia species (Machimbirike et al. 2019). The most important viral diseases in cultured tilapia are arranged in Table 2.2.

Tilapia lake virus (TiLV) is one of the most devastating viral infections negatively affecting tilapia culture worldwide. It has been associated with mass kills of

Table 2.2 Summary of tilapia viral diseases (Machimbirike et al. 2019)

Agent	Family/ genus	Host	Clinical signs	Distribution
Tilapia lake virus (TiLV)	Tilapinevirus (Orthomyxovirus-like)	*Oreochromis niloticus*, hybrid tilapia, red tilapia, red hybrid tilapia, wild black tilapia, *Oreochromis aureus*	Lethargy, skin erosions, darkening of the skin, loss of appetite, exophthalmia	In 13 countries in Asia, Africa, and South America
Viral nervous necrosis (VNN)	Nodaviridae/ Betanodavirus	*Oreochromis niloticus*	Erratic swimming, loss of balance	Indonesia, France, and Thailand
Infectious spleen and kidney necrosis virus (ISKNV)	Iridoviridae/ unknown	*Oreochromis niloticus*	Lethargy, ascites, pale gills	Canada
Infectious pancreatic necrosis virus (IPNV)	Birnaviridae/ aquabirnavirus	*Tilapia mossambica*, *Oreochromis niloticus*	Darkening of skin, ascites, pale gills, abnormal swimming	Taiwan, Kenya
Lymphocystis disease virus (LCDV)	Iridoviridae/ lymphocystivirus	*Tilapia amphimelas, T. esculenta, T. variabilis, Haplochromis* sp.	Wart-like growths on the tails of fish	North Tanzania

tilapia species in Ecuador, Colombia, Egypt, and Malaysia (Abdel-Latif et al. 2020). Tilapia lake virus can cause disease in various species by natural or experimental infection (Surachetpong et al. 2020). Similar to natural infection, tilapia lake virus disease could be reproduced in different species of tilapia and giant gourami under laboratory-challenged conditions (Behera et al. 2018). However, most warm-water fish are resistant to the virus as determined by experimental studies (Jaemwimol et al. 2018); the labeorohita also appear to be resistant to TiLV since experimentally challenged fish displayed no clinical signs or mortality, and no viral RNA was detected (Pradhan et al. 2020). The reason behind the insusceptibility of these fish species to TiLV could be due to the absence of viral receptors of mechanisms that allow the virus to replicate in these non-natural species. Therefore these fish are unlikely to be the carrier of TiLV.

Furthermore, *Iridovirus* is another important virus that has been reported to induce heavy mortalities in Nile and Mozambique (*O. mossambicus*) tilapias (Ariel and Owens 1997). In addition, infectious spleen and kidney necrosis virus (ISKNV), first reported in Africa, has been linked with mass mortalities in farmed tilapia (Ramírez-Paredes et al. 2019).

2.2.3 Parasitic Diseases

Parasites are important components of host biology and can be found in any fish species. Numerous parasites, especially monogenetic and digenetic trematodes, and ciliated protozoans have been identified in tilapia. Moreover, ectoparasite infestation with ciliated protozoans, such as *Trichodina* sp. and *Ichthyophthirius multifiliis*, has been investigated and described in tilapia (Pantoja et al. 2012). The monogenetic trematode *Cichlidogyrus* species have been described from tilapia such as *C. sclerosus* and *C. tilapiae* (Le Roux and Avenant-Oldewage 2010). In addition, gyrodactylid parasites such as *Gyrodactylus cichlidarum* are one of the main *Monogenean* species infecting juvenile Nile tilapia and could lead to severe mortalities of intensively farmed fish worldwide (García-Vásquez et al. 2010).

Among the digenetic trematodes that infect tilapia is *Clinostomum* and *Euclinostomum*. *Clinostomum tilapiae* and *C. complanatum* have been identified in the intestine of Nile tilapia and *Sarotherodon galilaeus* (Ukoli 1966). These digenetic metacercariae are encysted in tilapia tissues such as the sub-mucosa of the mouth cavity, the gill chamber, the operculum, the muscles, the abdominal cavity, mesentery, and viscera (Chung et al. 1995). The encysted metacercariae of *Euclinostomum heterostomum* were also recorded in tilapia in Egypt (Eissa et al. 2011).

A particularly common instance of concurrent infections may be attributed to colonization by ectoparasites, which act as a portal of entry for the second co-infecting pathogen. Suppression of the host immune response resulting in an enhanced effect of the secondary invaders can also occur, and both co-infecting pathogens act synergistically to induce disease with greater economic losses for the affected tilapia (Abdel-Latif et al. 2020).

2.3 Control and Treatment Approach of Diseases

As tilapia culture has become more intensive, the need for environmental control, water quality stability through management, stress reduction, and other sound health maintenance procedures are more essential.

2.3.1 *Management*

Most management procedures should be applied to regularly eliminate the accumulation of detritus, waste, and dead fish. Water quality should be maintained at the highest possible level, and the reduction of environmentally induced stress is a top urgency. Prudent stocking densities, adequate water exchange in intensive culture units, removal of metabolites and faecal waste from recirculating water, supplemental aeration, and feeding high-quality diets are a means to that end. Water treatment with ultraviolet (UV) light and ozone will help reduce bacterial populations in recirculating water or open water supplies. Disinfection is accomplished by dipping utensils and boots into 200 mg/L of chlorine, 100 mg/L of a quaternary ammonium compound (Roccal), or Iodine at 1000 mg/L.

2.3.2 *Chemotherapy*

Although not FDA-approved, some drugs and chemicals have been successfully used prophylactically or in chemotherapy for bacterial infections of tilapia. Prophylaxis includes salt baths (NaCl or $CaCl_2$ at 0.5–3%) for a dip or prolonged treatments and/or potassium permanganate (5–10 mg/L) for 1 h or 2–5 mg/L indefinitely. Erythromycin is effective against Gram-positive bacteria, and its use is currently under FDA consideration for use in *Streptococcus* spp.-infected tilapia. If approved, the medication level would probably be 50 or 100 mg/kg of fish per day for 10 days.

2.3.3 *Vaccination*

Few immunological or vaccination experiments involved tilapia until the last decade, when significant strides were made in vaccinating them against *S. iniae* and *S. agalactiae* (Klesius et al. 2008). The first vaccines used experimentally to prevent *Streptococcus* infections in tilapia were in Japan (Sakata 1988), but since then, numerous vaccination studies have been carried out. A formalin-killed vaccine of two *S. iniae* isolates (ARS-10 and ARS-60), injected IM and IP individually and in combination, protected against the respective antigens (Klesius et al. 2008). These data as a whole indicate that vaccines against *S. iniae, S. agalactiae*, or *E. tarda* provide a feasible and effective proactive approach to prevent serious infections in tilapia by these pathogenic bacteria.

References

Abdel-Latif HMR, Dawood MAO, Menanteau-Ledouble S, El-Matbouli M (2020) The nature and consequences of co-infections in tilapia: a review. J Fish Dis 43(6):651–664

Ariel E, Owens L (1997) Epizootic mortalities in tilapia *Oreochromis mossambicus*. Dis Aquat Org 29(1):1–6

Behera BK, Pradhan PK, Swaminathan TR, Sood N, Paria P, Das A, Jena JK (2018) Emergence of tilapia lake virus associated with mortalities of farmed Nile Tilapia *Oreochromis niloticus* (Linnaeus 1758) in India. Aquaculture 484:168–174

Boyd CE (1990) Water quality in ponds for aquaculture. Alabama Agricultural Experiment Station, Auburn University, Alabama

Chang PH, Plumb JA (1996) Effects of salinity on Streptococcus infection of Nile tilapia, *Oreochromis niloticus*. J Appl Aquac 6:39–45

Chung DI, Kong HH, Moon CH (1995) Demonstration of the second intermediate hosts of *Clinostomumcom planatum* in Korea. Korean J Parasitol 33(4):305–312

Eissa I, Gado M, Laila A, Zaki M, El-Deen NA (2011) Field studies encysted metacercariae infested natural male tilapias and monosex tilapias in Kafr El-Sheikh governorate fish farms. Life Sci 8(2):7–12

Evans JJ, Klesius PH, Gilbert PM, Shoemaker CA, Al Sarawi MA, Landsberg J, Duremdez Al Marzouk A, Al Zenki S (2002) Characterization of β-haemolytic Group B *Streptococcus agalactiae* in cultured seabream, *Sparus auratus* L., and wild mullet, *Liza klunzingeri* (Day), in Kuwait. J Fish Dis 25:505–513

FAO (2017) World aquaculture 2015: a brief overview, by Rohana Subasinghe. FAO Fisheries and Aquaculture Circular No. 1140. Rome, Italy: Food and Agriculture Organization of the United Nations

García-Vásquez A, Hansen H, Christison K, Rubio-Godoy M, Bron J, Shinn A (2010) Gyrodactylids (Gyrodactylidae, Monogenea) infecting *Oreochromis niloticus* niloticus (L.) and *O. mossambicus* (Peters) (Cichlidae): a pan-global survey. Acta Parasitol 55:215

Hoshina T, Sano T, Morimoto Y (1958) A streptococcus pathogenic to fish. J Tokyo Univ Fish 44: 57–68

Hsieh CY, Tung MC, Tu C, Chang CD, Tsai SS (2006) Enzootics of visceral granulomas associated with Francisellalike organism infection in tilapia (*Oreochromis spp.)*. Aquaculture 254:129–138

Jaemwimol P, Rawiwan P, Tattiyapong P, Saengnual P, Kamlangdee A, Surachetpong W (2018) Susceptibility of important warm water fish species to tilapia lake virus (TiLV) infection. Aquaculture 497:462–468

Khalil RH, Aly SM (2008) Photobacteriosis in some wild and cultured freshwater fishes in Egypt. Eighth International Symposium on Tilapia in Aquaculture. Cairo, Egypt, pp. 1211–1228

Klesius PH, Shoemaker CA, Evans JJ (2008). Streptococcus: a worldwide fish health problem. International Symposium on Tilapia Aquaculture, Cairo 1:83–107

Le Roux L, Avenant-Oldewage A (2010) Checklist of the fish parasitic genus Cichlidogyrus (Monogenea), including its cosmopolitan distribution and host species. Afr J Aquat Sci 35(1): 21–36

Machimbirike VI, Jansen MD, Senapin S, Khunrae P, Rattanarojpong T, Dong HT (2019) Viral infections in tilapines: more than just tilapia lake virus. Aquaculture 503:508–518

Marzouk MSM, Bakeer A (1991) Effect of water pH on *Flexibacter columnaris* infection in Nile tilapia. Egypt J Comp Pathol Clin Pathol 4:227–235

Mauel MJ, Miller DL, Frazier K, Liggett AD, Styer L, Montgomery-Brock D, Brock J (2003) Characterization of a piscirickettsiosis-like disease in Hawaiian tilapia. Dis Aquat Org 53:249–255

Mauel MJ, Soto E, Moralis JA, Hawke J (2007) A piscirickettsiosis-like syndrome in cultured Nile tilapia in Latin America with Francisella spp. as the pathogenic agent. J Aquat Anim Health 19: 27–34

Pantoja MFW, Neves LR, Dias MDR, Marinho RGB, Montagner M, Tavares-Dias M (2012) Protozoan and metazoan parasites of Nile tilapia Oreochromisniloticus cultured in Brazil. Revista MVZ Córdoba 17:2812–2819

Plumb JA, Hanson AL (2011) Health maintenance and principal microbial diseases of cultured fishes, 3rd edn. Blackwell Publishing Ltd., Iowa, USA

Pradhan PK, Paria A, Yadav MK, Verma DK, Gupta S, Swaminathan TR, Lal KK (2020) Susceptibility of Indian major carp *Labeo rohita* to tilapia lake virus. Aquaculture 515:734567

Ramírez-Paredes JG, Paley RK, Hunt W, Feist SW, Stone DM, Field T, Verner-Jeffreys DW (2019) First detection of infectious spleen and kidney necrosis virus (ISKNV) associated with massive mortalities in farmed tilapia in Africa. bioRxiv:680538

Sakata T (1988) Characteristics of Vibrio vulnificus isolated from diseased tilapia (*Saratherodonniloticus*). Fish Pathol 23:33–40

Snieszko SF (1973) Recent advances of scientific knowledge and development pertaining to diseases of fishes. Adv Vet Sci Comp Med 17:291–314

Soto E, Hawke JP, Fernandez D, Morales JA (2009) Francisella sp., an emerging pathogen of tilapia, *Oreochromisniloticus* (L.) in Costa Rica. J Fish Dis 32:713–722

Surachetpong W, Roy SRK, Nicholson P (2020) Tilapia lake virus: the story so far. J Fish Dis 43(10):1115–1132

Ukoli F (1966) On the life history, growth and development from the metacercarial stage to adulthood of *Clinostomumtilapiae* Ukoli. J Helminthol 40(1–2):215–226

Wedemeyer GA (1996) Physiology of fish in intensive culture systems. Chapman & Hall, New York

Wood EM, Snieszko SF, Yasutake WT (1955) Infectious pancreatic necrosis in trout. AMA Arch Pathol 60:26–28

Wu SY (1970) New bacterial disease of tilapia. FAO Fish Culture Bull 2:14

Chapter 3
Effects of Feed Additives on Tilapia Reproduction

Giorgia Gioacchini, Ike Olivotto, Ghasem Ashouri, and Oliana Carnevali

Abstract Reproduction in fish is an energy-demanding process. The successful development of gonads requires adequate resources to sustain the high-energy demand. Nutrition can affect gametogenesis, fecundity, and consequently, gamete and offspring quality in gravid females. Therefore, an artificial diet for brood stock is formulated to improve nutritional status during this important stage of their life cycle. A detailed reproductive physiology and breeding behaviour of tilapia is discussed in this chapter, with special emphasis on the use of feed additives to control tilapia reproduction, sexual maturation, and sexual determination.

Keywords Reproduction · Physiology · Gametogenesis · Feed additives

3.1 Introduction

Nutrition is known to have a considerable effect on the reproductive performance and gonadal development of fish (Izquierdo et al. 2001; Volkoff and London 2018; Saborido-Rey and Kjesbu 2005). Successful reproduction requires adequate resources to sustain the high-energy demand for gametes production and reproductive behaviours, and, in most cases, negative energy balance and low-food consumption adversely affect reproductive success. In particular, in reproductive females, nutrition can affect gametogenesis, fecundity, and, consequently, gamete and offspring quality (Izquierdo et al. 2001; Lubzens et al. 2017; Bombardelli et al. 2017). Consequently, over the last decades, for most cultured fish species, several commercial broodfish diets have been formulated to improve their nutritional status during this important stage of their life cycle.

G. Gioacchini (✉) · I. Olivotto · O. Carnevali
Dipartimento di Scienze della Vita e dell'Ambiente, Università Politecnica delle Marche, Ancona, Italy
e-mail: giorgia.gioacchini@staff.univpm.it

G. Ashouri
Department of Sciences and Technologies (DST), University of Sannio, Benevento, Italy

© The Author(s), under exclusive license to Springer Nature Switzerland AG 2023

S. H. Hoseinifar, H. Van Doan (eds.), *Novel Approaches Toward Sustainable Tilapia Aquaculture*, Applied Environmental Science and Engineering for a Sustainable Future, https://doi.org/10.1007/978-3-031-38321-2_3

Nile tilapia, *Oreochromis niloticus,* is one of the most cultured fish species worldwide. Thanks to its rapid growth rate and great adaptation to intensive culture conditions, its farming has been intensified; however, there is a strong need for improving the rearing techniques which are essential to guarantee a more effective and sustainable production. Despite the great potential of tilapia culture, there are still problems related to their reproduction, causing an imbalance between seed production and increased global demand representing the main limitation to its further intensive culture expansion (Bhujel 2000). These limitations are related to the low number of eggs released per spawn, the asynchronous spawning behaviour of tilapia, the early sexual maturation in farming conditions which cause overcrowding and fish stunting, and finally, the lower growth rate of females with respect to males, which induce to move toward a mono-sex (only males) culture (Tsadik Getinet 2008; El-Sayed 2019). Therefore, the ability to control sexual determination, sexual maturation, reproduction, and spawning to produce high-quality fry is essential for successful tilapia production (El-Sayed and Kawanna 2008a, b; Migaud et al. 2013). In this light, the optimisation of tilapia reproductive performances using feed additives has been exploited to maximise hatchery production efficiency, sustainability of seed production, mono-sex culture and in turn, profitability.

In this chapter, a general overview of fish reproduction and a focus on the reproductive behaviour and physiology of tilapia, besides a description of the most common systems for seed production, will be done. Particular emphasis will be done on the use of feed additives to control tilapia reproduction, sexual maturation, and sexual determination.

3.2 Fish Reproduction

3.2.1 Reproductive Strategies

The great diversification in reproductive strategies of fish is a key issue supporting their success during evolution (reviewed by Murua and Saborido-Rey (2003), Lowerre-Barbieri et al. (2011), Serrat et al. (2019)). Reproductive strategies could differ in several characteristics. For example, gonochorism, in which sexes are separated, is the most common condition in fish, although hermaphroditism, which could be either simultaneous or sequential, is also observed. Another diversifying factor consists in the choice of breeding dynamics: the most common situation is represented by iteroparity, i.e. fish have multiple reproductive cycles during their lifetime. However, some cases of semelparity, in which the spawning is followed by death, are observed. Reproductive strategy diversification can also depend on fertilisation type and embryo development. External fertilisation is the widespread strategy in fish, in which haploid gametes are released in the aquatic environment where fertilisation occurs. Anyway, several species of fish show internal fertilisation. Regarding embryo development, in oviparous species, it occurs within the eggshell outdoor the mother's body. On the contrary, in the case of viviparity, embryo development occurs within the uterus or the ovary. Finally, parental care can be considered a diversifying factor for reproductive strategies. Many fish species take care of their offspring by applying a high diversity of strategies, ranging from hiding to guarding eggs. They may include nest construction (e.g. *tilapia zillii*), mouthbrooding (e.g. *Oreochromis* spp.), or carrying embryos in specialised body structures (Murua and Saborido-Rey 2003; Lowerre-Barbieri et al. 2011; Serrat et al. 2019).

3.2.2 Endocrine Control of Reproduction

In sexually mature fish, reproduction is controlled by the brain–pituitary–gonad (BPG) axis, which regulates sexual maturity, gametogenesis, reproductive behaviour, integrating external stimuli, both environmental (e.g., photoperiod and water temperature), and social (e.g. courtship), and internal ones (e.g. nutritional status) (Muñoz-Cueto et al. 2020). Despite the wide spectrum of reproductive strategies, the endocrine system which controls reproduction is similar among teleost fish (Zohar 2021).

The BPG axis involves the hypothalamic gonadotropin-releasing hormone (GnRH), the pituitary hormones, the follicle-stimulating hormone (FSH) and the luteinizing hormone (LH), and the gonadal sexual steroid hormones (progestin, oestrogen, and androgen hormones) which are altogether key factors in the endocrine control of reproduction. GnRH stimulates the pituitary gland to produce and release FSH and LH into the bloodstream, which in turn induces gonadal steroidogenesis hormones (oestrogens, androgens, and progestogens) for the control of gonadal development and gametogenesis (Muñoz-Cueto et al. 2020; Zohar 2021).

3.2.3 Gametogenesis

Gametogenesis is the process that includes the formation, growth, and maturation of gametes in females and males, named oogenesis and spermatogenesis, respectively (Lubzens et al. 2010; Kagawa et al. 2013).

3.2.3.1 Oogenesis

Oogenesis is a dynamic process that occurs continuously in the ovary and consists of the development of female gametes from oogonia to eggs (reviewed by Le Menn et al. (2007), Babin et al. (2007), Lubzens et al. (2010), Kagawa et al. (2013), Carnevali et al. (2020)). Oogenesis is commonly divided into primary and secondary growth phases which are gonadotropin independent and dependent, respectively.

The first is characterised by oogonia proliferation and transformation into a primary oocyte (PO). Oogonia are diploid cells that enter meiosis but arrest it in the diplotene stage of the first meiotic division. Once enter meiosis and reach the PO stage, they could last at this stage for several years, at least the entire juvenile period, although in some species (e.g. iteroparous species), they are still present in mature fish. The secondary growth gonadotropin-dependent overlaps with three different phases. During the first one PO develop into the cortical alveoli stage (CA), characterised by the accumulation of alveolar vesicles in the cortical zone of the cytoplasm; during this phase, an accumulation in the form of oil droplets of neutral fatty acids synthesised by the liver may occur however not all species develop this phase. The second phase is represented by vitellogenesis, the main feature of which is yolk/lipid accumulation. Components derived from vitellogenins (Vgs) represent a huge portion of the yolk mass. The Vgs are phospholipoglycoproteins commonly found during vitellogenesis in the blood of oviparous vertebrate females. Teleost Vgs circulate as a large phosphoglycolipoprotein dimer, which is taken up by growing oocytes via receptor-mediated endocytosis. They are synthesised mainly in the liver, under the regulation of estradiol. During this phase, indeed, the oocyte increases in size as it sequesters vitellogenin from blood and increasingly fills the cytoplasm with yolk granules or globules; vitellogenesis is the more protracted oogenesis stage, ranging from days to several months. The final phase consists of the final oocyte maturation characterised by the resumption of meiosis, formation of large oil droplets or lipid coalescence (only in some species), yolk coalescence, and hydration. The maturation phase is fast (less than 1 day) and ends with eggs spawning. Once the oocyte reaches maturation, the energetic investing becomes irreversible. Indeed, vitellogenic oocytes can be reabsorbed before reaching maturation via a degenerative process named follicular atresia, although it has been observed in some oocytes before vitellogenesis. High rate of atresia is usually associated with inadequate fish conditions such as nutritional deficiency. Starting from proliferation, the oocyte is surrounded by a monolayer of somatic granulosa cells which in turn are surrounded by a somatic theca cells layer. These two somatic components represent the follicular layer of oocytes and are responsible for steroidogenesis. After spawning, the empty follicle cell layer degenerates into a post-ovulatory follicle (POF) (reviewed by Le Menn et al. (2007), Babin et al. (2007), Lubzens et al. (2010), Kagawa et al. (2013), Carnevali et al. (2020)).

3.2.3.2 Spermatogenesis

Spermatogenesis is a dynamic process that leads to the formation of mature sperm from spermatogonia and occurs in the testis (reviewed by Schulz et al. 2010; Schulz and Miura 2002; Miura and Miura (2003); Uribe et al. (2014)). In males, spermatogenesis can be categorised into three phases: (1) the spermatogonial phase (SG), during which spermatogonia undergoes several mitotic divisions, followed by Volkoff and London (2018) (2) the meiotic phase, including primary and secondary spermatocytes and finally Saborido-Rey and Kjesbu (2005) the spermiogenic phase consisting in the production of mature spermatozoa. Spermatogenesis takes place in the testis, in which a germinal epithelium, corresponding mainly to somatic Sertoli cells, surrounds germ cells giving rise to spermatogenic cysts formation. Sertoli cells, continuously interacting with germ cells, are essential for the support of spermatogenic cell growth, development and survival. Inside cysts, germ cells develop synchronously among them and with respect to Sertoli cells, until mature spermatozoa are released into the lumen of the testicular lobule. Somatic Leydig cells, responsible for androgen synthesis, are present among cysts in interstitial compartments containing also blood and lymphatics vessels) (reviewed by (Schulz and Miura 2002; Miura and Miura 2003; Schulz et al. 2010; Uribe et al. 2014)).

Mature gametes are the successful completion of gametogenesis, and their quality consists in the ability of spermatozoa to fertilise an egg, or an egg to be fertilised, resulting in the formation of a viable embryo (Bobe and Labbé 2010).

3.2.4 Spawning Dynamics

The spawning dynamics are crucial in reproductive success, as they must give suitable conditions for offspring survival (Saborido-Rey 2019). The size of the fish and hence energy availability is a key factor affecting spawning frequency, as well as the total number of batches produced during the spawning season. Although several species reproduce once per year, some species exhibit shorter cycles, more than one reproductive cycle per year or other species even spawn nearly continuously, i.e. tropical and subtropical species (e.g. *Oreochromis niloticus*). However, reproductive cycles can be interrupted prior to the spawning entering into a regressing phase until the onset of the next breeding season. This event, termed *skipped spawning*, can be caused by inadequate diet and nutritional condition, and it is characterised by a high rate of atresia, clearly indicating that female spawning success is strictly related to energy availability (Carnevali et al. 2017; Saborido-Rey 2019; Yoshida 2020; Alix et al. 2020).

3.2.5 Puberty

Puberty is considered the developmental period covering the transition from a juvenile (immature) to an adult (mature) fish (Taranger et al. 2010; Okuzawa 2002). The main feature of puberty is the associated initiation of the two main gonadal functions: the synthesis of reproductive hormones (oestrogens, androgens, and progestogens) and the production of mature gametes (Taranger et al. 2010). Sex hormones are required during gametogenesis in both males (androgens for spermatogenesis) and females (oestrogens for vitellogenesis). According to Okuzawa (2002), the successful production of the first batch of fertile gametes determines puberty's end.

The timing of puberty onset in fish is mostly determined by several internal (such as genetic) and external factors (including food availability and somatic growth), all of which could be manipulated under farming conditions. Sexual maturation includes physiological, behavioural, and energetical changes and represents a critical shift in the life of an individual (Taranger et al. 2010). Before sexual maturation, assimilated energy is fully expended for growth and survival. When immature fish reaches a determined threshold of the energetic state, sexual maturation can occur. Therefore, puberty involves several endocrine and metabolic changes which lead to the mobilisation and reallocation of energy and depends on the achievement of a critical size/age (size/age of first reproduction) corresponding to a determined physiological state which could support reproductive processes (Taranger et al. 2010).

3.3 Tilapia Reproductive Behaviour and Physiology

3.3.1 Tilapia Reproductive Strategies

Tilapia species are gonochoristic and oviparous with external fertilisation, but the trademark of their reproduction is parental care, which is not typical of most teleost fish (Rana 1988; Bhujel 2000; Coward and Bromage 2000; Dieudonne et al. 2017).

Tilapia are classified into two main generic groups based on their mode of parental care:

1. 'Substrate spawners' (guarders), which contain the fish belonging to the genus Tilapia (e.g. *T. zillii*, *T. rendalli*). In this case, both males and females contribute to the defence of the nest containing fertilised eggs until hatched larvae become independent of their parents (Coward and Bromage 2000).
2. 'Mouthbrooders', which include the genus *Sarotherodon* and *Oreochromis* in which fertilised eggs are incubated in the parents' buccal cavities. In particular, 'mouthbrooders' could be: biparental (both males and females brood and defend the fertilised eggs and hatched fry) such as *S. galilaeus*, or mono-parental which in turn could be paternal mouthbrooders (only males brood the eggs) such as *S. melanotheron* Coward and Bromage (2000); Rana (1988); Turner and Robinson (2000), or maternal mouthbrooders (only females brood the eggs) such as the most cultured tilapia *O. niloticus*, *O. aureus,* and *O. mossambicus.*

Oreochromis males build and defend the spawning territory; courtship between a visiting female and a resident male takes only a few hours and ends with egg release and fertilisation. The Nile tilapia female releases a string of about 20–50 eggs which will be immediately fertilised by the male. After fertilisation, the female takes the fertilised eggs into her mouth and moves to a safe place to incubate. The male no longer participates in incubation and breeding while lures further females to the nest. During egg incubation, the female, holding the eggs in her mouth and continuing to turn them continuously, is unable to eat. After about 5–10 days, egg hatching occurs, but the newly hatched fry continues to develop in the mouth. After yolk-sac absorption, the fry leaves the mouth of the mother and returns to it in the presence of danger. (Rana 1988; Coward and Bromage 2000).

Tilapia are multiple spawners, releasing eggs several times per year, if environmental and/or farming conditions are favourable. At the end of the spawning, the tilapia ovary regenerates very rapidly, and previtellogenic oocytes start vitellogenesis within a week (Coward and Bromage 2000; Elizabeth Cruz Suárez et al. 2006). Parental care characterising tilapia reproduction has led to an increase in egg size and a corresponding reduction of fecundity, which in turn, are directly related to fish size, length, or age. Larger females usually produce a higher number of eggs per clutch with respect to smaller ones. Conversely, "mouthbrooding tilapia" produce a lower number of bigger eggs with respect to "substrate spawners" (Rana 1988; Coward and Bromage 2000). It was also demonstrated that bigger eggs contained more yolk and led to larger fry, more resistance to unfavourable environmental conditions and adverse starvation (Rana 1988).

3.3.2 Tilapia Seed Production Systems

In tilapia farming, three seed production systems are commonly used. The first and most widely used is seed production in earth ponds. In these ponds, a limited number of tilapia grow and reproduces naturally. The seed remains with the parent fish until the time of the seasonal or annual harvesting of the entire pond. During harvesting, new broodfish are selected and restocked into new ponds for the next production cycle. Alternatively, the seed may be harvested separately and restocked in fattening or on-growing ponds.

The second system of seed production is based on managed spawning ponds. In this system, seed production is improved by enhancing the quality and quantity of the broodfish diet. Feed quality improvement could consist of organic and/or inorganic fertilisers or supplementary feeding with leaves, rice bran, oil seed cake, or commercial diets containing about 20% protein. Fertilisers and feed availability and costs determine the combined rate of the two methods.

Finally, the third system for seed production is represented by ponds of special design. In this system, there is an upper egg-laying area from which the eggs drain by gravity into an underlying nursery area through a series of graduated screens. In this way, the broodfish remain confined to the spawning area while the fry enters the nursery. The spawning pond is soon available for the next spawning cycle.

In some tilapia farming situations, land for ponds may not be available due to environmental restrictions or competition with other farming sectors. In these cases, seed production occurs in net enclosures suspended in nutritionally rich shallow water areas of lakes and lagoons, traditionally known as hapas. In recent years, the usage of ad-hoc manual or semi-automated seed production tanks has been developed.

3.4 Feed Additives Administration to Control Tilapia Reproduction

3.4.1 Reproduction Improvement by Using Feed Additives

The fast breeding of tilapias in various tropical and sub-tropical environments makes them ideal farming species; however, their extensive cultivation is still challenged by their low reproductive efficiency together with advanced sexual maturation (as early as 3 months) (Coward and Bromage 2000). As stated above, the complex reproductive strategy, together with the asynchronous spawning and low fecundity, require rigorous management of large numbers of broodfish fish. Feeding management of broodfish tilapia can significantly improve reproductive performance enhancing eggs quality and, in turn, hatchery efficiency and profitability (Bhujel 2000; Izquierdo et al. 2001; Migaud et al. 2013; Bombardelli et al. 2017). Female nutrition is extremely important because dietary constituents are employed for vitellogenesis Izquierdo et al. (2001) and affect gonadal maturation and the quality of yolk (Bobe and Labbé 2010). This determines the quality and survival of the offspring Bombardelli et al. (2017); El-Sayed and Kawanna (2008a, b) because the embryos and larvae, in the initial developmental stages, solely depend on the yolk for sustenance (Fernández-Palacios et al. 2011). Limiting the feed during this period reduces the percentage of eggs that finally mature and ovulate, resulting in a reduction in fry production. Furthermore, nutrition controls the health and survival of females and establishes their capability to allow metabolic stress imposed by the intense reproduction process and rearing characteristics of tilapia (Lupatsch et al. 2010).

The quality and quantity of broodfish diet depend on the type of seed production system. Fertilisation of ponds led to reducing costs and providing a well-balanced diet. Instead, clear water systems require a completely well-balanced diet. As for fry management, diets containing 35–45% protein allow for excellent growth rates. It is also known that the quality of the protein is important. Fishmeal is the best source of high-quality protein, but it is expensive and often unavailable. Therefore, optimising proper feeding management for tilapia broodfish becomes necessary for maximising seed production sustainability and culture productivity (Bhujel 2000).

Up today there are several live and non-living feed additives available to improve tilapia reproduction which include probiotics, phytobiotic, vitamins, and other organic compounds.

3.4.1.1 Probiotics

Probiotics are live microorganisms which confer a health benefit on the host when administered in adequate amounts. Nowadays, since the growing demand for environment-friendly aquaculture probiotics is considered an integral part of the aquaculture systems to increase production also in tilapia (Nayak 2010; Hai 2015; Standen et al. 2015).

During the last decade, several studies evaluated the efficacy of probiotics on the reproductive performance of different teleost (Carnevali et al. 2017; Aydın and Çek-Yalnız 2019). Regarding tilapia farming, several papers showed the positive effects of probiotics on *O. niloticus* fry and fingerlings (Gioacchini et al. 2018; Lara-Flores et al. 2003; Standen et al. 2013). Recently, Mehrim and collaborators (Mehrim et al. 2015) evaluated the effects of graded levels (0, 5, 10, and 15 g/kg diet) of a commercial probiotic, Hydroyeast Aquaculture®, on different reproductive parameters of both male and female Nile tilapia *O. niloticus* adults. In particular, the authors reached the best results using a 15 g/kg diet and a 10 g/kg diet of Hydroyeast Aquaculture® probiotic. At these concentrations, in males, the probiotic treatment enhanced reproductive efficiency in terms of an increase of gonadosomatic index (GSI) and sperm quality parameters (such as total count, motility, abnormalities, and dead); in females, probiotic-enhanced GSI, egg weight numbers and diameters, and fecundity (Mehrim et al. 2015).

3.4.1.2 Phytobiotics

Phytobiotics have recently attracted significant attention in aquaculture as they showed very potent immunostimulant, antioxidant and antimicrobial functions (Jana et al. 2018). Among them, Wahbi and Sangak Wahbi and Sangak (2017) demonstrated the beneficial effect of dietary inclusion of graded levels of *Spirulina platensis* (0.5, 7.5, 10, and 20 g kg^{-1} diet) in Nile Tilapia *Oreochromis niloticus* reproductive performances. In particular, they found that Spirulina supplementation increased total fish weight, GSI, spawning activity and spawned egg number, hatching rate, fry length, and weight (Wahbi and Sangak 2017).

3.4.1.3 Vitamin C

Many authors highlighted the influence of vitamin C in fish reproduction(reviewed by (Dabrowski and Ciereszko 2001). In those studies, it was evidenced that a diet free from or deficient in ascorbic acid reduces in females the availability of vitamin C in the ovary, impairing reproduction, decreasing fecundity and egg quality, and increasing larval deformities and mortality (Dabrowski and Ciereszko 2001).

Martins and collaborators showed (Martins et al. 2016) that supplementation of 150 and 300 mg kg^{-1} vitamin C to the diet given to Nile tilapia positively increased GSI. Similar results have been obtained by Sarmento and collaborators (Sarmento et al. 2018) after vitamin C dietary supplementation (0, 261, 599, and 942 mg/kg of Nile tilapia (*Oreochromis niloticus*) broodfish. In this study, the best results in terms of GSI, egg weight, and larval measurements have been obtained by using 599 and 942 mg vitamin C/kg diet supplementations. A higher total and relative fecundity were found for females receiving a 599 mg vitamin C/kg diet. An increase in hatching rate and egg and larvae production per female was found for females receiving all doses of vitamin C. Interestingly, the authors also performed two stress tests on fertilised eggs from treated females, and they demonstrated that when eggs were incubated at a different rate of salinity, the worst survival rate was achieved with the 0 mg vitamin C/kg diet at the end of 120 h of incubation. While the tests of exposure to air for 40 and 50 min, showed that females receiving the 942 mg vitamin C/kg diet produced newly hatched larvae showing the highest resistance to the stress, while females receiving 0 and 261 mg vitamin C/kg diets produced newly hatched larvae showing the lowest stress tolerance (Sarmento et al. 2018).

3.4.1.4 Other Chemical Compounds

Taurine (2-aminoethanesulfonic acid) is an end product of the metabolism of sulphur-containing amino acids, and its feed supplementation significantly affects growth rates and physiological functions of marine and freshwater fishes (reviewed by El-Sayed El-Sayed (2014)). Previous studies revealed that taurine plays a positive role in Nile tilapia (*Oreochromis niloticus*) performance (Al-Feky et al. 2016a, b). Al-feki and collaborators (Al-Feky et al. (2016a, b)) demonstrated that dietary taurine (8 g kg^{-1}) improved the reproductive performance of Nile tilapia (*Oreochromis niloticus*) broodfish. In particular, taurine supplementation reduced the size at first maturation and the time to first spawning. At the same time, dietary taurine significantly increased spawning performances, including spawning frequencies, the total number of spawning per tank, the number of spawning per female, and absolute fecundity. Finally, eggs produced from broodfish fed on taurine exhibited significantly higher hatchability, reduced time for hatching and yolk-sac absorption.

Ahamed and co-workers (Ahamed et al. 2020) demonstrated the effects of a diet supplemented with Azomite, a natural mineral of volcanic ash, on gonadal steroid hormone levels and milt quality in *Oreochromis mossambicus*. The Azomite administration enhanced the quantity of milt, the percentage of sperm count, the duration of sperm motility, as well as gonadal sex-specific steroid hormones, such as testosterone (T), 11-ketotestosterone (11-KT), and 17β-estradiol (E2) (Ahamed et al. 2020).

De Lima and collaborators (de Lima et al. 2020) assessed the effects of diet supplementation with purified nucleotides (NT) on the health, growth, and reproductive performance of Nile tilapia females. In particular, they found that NT supplementation induced an increase in the percentage of secondary growth oocytes and egg number production, together with a decrease in lipid peroxidation in the ovaries (de Lima et al. 2020).

Finally, Abdelhamid and collaborator (Abdelhamid et al. 2010) investigated the effects of three commercial feed additives (Therigon®, Nuvisol Hatch P®, and Gibberellic acid) on females and one (L-carnitine) on males tilapia broodfish performances. Results indicated that all treatments positively affected reproduction inducing, in females, an increase in ovaries weight and egg number per fish, while in males, an increase of quality sperm parameters such as motility, forward, sluggish and dead percentages.

3.4.2 Mitigation of Tilapia Reproduction by Using Feed Additives

Even in farming conditions, tilapias reach sexual maturity and easily spawn, and the simplicity by which they breed is one of the reasons why tilapia farming has spread throughout the world. However, their plasticity and readiness for sexual maturation acquiring can trigger difficulties associated with overpopulation and reduced growth (El-Sayed 2019). The age and size at which tilapias mature and breed vary with environmental conditions. Farmed Tilapia females mature at an earlier age and smaller size compared to those raised in natural conditions. Generally, natural or wild populations breed at an age of between 2 and 3 years. Farmed stock in ponds and similar shallow water bodies reach sexual maturation at an age of only 2–-3 months, and sexually mature fish breed at 5–6 months. Among different factors, it was evidenced that feed quantity and quality could affect size at first maturity. Many authors found that the use of feed additives could control tilapia reproduction by impairing fertility through gonads (testes and ovaries) damaging. Among them, *Aspilia mossambicensis* and *Azadirachta indica* have recently been used to mitigate the prolific breeding of *O. niloticus* (Kapinga et al. 2019). *A. mossambicensis* and *A. indica* can mitigate the prolific breeding of *O. niloticus* thanks to the presence of phytocompounds which can affect steroidogenesis reproductive organs structures. Recently Kapinga and collaborators demonstrated that *A. mossambicensis* and *A. indica* leaf powder administration impair reproduction by decreasing hatchlings production in *O. niloticus (*Kapinga et al. 2019). The same authors also studied the effect of *A. mossambicensis* or *A. indica* leaf powders supplemented with four doses (1.0, 2.0, 4.0, and 8.0 g/kg diet) on a juvenile at 3% of their body weight daily for 90 days. All doses of *A. indica* and *A. mossambicensis* used significantly lowered absolute fecundity and GSI values, in a dose-dependent manner, starting from doses of 2.0 g/kg in males and 4.0 g/kg in females. In males fed at a lower dose (2.0 and 4.0 g/kg for both *A. mossambicensis* and *A. indica*) were observed histological alterations such as the degeneration of seminiferous tubules presenting a dilatated lumen and containing very few spermatozoa, and the presence of inflammatory cells scattered in the testicular parenchyma. The dose of 8.0 g/kg for both *A. mossambicensis* and *A. indica,* induced the loss of testicular architecture, a dramatic decrease in spermatozoa production, and the increased presence of squamous cells and inflammatory cells within the tubules. In females fed at higher doses (4.0 and 8.0 g/kg for both *A. mossambicensis* and *A. indica*), was evidenced the presence of numerous vacuoles in the ooplasm and alteration of granulosa cells nuclei and thecal layer thickness (Kapinga et al. 2019). A study by Jegede and Fagbenro (2008a, b) reported that incorporation of *Azadirachta indica* in *Tilapia zillii* basal diet at 2.0 g kg^{-1} induced swollen spermatids nuclei, increased interstitial cells and focal necrosis in testes and hydropic degeneration, ruptured follicles, granulomatous inflammation in the interstitium and necrosis ovaries. Similar findings were reported in *O. niloticus* fed Carica papaya Okomoda (2017); Jegede and Fagbenro (2008a, b), Hibiscus rosa-sinensis Jegede (2010), and Aloe vera

Kushwaha (2013) as well as in *O. mossambicus* fed dietary *Carica papaya* and *Moringa oleifera*, respectively (Ampofo-yeboah 2013).

3.4.3 Hormonal Sex Reversal by Feed Additives

Although the mixed-sex culture of tilapia represented a widespread practice in many countries for decades, nowadays, broad attention has been given to mono-sex culture. The production of mono-sex (all-male) tilapia represented a great improvement in productivity due to their fast growth rates, tolerance to different culture conditions, stress, and diseases, greater energy conservation, decreased aggressiveness, and better uniformity of size at harvest (Elizabeth Cruz Suárez et al. 2006). Today, mono-sex tilapia populations are obtained by applying several methods including manual separation of sex, sex reversal hormone, genetic manipulation, and hybridisation (Elizabeth Cruz Suárez et al. 2006). Among them, hormonal treatment is the most successful and widely used method responsible for the production of all male culture (Megbowon and Mojekwu 2014). Tilapia larvae are sexually undifferentiated up until 2 weeks after hatching, and at this time, there are no differences between sex hormone levels (Besseau et al. 2013). Therefore, the administration of exogenous steroid hormones such as androgen (male) or oestrogen (female) during gonadal development or before sexual differentiation would influence the larvae to become either male or female based on the hormone administered (Megbowon and Mojekwu 2014). Unfortunately, synthetic hormones lower fish immunity and survival and require human and environmental health concerns (Megbowon and Mojekwu 2014).

Consequently, alternative methods to obtain mono-sex tilapia culture that is safe and environmentally friendly have been evaluated. Among them, the use of medicinal plants in aquaculture to control reproduction represents a valid technology (Francis et al. 2002; Gabriel et al. 2017).

Recently, it was evidenced that several medicinal herbs containing phytochemicals (structurally similar to steroid hormones such as phytoestrogens or phytoandrogens) can induce tilapia masculinisation, feminisation, or impair fertility (Gabriel 2019). Saponin extracts from *Quillaja saponins* (QS) Francis et al. (2002, fenugreek) (*Trigonella foenum-graecum*) and soapbark tree (*Quillaja saponaria*) Stadtlander et al. (2013), and *Tribulus terrestris* B.O. Omitoyin et al. (2013), induced masculinisation in a dose-dependent manner in Nile tilapia when were incorporated in their diet, Similarly, masculinisation induction was also described in Nile tilapia fed dietary Aloe vera, *Mucuna pruriens* Mukherjee et al. (2015), *Butea superb* (Kiriyakit 2014). The same results have been achieved in other tilapia species, such *O. mossambicus* fed dietary moringa and paw crude extracts (Ampofo-yeboah 2013). Finally, El-Sayed and co-workers revealed that soybean meal significantly induced femininisation in Nile tilapia culture and further warned tilapia farmers to avoid using soybean as a source of protein during sex reversal (El-Sayed et al. 2012).

3.5 Conclusion

Constraints associated with the wide adoption of tilapia intensive aquaculture farming systems such as the increased request for proper seed quality and a monosex culture, among others, cannot be overemphasised. Moreover, the consequent use of drugs (i.e. sex reversal hormones) to improve production represents a prominent risk not only for animals but also for the environment and humans. The overview of the results achieved in recent years on the application of feed additives (often of natural origin and not of synthetic origin) for the control of Tilapia reproduction is presented in this chapter, showing that the use of feed additives could represent a valid alternative to drugs. However, the rate of implementation of this innovation is still too slow, despite the numerous advantages associated with it, and there is still the need to deepen knowledge to further validate the daily use of feed additives in tilapia farming.

References

Abdelhamid AM, Mehrim AI, El-Barbary MI, El-Sharawy MA (2010) An attempt to improve the reproductive efficiency of Nile tilapia brood stock fish. Fish Physiol Biochem 36:1097–1104

Ahamed HA, Mohamed MJ, Arunachalam KD, Raiyaan GID, Musthafa MS, Begum SS (2020) Effects of Azomite enriched diet on gonadal steroid hormone levels and milt quality indices in Oreochromis mossambicus. Aquac Rep 17:100341

Al-Feky SSA, El-Sayed AFM, Ezzat AA (2016b) Dietary taurine improves reproductive performance of Nile tilapia (Oreochromis niloticus) broodstock. Aquac Nutr 22:392–399

Al-Feky SSA, El-Sayed A-FM, Ezzat AA (2016a) Dietary taurine enhances growth and feed utilization in larval Nile tilapia (Oreochromis niloticus) fed soybean meal-based diets. Aquac Nutr 22:457–464

Alix M, Kjesbu OS, Anderson KC (2020) From gametogenesis to spawning: how climate-driven warming affects teleost reproductive biology. J Fish Biol 97:607–632

Ampofo-yeboah A (2013) Effect of phytogenic feed additives on Gonadal development in Mozambique Tilapia (Oreochromis mossambicus). PhD Thesis, 254

Aydın F, Çek-Yalnız Ş (2019) Effect of probiotics on reproductive performance of fish. Nat Eng Sci 4:153–162

Babin PJ, Carnevali O, Lubzens E, Schneider WJ (2007) Molecular aspects of oocyte vitellogenesis in fish, in The Fish Oocyte, pp 39–76. Springer Netherlands, Dordrecht

Besseau L, Fuentès M, Sauzet S, Beauchaud M, Chatain B, Covès D (2013) Somatotropic axis genes are expressed before pituitary onset during zebrafish and sea bass development. Gen Comp Endocrinol 194:133–141

Bhujel RC (2000) A review of strategies for the management of Nile tilapia (Oreochromis niloticus) broodfish in seed production systems, especially hapa-based systems. Aquaculture 181:37–59

Bobe J, Labbé C (2010) Egg and sperm quality in fish. Gen Comp Endocrinol 165:535–548

Bombardelli RA, dos Reis Goes ES, de Negreiros Sousa SM, Syperreck MA, Goes MD, de Oliveira Pedreira AC (2017) Growth and reproduction of female Nile tilapia fed diets containing different levels of protein and energy. Aquaculture 479:817–823

Carnevali O, Forner-Piquer I, Gioacchini G (2020) Structure of mature oocytes, in Reproduction in aquatic animals: from basic biology to aquaculture technology, pp 93–111. Springer Singapore

Carnevali O, Maradonna F, Gioacchini G (2017) Integrated control of fish metabolism, wellbeing and reproduction: the role of probiotic. Aquaculture 472

Coward K, Bromage NR (2000) Reproductive physiology of female tilapia broodstock. Rev. Fish Biol. Fish. Springer

Dabrowski K, Ciereszko A (2001) Ascorbic acid and reproduction in fish: endocrine regulation and gamete quality. Aquac Res 32:623–638

Dieudonne A, Kouakou Y, Frederic C, Paul L, François J, Beatrice C (2017) Reproductive behaviour of two tilapia species (Oreochromis niloticus , Linné, 1758; Sarotherodon melanotheron, Rüppel, 1852) in freshwater intra and interspecific pairing context. Appl Anim Behav Sci 193:104–113

Elizabeth Cruz Suárez L, Ricque Marie D, Tapia Salazar M, Nieto López MG, Villarreal Cavazos DA, Cruz P García A Ortega AC (2006) Tilapia culture in salt water: environmental requirements, nutritional implications and economic potentials. Av. en Nutr. Acuicola

El-Sayed A-FM (2014) Is dietary taurine supplementation beneficial for farmed fish and shrimp? A comprehensive review. Rev Aquac 6:241–255

El-Sayed A-FM (2019) Tilapia culture, 2nd edn. Acadey Press

El-Sayed AFM, Abdel-Aziz ESH, Abdel-Ghani HM (2012) Effects of phytoestrogens on sex reversal of Nile tilapia (Oreochromis niloticus) larvae fed diets treated with 17-α-Methyltestosterone. Aquaculture 360–361:58–63

El-Sayed AFM, Kawanna M (2008a) Effects of dietary protein and energy levels on spawning performance of Nile tilapia (*Oreochromis niloticus*) broodstock in a recycling system. Aquaculture 280:179–184

El-Sayed A-FM, Kawanna M (2008b) Optimum water temperature boosts the growth performance of Nile tilapia (Oreochromis niloticus) fry reared in a recycling system. Aquac Res 39:670–672

Fernández-Palacios H, Norberg B, Izquerdo M, Hamre K (2011) Effects of broodstock diet on eggs and larvae, in Larval fish nutrition, pp 151–181. Wiley-Blackwell, Oxford, UK

Francis G, Levavi-Sivan B, Avitan A, Becker K (2002) Effects of long term feeding of Quillaja saponins on sex ratio, muscle and serum cholesterol and LH levels in Nile tilapia (Oreochromis niloticus (L.)). Comp Biochem Physiol C Toxicol Pharmacol 133:593–603

Gabriel NN (2019) Review on the progress in the role of herbal extracts in tilapia culture. Cogent Food Agric 5

Gabriel NN, Qiang J, Ma XY, He J, Xu P, Omoregie E (2017) Sex-reversal effect of dietary Aloe vera (Liliaceae) on genetically improved farmed Nile Tilapia Fry. N Am J Aquac 79:100–105

Gioacchini G, Ciani E, Pessina A, Cecchini C, Silvi S, Rodiles A (2018) Correction to: effects of Lactogen 13, a new probiotic preparation, on gut microbiota and endocrine signals controlling growth and appetite of Oreochromis niloticus juveniles. Microb Ecol 76:1063–1074

Hai NV (2015) Research findings from the use of probiotics in tilapia aquaculture: a review. Fish Shellfish Immunol 45:592–597

Izquierdo MS, Fernandez-Palacios HT, Tacon AGJ (2001) Effect of brood stock nutrition on reproductive performance of fish. Aquaculture 147:25–42

Jana P, Karmakar S, Roy U, Paul M, Singh AK, Bera KK (2018) Phytobiotics in aquaculture health management: a review. ~ 1422 ~. J Entomol Zool Stud 6:1422–1429

Jegede T (2010) Control of reproduction in Oreochromis niloticus (Linnaeus 1758) using Hibiscus Rosa-sinensis (Linn.) leaf meal as reproduction inhibitor. J Agric Sci 4:149–154

Jegede T, Fagbenro O (2008a) Dietary neem (*Azadirachta indica*) leaf meal as reproduction inhibitor in redbelly tilapia, tilapia zillii. From the pharaohs to the future. Eighth International Symposium on Tilapia in Aquaculture. Proceedings. Cairo, Egypt, 12–14 October, 2008

Jegede T, Fagbenro O (2008b) Histology of gonads in Oreochromis niloticus (Trewavas) Fed Pawpaw (Carica Papaya) seed meal diets. Eighth International Symposium on Tilapia in Aquaculture 2008, pp 1135–1141

Kagawa H, Sakurai Y, Horiuchi R, Kazeto Y, Gen K, Imaizumi H (2013) Mechanism of oocyte maturation and ovulation and its application to seed production in the Japanese eel. Fish Physiol Biochem 39:13–17

Kapinga IB, Limbu SM, Madalla NA, Kimaro WH, Mabiki FP, Lamtane HA (2019) Dietary Aspilia mossambicensis and Azadirachta indica supplementation alter gonadal characteristics and histology of juvenile Nile tilapia (Oreochromis niloticus). Aquac Res 50:573–580

Kiriyakit A (2014) Efficacy of red kwao krua (Butea superb Roxb) crude extract for all male production of Nile tilapia (*Oreochromis niloticus*). J Agric Technol 10(2):391–398

Kushwaha MP (2013) Effects of Aloe vera (Liliaceae) on Gonad development in Nile tilapia, *Oreochromis niloticus* (L.) during intensive aquaculture. Int J Fish Aquat Stud 1(2):56–60

Lara-Flores M, Olvera-Novoa MA, Guzmán-Méndez BE, López-Madrid W (2003) Use of the bacteria Streptococcus faecium and lactobacillus acidophilus, and the yeast Saccharomyces cerevisiae as growth promoters in Nile tilapia (Oreochromis niloticus). Aquaculture 216:193–201

Le Menn F, Cerdà J, and Babin PJ (2007) Ultrastructural aspects of the ontogeny and differentiation of ray-finned fish ovarian follicles, in The Fish Oocyte, pp 1–37. Springer Netherlands, Dordrecht

de Lima SA, Pedreira ACDO, de Freitas JMA, Dalmaso ACS, Chiella RJ, Meurer F (2020) Diets containing purified nucleotides reduce oxidative stress, interfere with reproduction, and promote growth in Nile tilapia females. Aquaculture 528:735509

Lowerre-Barbieri SK, Ganias K, Saborido-Rey F, Murua H, Hunter JR (2011) Reproductive timing in marine fishes: variability, temporal scales, and methods. Mar Coast Fish 3:71–91

Lubzens E, Bobe J, Young G, Sullivan CV (2017) Maternal investment in fish oocytes and eggs: the molecular cargo and its contributions to fertility and early development. Aquaculture 472:107–143

Lubzens E, Young G, Bobe J, Cerdà J (2010) Oogenesis in teleosts: how fish eggs are formed. Gen Comp Endocrinol 165:367–389

Lupatsch I, Deshev R, Magen I (2010) Energy and protein demands for optimal egg production including maintenance requirements of female tilapia Oreochromis niloticus. Aquac Res 41: 763–769

Martins TPA, Gomides PFV, Pereira Navarro FKS, Navarro RD (2016) Suplementação de vitamina C sobre o desempenho e desenvolvimento gonadal de tilápiado-Nilo. Acta Sci Technol 38:477–481

Megbowon I, Mojekwu TO (2014) Tilapia sex reversal using methyl testosterone (MT) and its effect on fish, man and environment. Biotechnology 13:213–216

Mehrim AI, Khalil FF, Hassan ME (2015) Hydroyeast Aquaculture® as a reproductive enhancer agent for the adult Nile tilapia (Oreochromis niloticus Linnaeus, 1758). Fish Physiol Biochem 41:371–381

Migaud H, Bell G, Cabrita E, McAndrew B, Davie A, Bobe J (2013) Gamete quality and broodstock management in temperate fish. Rev Aquac 5:S194–S223

Miura T, Miura CI (2003) Molecular control mechanisms of fish spermatogenesis. Fish Physiol Biochem 28:181–186

Mukherjee D, Ghosal I, Chakraborty SB (2015) Production of Monosex Nile tilapia, Oreochromis niloticus using seed of Mucuna pruriens. IOSR J Pharm Biol Sci (IOSR-JPBS) 10:55–59

Muñoz-Cueto JA, Zmora N, Paullada-Salmerón JA, Marvel M, Mañanos E, Zohar Y (2020, May 15) The gonadotropin-releasing hormones: lessons from fish. Gen. Comp. Endocrinol. Academic Press Inc.

Murua H, Saborido-Rey F (2003) Female reproductive strategies of marine fish species of the North Atlantic. J Northwest Atl Fish Sci 33:23–31

Nayak SK (2010) Probiotics and immunity: a fish perspective. Fish Shellfish Immunol 29(1):2–14

Okomoda VT (2017) Haematology and gonad histology of Oreochromis niloticus (Linnaeus, 1758) fed Carica papaya seed meal. Artic. Brazilian J. Aquat. Sci. Technol. 21

Okuzawa K (2002) Puberty in teleosts. Fish Physiol. Biochem. Springer

Omitoyin BO, Ajani EK, Sadiq HO (2013) Preliminary investigation of Tribulus terrestris (Linn., 1753) extracts as natural sex reversal agent in Oreochromis niloticus (Linn., 1758) Larvae. Int J Aquac 3:133–137

Rana K (1988) Reproductive biology and the hatchery rearing of tilapia eggs and fry, in Recent advances in aquaculture, pp 343–406. Springer Netherlands
Saborido-Rey F (2019) Fish reproduction, in encyclopedia of ocean sciences, pp 232–245. Elsevier
Saborido-Rey F, Kjesbu (2005) OS Growth and maturation dynamics. http://hdl.handle.net/10261/47150
Sarmento NLAF, Martins EFF, Costa DC, Mattioli CC, da Costa Julio GS, Figueiredo LG (2018) Reproductive efficiency and egg and larvae quality of Nile tilapia fed different levels of vitamin C. Aquaculture 482:96–102
Schulz RW, de França LR, Lareyre JJ, LeGac F, Chiarini-Garcia H, Nobrega RH (2010) Spermatogenesis in fish. Gen Comp Endocrinol 165:390–411
Schulz RW, Miura T (2002) Spermatogenesis and its endocrine regulation. Fish Physiol. Biochem. Springer
Serrat A, Saborido-Rey F, Garcia-Fernandez C, Muñoz M, Lloret J, Thorsen A (2019) New insights in oocyte dynamics shed light on the complexities associated with fish reproductive strategies. Sci Rep 9:1–15
Stadtlander T, Levavi-Sivan B, Kerem Z, Dweik H, Qutob M, Abu-Lafi S (2013) Effects of a saponin fraction extracted from Trigonella foenum-graecum L. and two commercially available saponins on sex ratio and gonad histology of Nile tilapa fry, Oreochromis niloticus (L.). J Appl Ichthyol 29:265–267
Standen BT, Rawling MD, Davies SJ, Castex M, Foey A, Gioacchini G (2013) Probiotic Pediococcus acidilactici modulates both localised intestinal- and peripheral-immunity in tilapia (Oreochromis niloticus). Fish Shellfish Immunol 35
Standen BT, Rodiles A, Peggs DL, Davies SJ, Santos GA, Merrifield DL (2015) Modulation of the intestinal microbiota and morphology of tilapia, Oreochromis niloticus, following the application of a multi-species probiotic. Appl Microbiol Biotechnol 99:8403–8417
Taranger GL, Carrillo M, Schulz RW, Fontaine P, Zanuy S, Felip A (2010) Control of puberty in farmed fish. Gen Comp Endocrinol 165:483–515
Tsadik Getinet G (2008) Effects of maternal age on fecundity, spawning interval, and egg quality of Nile tilapia, Oreochromis niloticus (L.). J World Aquac Soc 39:671–677
Turner GF, Robinson RL (2000) Reproductive biology, mating systems and parental care, in Tilapias: biology and exploitation, pp 33–58. Springer Netherlands
Uribe MC, Grier HJ, Mejía-Roa V (2014) Comparative testicular structure and spermatogenesis in bony fishes. Spermatogenesis 4:e983400
Volkoff H, London S (2018) Nutrition and reproduction in fish. Encycl Reprod:743–748
Wahbi OM, Sangak Y (2017) Enhancement of reproductive performance of nile tilapia Oreochromis niloticus using phytobiotic Spirulina platensis. J Biol Sci 17:305–311
Yoshida M (2020) Overview: reproductive systems in aquatic animals, in Reproduction in aquatic animals: from basic biology to aquaculture technology, pp 13–21. Springer Singapore
Zohar Y (2021) Fish reproductive biology—reflecting on five decades of fundamental and translational research. Gen Comp Endocrinol:300

Chapter 4
Intestinal Microbiota of Tilapia: Characterization and Modulation

Zhen Zhang, Tsegay Teame, Anran Wang, Qianwen Ding, Chao Ran, Yalin Yang, and Zhigang Zhou

Abstract Tilapia (*Oreochromis* sp.) are cultured in many countries of the world. Its production is increasing from time to time to fulfill the demand for fish for the fast-growing population of the world. The significant importance of intestinal microbiota in several fish conditions including growth, digestion, energy homeostasis, immunity, and gut-brain axis were reported in a plethora of studies. Characterization of the intestinal microbiota is important to deal with the detailed function of these microbes. The majority of the bacteria species found in the intestine of fish are not culturable and to date, due to the advancement of molecular techniques including denaturing gradient gel electrophoresis (DGGE), temperature gradient gel electrophoresis (TGGE), and next-generation sequencing (NGS) has been employed to investigate and characterize intestinal microbiota. Researchers have been working to increase the production and productivity of tilapia in aquaculture by altering the different microorganisms found in the intestinal of the fish. Studies showed types of diets and their ingredient compositions have a significant role in improving and modulation of the intestinal microbiota of tilapia. Furthermore, the role of several "-biotics" including prebiotics, probiotics, paraprobiotics, and postbiotics in the improvement of the intestinal microbiota was described in many studies. The significance of dealing with intestinal microbiota is to provide a scientific basis for

Zhen Zhang and Tsegay Teame contributed equally to this work.

Z. Zhang · Q. Ding · C. Ran · Y. Yang
Key Laboratory for Feed Biotechnology of the Ministry of Agriculture and Rural Affairs, Institute of Feed Research, Chinese Academy of Agricultural Sciences, Beijing, China

T. Teame
China-Norway Joint Lab on Fish Gut Microbiota, Institute of Feed Research, Chinese Academy of Agricultural Sciences, Beijing, China

Tigray Agricultural Research Institute, Mekelle, Tigray, Ethiopia

A. Wang · Z. Zhou (✉)
China-Norway Joint Lab on Fish Gut Microbiota, Institute of Feed Research, Chinese Academy of Agricultural Sciences, Beijing, China
e-mail: zhouzhigang03@caas.cn

© The Author(s), under exclusive license to Springer Nature Switzerland AG 2023
S. H. Hoseinifar, H. Van Doan (eds.), *Novel Approaches Toward Sustainable Tilapia Aquaculture*, Applied Environmental Science and Engineering for a Sustainable Future, https://doi.org/10.1007/978-3-031-38321-2_4

the establishment of useful approaches for the effective manipulation of these microorganisms to improve fish health and increase production. The intimate knowledge of the usefulness of intestinal microbiota on the overall fitness of tilapia is necessary. Therefore, the objective of this review was to entirely understand the importance and function of these microorganisms invading the intestine of tilapia on the growth, development, metabolism, immunity, and overall fitness of tilapia. Moreover, we highlight the existing information on the strategies for changing the structure of the microbiota community of tilapia via diets or/and "-biotics" and recommendations for future research with the final objective of the applications of these approaches in the tilapia aquaculture industry.

Keywords Diet · Intestinal microbiota modulation · Probiotics · Prebiotics · Tilapia

4.1 Introduction

Next to carp, tilapia, which consists of several species in the family of *Cichlidae*, is the most common cultured fish in many countries of the world (Watanabe 2002). Tilapia are hardy, prolific, and fast-growing tropical fishes. Tilapia is the most popular culture fish in the world and is the second most important finfish species group cultured worldwide in terms of production. Its production has been increasing fourfold in the past decade due to its appropriateness for aquaculture in different environments, preferable for consumers, and stable market prices (Wang and Lu 2016). Studies demonstrated that many species of tilapia have been produced in more than 140 countries, and this data showed that this fish species is one of the most useful aquatic animals in the fish farming sector throughout the world (Fitzsimmons 2016; Junning et al. 2018). Even though tilapia are produced in several developed and developing countries of the world, most producers of the fish are from developing countries such as China, Indonesia, the Philippines, Thailand, Ecuador, and Egypt are the major producers (Yue et al. 2016). FAO's (2020) fish production data showed that the production of tilapia in 2018 accounted for more than 10% of overall cultured fish species in the world. The increasing trends in the production of tilapia were because of the improvement of the production performances of several breeds and strains of the fish via the development of new breeding strategies (Ponzoni et al. 2011; Nguyen 2015; Fitzsimmons 2016).

Several studies demonstrated that tilapia farming is common practice in these countries with low income and has a great contribution to the improvement of the nutritional security status and increase the economic status of the people. Because of this, researchers have been recommended to promote the production of tilapia. Furthermore, they have been working to improve the production of tilapia via several strategies, and one of the strategies was through modulation of the intestinal microbiota using different types of modulators. Identification and characterization of these microorganisms found in the intestinal of tilapia were done through culture-dependent methods or using modern technologies including high throughput.

To date, researchers performed culture-dependent methods of exploration of the intestinal microbiota of tilapia, and they identified fewer microbiota species, but later due to molecular improvement in high-throughput sequencing technology and advancement of bioinformatics have allowed more accurate data on identification and in-depth analysis of intestinal bacterial communities of fish (Opiyo et al. 2019). Especially, the discovery of new and sophisticated molecular technologies and techniques to identify and characterize these via sequencing genes of these microorganisms has widened our understanding of intestinal microbial community structure and their specific function in fish (Zarkasi et al. 2014; Falcinelli et al. 2015). Using high-throughput sequencing identified that the dominant phyla found in the intestinal microbiota of tilapia include *Proteobacteria, Cyanobacteria, Fusobacteria, Actinobacteria,* and *Bacteroidetes* (Standen et al. 2015). The advancement of molecular techniques for the isolation and identification of microorganisms was not only important for investigation but also for appreciating the significant role of the microbiota that occupy the tilapia intestine. Accordingly, a plethora of studies revealed that the intestinal microorganisms of fish have a significant role in the physiology, growth, and immunity of the fish (Burns et al. 2016; Yan et al. 2016; Nie et al. 2017; Smith et al. 2017). Consequently, these and other studies showed these microorganisms found in the intestine of tilapia play a vital role in overall fish fitness. Currently, the advancement of bacteria identification and characterization methods including next-generation sequencing plays a significant role in the detailed characterization of these microorganisms isolated from the intestine of tilapia. Though, the information on the composition, community structure, and role of the intestinal microbiota of tilapia is not fully understood. Therefore, here below, we provide detailed information on the microorganisms found in the intestinal of tilapia to permit their importance on the overall fitness of tilapia to increase the production and productivity of the fish.

4.1.1 Characterization of the Intestinal Microbiota of Tilapia

To date, our knowledge of the characterization of the microorganisms originating from the intestinal of fish results from the collection of information in several studies; the majority of them were focused on cultured fish species, and these tilapia species have gained much consideration and attention. Characterization of the intestinal microbiota of fish has a significant role in understanding the function of this microbiota of the fish. Studies indicated that the intestine of fish was a harbor with approximately 10^8 bacterial cells belonging to more than 500 beneficial and harmful bacteria species (Rawls et al. 2004; Romero and Navarrete 2006). Colonization of the intestinal of fish starts at the early larval stage and it continued with the fish growth and reaches its complex assemblage of microbes (Nayak 2010). Yet, the compositions of intestinal microbiota were changed rapidly at the fish developmental stages (Del'duca et al. 2013). The intestinal microbiota has multiple functions in the fish host physiology, metabolism, and immunology. Previous studies provide a number of evidence on the significant role of intestinal microbiota using germ-free fish and gnotobiotic models. Oyarbide et al. (2015) demonstrated that gnotobiotic zebrafish *(Danio rerio)* (Zebrafish developed in germ-free conditions and inoculated with few known microorganisms) *is* more affected by pathogenic bacteria compared with the conventional zebrafish. The specific taxa of bacteria found in the intestine as well as the diversity of these microorganisms have been applied as an indicator (biomarker) of the metabolism and health status of the fish (Clarke et al. 2014). Studies revealed that most of the time low diversity of these microorganisms isolated from the intestine of fish was associated with abnormal health of the fish (He et al. 2017; Nie et al. 2017). Several external factors' results reduced the diversity of intestinal microbiota these studies indicated that the application of antibiotics resulted in a reduction in the diversity of intestinal microbiota and affected fish health via facilitating colonization by pathogenic microbes (He et al. 2017).

Most of the time, more diversified intestinal microbiota was an indicator of good fish health because they provide greater protective effects against pathogenic bacteria (De Schryver and Vadstein 2014; Zhu et al. 2016). Therefore, maximizing the diversity of intestinal microbiota is useful since tight mutual interaction between the fish and intestinal microbiota is vital to reduce the effect of harmful microbiota on the host, and increase the production of fish in aquaculture (McFall-Ngai et al. 2013). Analysis of OUT is a good indicator of the diversity formed within the sample (alpha diversity) and the other diversity observed among different samples (beta diversity) while interpreting the composition of the intestinal microorganisms. Recently de Bruijn et al. (2017) demonstrated the relationship between intestinal microbiota diversity, function, and fish health status. Another study also revealed the interaction between intestinal microorganisms and fish innate immunity (Nie et al. 2017).

Studies revealed that the intestinal microorganisms of fish were not easily identified using culture-based techniques, and less the 1% of the total intestinal microbiota of fish was culturable, and this does not show an accurate image (Ray et al. 2012; Banerjee and Ray 2017). Culture-based techniques of identification and isolation of intestinal microbiota were used in many specific media, including nutrient agar, tryptone soya agar, peptone gelatin agar, cellulose agar, and starch agar (Ray et al. 2012). Nowadays, many researchers tried to identify and characterized the intestinal microbiota of tilapia using advanced technologies such as DGGE, TGGE, and NGS (Banerjee and Ray 2017). The function and factors that affect the intestinal microbiota of tilapia were summarized (Fig. 4.1).

4.1.1.1 Common Intestinal Microbiota in Tilapia

Understanding the usefulness of microorganisms found intestines of fish in improving the development and well-being of tilapia was explained by many researchers. A plethora of studies identify and characterize the microorganisms isolated from the intestine of tilapia via culture-dependent and independent methods. The culture-

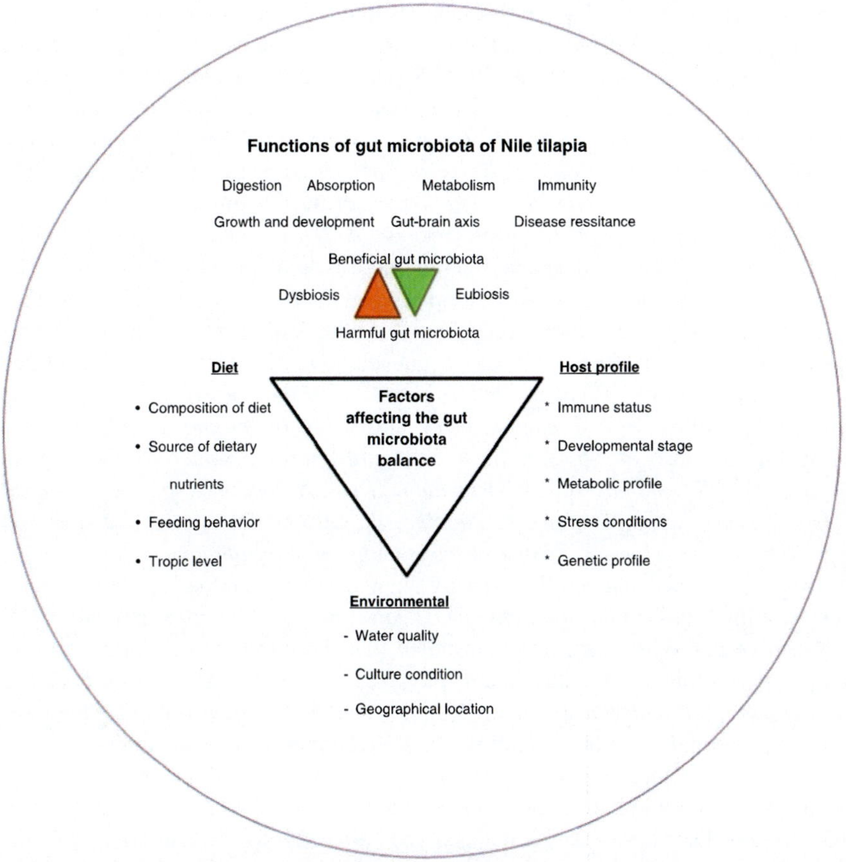

Fig. 4.1 Environmental, diet-associated, or host—associated factors affect the structure and function of the intestinal microbiota of tilapia. These factors can result in either a healthy state (normobiosis) or altered microbiota (dysbiosis) of Tilapia

dependent technique of identification of bacteria from the tilapia intestine was accomplished by Cahill (1990). Before three decades, the identification and characterization of the intestinal microbiota of fish were restricted to the most common and easily culturable bacteria such as *Pseudomonas* sp., *Vibrio* sp., *Aeromonas* sp., and other unidentified species (Cahill 1990). Molinari et al. (2003) identified several bacteria including *Aeromonas hydrophila, A. veronii, Citrobacter freundii, Escherichia coli,* and *Plesiomonas shigelloides* from the intestine of tilapia. The advancement of molecular technologies in the identification of microbiota has increased our understanding of fish microbiota (Nayak 2010). The intestinal microbiota of tilapia was largely affected by several external and internal factors and has dynamic composition and community structure (Wang et al. 2016).

Studies demonstrated that the intestinal of tilapia harbor a complex microbiota, and the main phyla are *Proteobacteria, Fusobacteria, Firmicutes,* and *Bacteroidetes* (Fan et al. 2017; de Souza et al. 2020; Souza et al. 2020). Foysal et al. (2019) identified that the most common intestinal bacteria of tilapia fed on the control diet were *Proteobacteria*, *Bacteroidetes*, *Planctomycetes*, *Fusobacteria*, *Verrucomicrobia*, and *Acidobacteria*. *Fusobacteria*, *Firmicutes*, *Proteobacteria*, *Actinobacteria*, *Bacteroidetes*, and *Planctomycetes* were identified from the intestinal of farmed Nile tilapia (Melo-Boliver et al. 2019). Furthermore, Bereded et al. (2020) identified the core intestinal microbiota of Nile tilapia were *Proteobacteria*, *Firmicutes*, *Cyanobacteria*, *Fusobacteria*, and *Actinobacteria*. In addition to this, Baldo et al. (2015) assessed the composition of intestinal microbiota from different species of tilapia using 16S rRNA pyrosequencing DNA sequencing and identified seven bacteria phyla such as *Proteobacteria, Actinobacteria, Bacteroidetes, Firmicutes, Fusobacteria, Planctomycetes,* and *Verrucomicrobia.* Besides, several major typical taxa of bacteria such as *Cetobacterium somerae, Clostridium perfringens, Plesiomonas* spp., *Clostridium XI* sp., *Aeromonas* sp., *Neisseriacea, Clostridiaceae, Acromobacter* sp., *Bacillus* sp., *Citrobacter koseri, Edwardsiella* spp., *Enterobacter cloacae, Pasteurella pneumotropica, Photobacterium damselae, Shewanella putrefaciens, Staphyloccocus* sp., *Vibrio* spp., and unidentified species were identified for the intestine of tilapia (Baldo et al. 2015; Pakingking et al. 2015).

All these and other studies demonstrated that the intestinal of tilapia harbor a complex microbiota, and the main phyla are *Proteobacteria, Fusobacteria, Firmicutes,* and *Bacteroidetes*. Figure 4.2 indicated that summarization of the most common intestinal bacteria phyla of tilapia was reported in several studies.

There are a number of factors that can influence the composition and diversity of the intestinal microbiota of tilapia. Administration of antibiotics such as oxytetracycline for the long-term has been negatively affecting the intestinal microbiota, especially *Bacteroidetes* and *Firmicutes* phyla of tilapia, and this, in turn, affects the growth and immunity (Limbu et al. 2018). Numerous studies reported that the characterization and modulation of microorganisms found in the intestinal of tilapia

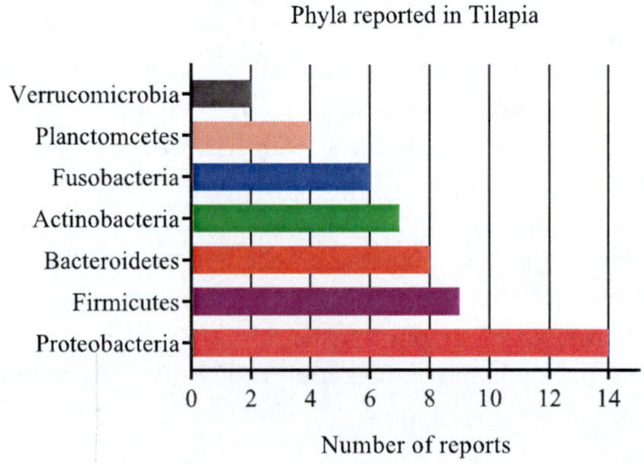

Fig. 4.2 Bacteria phyla observed in the intestinal of tilapia

have a great role and are used as a model for other commercially important fish species.

4.1.2 Function

Studies demonstrated that the intestinal microbiota of fish has several functions including digestion, absorption, development of the mucosal system, and disease (Rawls et al. 2004). In the intestinal of fish diverse microbiota including archaea, protoctista, fungi, yeast, viruses, and bacteria were identified. The intestinal microbiota of fish was dominated by *Proteobacteria* and *Fusobacteria* bacteria phyla. The different sections of the intestine of fish consist of various species of microbiota, and these bacteria were different in density, composition, and function. The bacteria communities were grouped into allochthonous (free-living, transient microbiota associated with the digesta) and autochthonous (colonize the mucosal surface of the digestive tract), and these autochthonous bacteria are considered the core bacteria species of the fish intestine (Banerjee and Ray 2017).

The intestinal microbiota of fish has several functions and regulates many genes. Rawls et al. (2004), confirmed that fish intestinal microbiota regulates the expression of 212 genes, and these genes were related to nutrient metabolism, innate immunity, and stimulation of epithelial proliferation and growth. Studies demonstrated that the specificity of the host response depended on the bacterial species that colonize the digestive tract (Rawls et al. 2004). The flagella of *Pseudomonas* were able to interact with the host cells, and the motility of *Pseudomonas* spp. was also important to stimulate inflammatory signals in zebrafish (Rawls et al. 2004). Furthermore, they demonstrated that the interaction between *Pseudomonas* and the host epithelium was improved by the flagella-dependent swimming motility of the bacteria.

Moreover, Bates et al. (2006) demonstrated the significant role of intestinal microbiota on fish digestive tract differentiation and development, and function. They also revealed that in the absence of intestinal microbiota, the intestine was developed without brush-border intestinal alkaline phosphate activity, immature patterns of glycan expression, and a reduction of goblet cells and enteroendocrine cells. Finally, these effects lead to the abnormal function of the intestine, which proves how much intestinal microbiota is vital for the development and normal function of the intestine of fish. Remarkably, yet, the normal function and structure of the intestine were reversed by the reintroduction of the common intestinal microorganisms of the fish. The intestinal microbiota of fish including *Proteobacteria, Bacteroidetes, Firmicutes,* and *Fusobacteria* phyla also involved in fish nutritional metabolism.

Some species of bacteria which colonize the intestine of fish can produce short-chain fatty acids and vitamins), which may have a beneficial effect on fish growth and health. A large quantity of vitamin B12 (cobalamin) was produced by *Cetobacterium somerae (C. somerae),* which colonized the intestine of several fish species including tilapia (Sugita et al. 1991; Tsuchiya et al. 2008). Some fish species such as carp and tilapia, due to the presence of *C. somerae* in their intestine, have no requirement for dietary vitamin B12.

Furthermore, studies demonstrated that fish intestinal microbiota other than *C. somerae* bacterium contribute a significant role in host metabolism via synthesizing several enzymes and co-enzymes nutrition by providing enzymatic activities corresponding to the fish host (Ray et al. 2012). As well, some bacteria species, *Microbacterium, Micrococcus*, unidentified anaerobes, and yeast have also affected the digestion of the host positively, since these and other microbiota were isolated, and identified from the intestine of fish and can produce a wide range of digestive enzymes including *amylase, protease, lipase, chitinase, cellulose,* and *phytase* (Ray et al. 2012). Although estimation of the contribution of the specific microorganism that colonizes the intestine of the fish is difficult, it is rational to think on the overall microenvironment would be strongly influenced by the predominant populations of microorganisms. To date, our knowledge of the function of a specific microorganism that colonizes the intestine of fish and its effect on the host was increasing due to the modernization and advancement of molecular approaches and new sequencing technologies.

4.1.2.1 The Contribution of the Intestinal Microbiota of Tilapia on Growth, Development Metabolism, and Immunity

The significant role of intestinal microbiota to host digestion, metabolism, and immunity was principally explained in humans (Fouhy et al. 2012) and then in economically valuable animal species including cattle (Li et al. 2019), pig (Jha and Berrocoso, 2015) and chicken (Brisbin et al. 2008). In the previous two decades with the advancement of molecular technologies, the function of intestinal microbiota fish in digestion and metabolism was studied. The intestinal microbiota of fish can break down nondigestible fibers and provides nutrients to the host, and metabolites of the microbes such as enzymes, vitamins, and short-chain fatty acids were absorbed and utilized by the host for growth and development.

In addition to growth and metabolism, the intestinal microbiota of fish can also control the gut-brain axis, innate immunity, and overall fitness of the fish. Intestinal microbiota contributes a significant role in the host health using mechanisms including the development and maintenance of the epithelial cell of the intestine, reducing the growth of pathogenic microbes, supporting the process of angiogenesis, and stimulating the expression of genes related to intestinal development and innate immune system. Besides, these intestinal beneficial bacteria compete for available nutrients with pathogenic bacteria and modify the ecological niche of the bacteria community to prevent the colonization and proliferation of arriving harmful microbes in the fish intestine (Kamada et al. 2013). Fukuda et al. (2012) demonstrated that the beneficial bacteria *Bifidobacterium* can inhibit the growth and invasion of pathogenic *Escherichia coli* using acidification of the fish's intestinal environs. Some other probiotic bacteria species can also synthesize bacteriocins (ribosomal synthesized antimicrobial peptides) and proteinaceous toxins that hinder the growth of members of the same or similar bacterial species. Hence, these and other studies provided evidence that the effect of pathogenic bacteria depended on the stability of the intestinal microbiota of the fish (Galindo-Villegas et al. 2012; He et al. 2017). However, the strong interaction of intestinal microbiota and host health can be affected via several factors, and disease was the main reason (Li et al. 2019; Nie et al. 2017) and mostly resulted in dysbiosis (imbalance of microbiota) of the microbes found in the intestine of the fish (He et al. 2017).

Currently, due to the advancement of molecular technologies, more investigation will require to understand the environmental interactions between members of the microorganisms found in the intestine and host intestinal microbiota interaction in terms of the function and importance of the intestinal microbiota on tilapia health. Development and improvements of techniques applied in genetics and molecular data analyses can provide detailed information on the ecological interaction of microbiota found in the intestinal of tilapia. Roh et al. (2010) used DNA microarrays to determine the regulation of the expression of microbial genes in different parts of the fish intestine. Furthermore, studies revealed that DGGE, TGGE, RNA-seq, and NGS have been used to quickly detection of much autochthonous intestinal microbiota of fish. Moreover, the development of digital transcriptomics techniques was applied to govern the expression level of several microbial genes (Ekblom and Galindo 2011).

4.1.3 Techniques of Modulation of the Intestinal Microbiota of Tilapia

It is clear that aquaculture contributes much to the world's food supply via aquatic products has been increasing these dates, and the main aim of aquaculture is to provide a high-quality product for human consumption. However, several factors negatively affect the production of tilapia in aquaculture, and the primary factor is diseases. Currently, several pathogenic bacteria such as *Streptococcus* sp. and *Francisella noatunensis* subsp. *orientalis*, was affecting the intestine microbiota and health of tilapia (Amal and Zamri-Saad 2011; Soto et al. 2013). These pathogenic bacteria symbolize a wide range of virulence and have different modes of infection. At present, for many bacteria diseases of tilapia, there is no effective treatment, and this results in huge economic loss in tilapia production. Researchers recommend using prevention methods and substitutions to antibiotics including probiotics, prebiotics, paraprobiotics, postbiotics, and feed additives. Understanding the functional significance of intestinal microbiota in many aspects of host physiology and modulation of this intestinal microorganism of fish is a reasonable approach to reducing the number of emerging infectious diseases in aquaculture. To attain this goal, a fundamental understanding of the interaction among fish health, beneficial intestinal microbiota, and pathogens microbes is a prerequisite. These clarifications focus on existing knowledge concerning the links between fish diseases, dysbiosis, and immune responses.

To accomplish the positive alteration of the intestinal microbiota of tilapia, researches used different types of modulators including prebiotics, probiotics, paraprobiotics, postbiotics, and optional feed ingredients in fish diets (Aguilar-Toalá et al. 2018; Wu et al. 2020). Due to the increasing number of fish pathogens, the application of "-biotics" an alternative means to antibiotics, has shown benefits in aquaculture. Improvement of the immunity of farmed fish species via the establishment of normal intestinal microbiota is important, as it affects various biological processes such development and assembly of gut-linked lymphoid tissue and the capability to compete for infectious disease caused by several microorganisms (Merrifield et al. 2011; Ringø et al. 2016).

Several studies indicated administration of probiotics plays a significant role in improving the intestinal microbiota balance and modulate the nonspecific immune system (Gibson & Roberfroid, 1995; Nayak 2010). Banerjee and Ray (2017) also demonstrated the importance of probiotics in the improvement of the immune response and resistance against infection. Furthermore, the probiotics administrated to the fish remain viable in the intestinal of the fish throughout the growth period, and this indicates that probiotics are important to the host via balancing the microbiota communities in the fish intestine (Li et al. 2019; Sayes et al. 2018). Other than probiotics, there are several "-biotics" which have a great contribution to the modulation of the gut microbiota and the improvement of the health of tilapia. Consequently, it is compulsory to start the incorporation of prebiotics, probiotics, or alternative feed ingredients to effectively modify the composition and diversity of the intestinal microbiota of tilapia.

4.1.3.1 Modulation by Watery Environments

The surrounding water contains several species of microbiota, and most of the time; the intestine of tilapia is colonized with these microbiota species at an early stage of growth. Studies revealed that biotic factors, abiotic factors including salinity, the trophic level of the host, and host phylogeny affect the composition and diversity of the intestinal microbiota of different fish species (Sullam et al. 2012). Connected with this, Del'duca et al. (2013) explained the type of different bacteria species isolated from the intestine of tilapia and demonstrated the similarity of the intestinal bacteria and the bacteria collected from the water and its sediments. These are valuable concerns in the experimental design during the exploration of tilapia intestine microorganisms. The type of aquaculture system and the water system are some of the factors that affect the intestinal microbiota of tilapia. Besides, rearing conditions also have a significant role in the intestine microbiota. Studies recommended that rearing conditions should be more similar to the stocking density and system design of the aquaculture since the environment has a great influence on the intestinal microbiota of the host.

Even though the composition of the intestinal microbiota of fish changes quickly at the time of transition from fry to fingerlings stages, many factors contribute to the change. The fish ingested the microbiota found in the water, contributes a lot to the modulation of the intestinal microbiota of the host (Giatsis et al. 2015; Del'duca et al. 2013). The different types of aquaculture systems affect the diversity and establishment of the intestinal microbiota of tilapia, which is explained how a type of aquaculture system (recirculating versus active suspension) influences the majority of the differences of microorganisms to colonize the intestine of the fish larvae (Giatsis et al. 2015). All these studies give evidence that the quality of the water used in the aquaculture system is the most important for the colonization of the gut microbiota of tilapia and is the primary indicator of host-microbial composition.

4.1.3.2 Modulation by Diet

Diet has a great potential to alter the fish intestinal microbiota (Ingerslev et al. 2014). A plethora of studies reported a strong link between the intestinal microbiota of fish with diet, yielding an understanding of the mutual interaction between some microbes and the host fish. These intestinal microbiota communities are also vital for several biological functions of the host including nutrition and metabolism by promoting nutrient uptake and utilization, increasing the production of enzymes, amino acids, short-chain fatty acids, and vitamins, and enhancing the digestion system of the fish (Merrifeld et al. 2010). Digestion enzymes such as carbohydrases, cellulase, phosphatases, esterases, lipases, and proteases were produced by the intestinal microbiota of fish which are responsible for the digestion of different types of diets (Ray et al. 2012; Wu et al. 2015). SCFAs are produced by gut microbiota from indigestible food components such as fiber, oligosaccharides, and polysaccharides via different metabolism channels (Layden et al. 2012; Tan et al. 2014). The SCFAs have a wide range of positive effects on the host, such as providing energy sources for colonic epithelium cells (Pascale et al. 2018), maintaining metabolic homeostasis (Canfora et al. 2015), regulating T regulatory cells (Smith et al. 2013; Furusawa et al. 2013), and anti-inflammatory effects (D'Souza et al. 2017; Ferrer-picón et al. 2019; Sun et al. 2018). Generally, they are indispensable for the growth and well-being of the host when present in sufficient amounts (Leblanc et al. 2017). Intimate relationships also exist between gut bacteria and teleost, and these gut bacteria have a role in supplying the host with volatile fatty acids (Ramirez and Dixon 2003) that provide energy for intestinal epithelial cells (Clements 1997). Furthermore, the gut microbiota of aquatic vertebrates can also produce vitamins and amino acids (Balcázar et al. 2006; Nayak 2010). Sugita et al. (1991) demonstrated that the amount of vitamin B_{12} was positively correlated with the abundance of *Bacteroides* and *Clostridium* in *Oreochromis niloticus*. Nile tilapia fed with a diet deficient in B12, was able to produce 11.2 ng vitamin B12/gram of body weight per day (Lovell and Limsuwan 1982). Tsuchiya et al. (2008) also demonstrated that strains of *Bacteroides* identified from the intestine of tilapia synthesized vitamin B12 at a rate of approximately 8 ng/mL culture in 48 hr.

Here, we discuss the modulation of the gut microbiota of tilapia using diet. Some studies revealed that the diet altered the intestinal microbiota of tilapia (Table 4.1).

4.1.3.3 Modulation by Prebiotics, Probiotics, Paraprobiotics, and Postbiotics

Currently, due to the increase in the number of reports on the negative effect of antibiotics, scientists have been working to develop alternative methods to antibiotics and examining the importance of the application of prophylactic measures such as probiotics, paraprobiotics, postbiotics, and prebiotics (Table 4.2). The word probiotics were defined by the Food and Agriculture Organization of the United Nations/World Health Organization (FAO/WHO) "as living microorganisms, which, once administered in appropriate amounts, confer a health profit on the host." With the advancement of molecular technologies and increasing our knowledge of the health benefits and their interaction with the host cells, the concept of paraprobiotics was proposed to indicate the use of inactivated microbial cells or cell fractions that confer a health benefit to the host (Taverniti and Guglielmetti 2011; Posadas et al. 2012). Moreover, postbiotics which are the soluble products or metabolites secreted by probiotics that have physiological benefits to the host as defined by (Aguilar-Toalá et al. 2018). The effect of the different types of "-biotics" on the modulation of the intestine microbiota has been reported (Nayak 2010; Wu et al. 2020). In this section, our focus is to review the ability of these compounds to modulate the function of the gut microbiota of tilapia and their role in host physiology and function through shifts in the gut microbiome. The role of different types of prebiotics and probiotics on growth, metabolism, composition, and diversity of gut microbiota communities, and the innate immune system of tilapia have been comprehensively studied and reviewed detailed these days (Goutam and Arun Kumar 2017). A plethora of studies confirmed that some probiotics such as *Bacillus* sp. and *Lactobacillus* sp. have a positive effect on the modulation of the gut microbiota of tilapia and their consequences on the host innate immune system and are recommended to provide an alternative strategy to governing many infectious diseases in aquaculture. The impact of live and dead probiotic cells on the nonspecific immune system of *O. niloticus* was investigated, and probiotics treatment increased the nonspecific immune parameters like lysozyme activity, migration of neutrophils, and plasma bacteriocidal activity, leading to improvement of resistance to *Edwardsiella tarda* infection (Taoka et al. 2006).

Manipulation of useful gut bacteria via improving the intestine of tilapia is one of the strategies applied in aquaculture. Studies reported these modulated intestinal microbes could exert positive effects on fish growth by altering the gut morphology, improving digestion and metabolism, and regulating the expression of microbial genes related to growth and appetite control (Rawls et al. 2004; Bates et al. 2006; Round and Mazmanian 2009). The role of intestinal microbes on the gut-brain axis was studied. Probiotic *Lactobacillus* bacteria within the intestinal microbiota of fish were associated with improved learning/memory capacity (Borrelli et al. 2016).

Table 4.1 Studies assessing the influence of diet on the microbiota and health of tilapia

Type of diet	Effect on gut microbiota	Effect on the host	References
Resveratrol	Increased the ratio of beneficial microbial taxa (*Acetobacteraceae* and Methylobacteriaceae), while the proportion of harmful microbial taxa decreased, e.g., Streptococcaceae deceased	Improve the growth and immunity of fish	Zheng et al. (2018)
Potassium diformate	Improved the relative richness of some intestinal allochthonous bacteria	Improved growth performance and feed conversion ratio	Zhou et al. (2009)
Garlic	Increased bacterial diversity, increase the abundance of *Proteobacteria* and *Tenericutes,* the phyla associated with healthy intestinal flora	Increase resistance against *streptococcus iniae*	Foysal et al. (2019)
Nucleotides	Moderate the intestinal microbiota	Improve the growth, feed utilization, intestinal growth, antioxidant status, and nonspecific immune response	Xu et al. (2015)
Betaine and the antibiotic florfenicol	Betaine can promote intestinal autochthonous bacteria	Improve immunity	He et al. (2012)
DL-methionyl-DL-methionine	Improve the diversity of gut microbiota	Increase growth performance and antioxidant ability	Guo et al. (2019)
Turmeric (Curcuma longa)	Total bacterial count and total lactobacillus count of intestinal content	Improve growth performance	Yusuf et al. (2017)
Low protein diet	Significantly decreased intestinal microbial diversity, *Bacteroidetes* were more abundant, and *Proteobacteria* and *Firmicutes* were less abundant	Restrict growth and weaken disease resistance	Zhu et al. (2020)
Sodium chloride	Increase in microbial diversity	Increased protein digestibility, dietary ions, and nutrient transporters	Hallali et al. (2018)

Administration of a combination of probiotic bacteria, *Bacillus subtilis*, *Saccharomyces cerevisiae*, and *Aspergillus* to tilapia showed a strong association among these microbes in the host intestine (Iwashita et al. 2015). Besides, this study demonstrated that the intestinal microbiota-host association permitted the upregulation of immune response and reduced the mortality rate of tilapia against *A. hyrdophila* and *S. iniae* challenge. Another study performed using a commercial probiotic Aquastart® Growout (a combination of *B. subtilis*, *E. faecium*, *Lactobacillus reuteri,* and *Pediococcus acidilactici*) were applied to tilapia and resulted improved growth

Table 4.2 Studies evaluating the influence of probiotic microbes on the function and composition of intestinal microbiota and the health of tilapia

Type of prebiotics or probiotics	Effect on gut microbiota	Effect on the host	
Rummeliibacillus stabekisii	Increased the abundance of *bacillus* and *lactobacillus* spp. and reduced abundances of pathogenic bacteria (*streptococcus* and *staphylococcus* spp.) in intestines	Improved weight gain, feed conversion ratio, and feed efficiency, increased intestinal digestive enzymes; improved immunity	Tan et al. (2019)
Lactobacillus rhamnosus JCM1136 and *Lactococcus lactis* subsp. *lactis* JCM5805	Decease the abundance of *Plesiomonas* and increased *rhizobium* and *Achromobacter*	Improved immunity of the host	Xia et al. (2018a, b)
Saccharomyces cerevisiae or *Bacillus subtilis*	Significant higher number of yeast and b*acillus* spp. cells counts in the intestinal of fish fed on a diet supplemented with probiotics	Improved immunity of the host	Opiyo et al. (2019)
Pediococcus acidilactici	Increase the abundance of lactic acid bacteria	Enhanced some aspects of the nonspecific immune response	Ferguson et al. (2010)
Lactobacillus reuteri, B. subtilis, enterococcus faecium, and *Pediococcus acidilactici*	Improve the growth of beneficial bacteria	Enhanced intestinal morphology by elevating the absorptive surface area	Standen et al. (2015)
Lactobacillus plantarum CCFM8610	Modulate gut microbiota	Promote growth performance and prevent the death of fish exposed to cd	Zhai et al. (2017)
Lactobacillus plantarum	Improve the growth of beneficial gut bacteria	Improve immunity	Foysal et al. (2019)
Lactobacillus plantarum	Lactic acid bacteria were increased	Improved growth and feed efficiency	Jatoba et al. (2011)
B. subtilis	Modulate the gut microbiota	Improve growth performance	Giatsis et al. (2015)
Lactococcus lactis subsp. *lactis* JCM5805	Modulate gut microbiota	Upregulated the expression of *IFNα* via the *TLR7/TLR9-Myd88* pathway and enhanced disease resistance of larvae	Xia et al. (2018a, b)
Bacillus cereus	Affected the autochthonous gut bacteria community of tilapia and	Enhance the immune status	Wang et al. (2016)

(continued)

Table 4.2 (continued)

Type of prebiotics or probiotics	Effect on gut microbiota	Effect on the host	
	stimulated various potentially beneficial bacteria		
Lacillus subtilis C-3102	The increased amount of gut bacteria altered the autochthonous gut bacterial communities	Improve fish immunity	He et al. (2013)
Schizochytrium species	Enhanced the richness (represented by the Chao index) of bacteria and increase the abundance of the bacterial phylum *Firmicutes*	Improve structure and integrity of the intestinal villi Modulatory effects on blood cells	de Souza et al. (2020)
Lactococcus lactis JCM5805	Altered the composition of host gut microbiota	Improve metabolic pathways including carbohydrate metabolism, nucleic acid metabolism, energy metabolism, and translation	Xia et al. (2020)
Saccharomyces cerevisiae or *B. subtilis*	Improved the growth of beneficial intestinal bacteria	Improves immunity and enhances fish flesh quality	Opiyo et al. (2019)
Lactobacillus brevis JCM 1170 and *Lactobacillus acidophilus* JCM 1132	Altered gut microbiota	Improve the immunity of the fish	Liu et al. (2013)
Clostridium butyricum	Improved the increased diversity of the intestinal microbiota and the relative abundance of beneficial bacteria (*bacillus*). Reduced the relative abundance of opportunistic pathogenic bacteria (*Aeromonas*)	Improved the specific growth rate and feed intake and increased resistance against *S. agalactiae*	Li et al. (2019)
Clostridium butyricum	Improve the growth of beneficial bacteria	Promoting growth, feed utilization, gut health	Poolsawat et al. (2019)
Mannan-oligosaccharide	Modulate the intestinal microbiota by increasing the number of beneficial bacteria species	Improve immunity	Levy-Pereira et al. (2018)
Oligochitosan	Improve the growth of beneficial gut bacteria	Improve the growth, and immunity and enhance resistance against *A. hydrophila* infection	Shi et al. (2020)

(continued)

Table 4.2 (continued)

Type of prebiotics or probiotics	Effect on gut microbiota	Effect on the host	
Lactogen 13	Modulate gut microbiota communities by increasing the proportion of *lactobacillus*. Decreasing the proportion of potential pathogens	Improved the expression of genes responsible for growth and appetite control and activating the endocrine system	Giorgia et al. (2018)
β-Glucan	Increased the Chao richness and changed the composition of gut microbiota	Modulating plasma glucose concentrations, decreasing mortality, improving immunity	de Souza et al. (2020)
Chito-oligosaccharides	Changed autochthonous gut bacteria,	Improve intestinal health, and improve resistance to infection by *A. hydrophila*	Qin et al. (2014)
Inulin	Improve the composition of α-diversity, and β-diversity of gut microbiota	Improve growth and alleviate oxidative stress o	Zhou et al. (2020)

and enhanced intestinal immunological status (Standen et al. 2016). Administration of a mixed probiotic *Bacillus licheniformis* and a prebiotic, yeast extract can improve the growth and feed utilization of Nile tilapia (*O. niloticus*) (Hassaan et al. 2014). Likewise, the administration of a mixture of exogenous enzymes phytase, protease, and xylanase with a probiotic containing *B. subtilis*, *B. licheniformis*, and *Bacillus pumilus* results in enhanced growth and fiber breakdown (Melo-Boliver et al. 2019). Correspondingly, this combination of probiotics and prebiotics increased the digestion of indigestible nonstarch polysaccharides and trypsin inhibitors, which may produce necrotic enteritis in Nile tilapia (Adeoye et al. 2016). Some microbes such as *Enterococcus*, *Myroides,* and *Exiguobacterium* isolated from the intestine of Nile tilapia were improved the survival rate of the fish after challenged with *Edwardsiella tarda* via enhancing the innate immunity (Villamil Diaz and Esguerra Rodriguez 2017).

Probiotics *L. rhamnosus* GG can improve the host's intestinal microflora balance and enhance the immune system and disease resistance of tilapia after being challenged by *Edwardsiella tarda* and *Streptococcus agalactiae* (Pirarat et al. 2006; Lahti et al. 2013; Pirarat et al. 2015). In addition to several species of bacteria, some yeast species including *Saccharomyces cerevisiae, Kloeckera apiculata, Candida* sp., *Metcschnikowia* sp., and *Rhodotorula* sp., have been used as a probiotic for tilapia (Ayyat et al. 2014). Besides these and other species of yeast have immunostimulants and play a significant role in the growth of fish (Gatesoupe 2007).

4.1.4 Conclusions and Future Perspective

Tilapia is one of the most cultivable fish species in the world. In the previous three decades, studies on the intestinal microbiota of tilapia increased with the aim of optimum manipulation of these microorganisms. The intestine of tilapia was harbored by different species of microbiota, and the dominant bacteria phyla were *Proteobacteria, Firmicutes, Fusobacteria,* and *Bacteroidetes.* The intestinal microbiota has been affecting the host in more than one way. Considering the importance of these bacteria phyla and other microbiota species in the overall fitness of tilapia in aquaculture, scientists have been characterizing and developing many techniques to manipulate the intestinal microbiota to increase the production and productivity of the fish. Conversely, the detailed mechanisms of these intestinal microbiota modulators are required to be fully explored. Presently, several infectious diseases of tilapia have been treated using antibiotics. However, a plethora of studies revealed that antibiotics affect the dynamics of fish intestinal microbiota and mostly cause dysbiosis. Furthermore, increasing the application of antibiotics has fortified antibiotic resistance microbiota, and these results in a condition difficult to fight against pathogens (Kalia et al. 2014). To date, the advancement of "forward microbiomics" was played a great role in manipulating the intestinal microbiota to improve the growth and health of tilapia. Modulation of the intestinal microbiota of tilapia using different diets and "-biotics" have been applied. Most of the time, probiotic bacteria including *Lactobacilli* species, have been used in a number of fish species including tilapia (He et al. 2017). But to date, the function of each intestinal microbiota taxa and modulation of these beneficial microorganisms is not fully understood. The research highlighted above is an important starting point for optimally manipulating the beneficial microbiota of tilapia.

Like other fish species, the intestinal microbiota of tilapia was influenced by many factors; however, the effect of each of these factors on the growth, metabolism, immunity, and physiology of fish remains poorly understood. For drawing meaningful conclusions, this review summarized the importance of intestinal microbiota and the modulation methods of microbiome structure and function. Furthermore, to create a comparative picture of the intestinal microbiota of tilapia needs a careful review of the studies employed using the different advanced molecular technologies such as DNA microarray, next-generation, and third-generation is required to increase our understanding of the role of intestinal microbiota to increase the production and productivity of tilapia. To date, scientists have been working to manipulate the intestinal microbiota to use as biomarkers for the health status of the fish and to assess the detail to uncover functional variation with diet or host factors. Conversely, the detailed mechanisms of these intestinal microbiota modulators are required to be fully explored. This review highlighted collecting information on the characterization and function of intestinal microbiota on tilapia growth, development, and immunity and its modulation methods. Gathering information from different studies would help to identify and characterize the intestinal microbiota of tilapia to use as biomarkers for the status of the fish in aquaculture. As a result, more study is required on the characterization and identification of specific taxa of intestinal microbiota, and detailed work is necessary to investigate the fundamental association between tilapia fitness and the function of each intestinal microbiota, which may allow us to enhance our knowledge on the importance of intestinal microbiota on improving tilapia aquaculture.

Acknowledgments This study was funded by National Natural Science Foundation of China (NSFC 31925038, 32061133004, 32172991, 32172958, 32122088, 32202959, 32102812, U21A20267 and 31972807); Agriculture Science and Technology Innovation Progam (ASTIP) of the Chinese Academy of Agricultural Sciences (CAAS-ZDRW202305 and CAAS-ASTIP-2023-IFR-05).

References

Adeoye AA, Yomla R, Jaramillo-Torres A, Rodiles A, Merrifield DL, Davies SJ (2016) Combined effects of exogenous enzymes and probiotic on Nile tilapia (*Oreochromis niloticus*) growth, intestinal morphology and microbiome. Aquaculture 463:61–70. https://doi.org/10.1016/j.aquaculture.2016.05.028

Aguilar-Toalá JE, Garcia-Varela R, Garcia HS, Mata-Haro V, González-Córdova AF, Vallejo-Cordoba B, Hernández-Mendoza A (2018) Postbiotics: an evolving term within the functional foods. Trends Food Sci Technol 75:105–114

Amal M, Zamri-Saad M (2011) Streptococcosis in tilapia (*Oreochromis niloticus*): a review. Pertanika J Trop Agric Sci 34:195–206

Ayyat MS, Labib HM, Mahmoud HK (2014) A probiotic cocktail as a growth promoter in Nile Tilapia (Oreochromis niloticus). J Appl Aquac 26(3):208–215. https://doi.org/10.1080/10454438.2014.934164

Balcázar JL, de Blas I, Ruiz-Zarzuela I, Cunningham D, Vendrell D, Múzquiz JL (2006) The role of probiotics in aquaculture. Vet Microbiol 114(3–4):173–186. https://doi.org/10.1016/j.vetmic.2006.01.009

Baldo L, Riera JL, Tooming-Klunderud A, Albà MM, Salzburger W (2015) Gut microbiota dynamics during dietary shift in eastern African cichlid fishes. PLoS One 10(5):e0127462

Banerjee G, Ray AK (2017) The advancement of probiotics research and its application in fish farming industries. Res Vet Sci 115:66–77

Bates JM, Mittge E, Kuhlman J, Baden KN, Cheesman SE, Guillemin K (2006) Distinct signals from the microbiota promote different aspects of zebrafish gut differentiation. Dev Biol 297(2):374–386. https://doi.org/10.1016/j.ydbio.2006.05.006

Bereded NK, Curto M, Domig KJ, Abebe GB, Fanta SW, Waidbacher H, Meimberg H (2020) Metabarcoding analyses of gut microbiota of Nile tilapia (*Oreochromis niloticus*) from Lake Awassa and Lake Chamo, Ethiopia. Microorganisms 8(7):1040

Borrelli L, Aceto S, Agnisola C, De Paolo S, Dipineto L, Stilling RM, Dinan TG, Cryan JF, Menna LF, Fioretti A (2016) Probiotic modulation of the microbiota-gut-brain axis and behaviour in zebrafish. Sci Rep 6:30046. https://doi.org/10.1038/srep30046

Brisbin JT, Gong J, Sharif S (2008) Interactions between commensal bacteria and the gut-associated immune system of the chicken. Anim Health Res Rev 9(1):101–110. https://doi.org/10.1017/S146625230800145X

Burns AR, Stephens WZ, Stagaman K, Wong S, Rawls JF, Guillemin K, Bohannan BJ (2016) Contribution of neutral processes to the assembly of gut microbial communities in the zebrafish over host development. ISME J 10(3):655–664

Cahill MM (1990) A review virulence factors in motile Aeromonas species. Journal App Biotechnol 69(1):1–16

Canfora EE, Jocken JW, Blaak EE (2015) Short-chain fatty acids in control of body. Nat Publ Gr 11:577. https://doi.org/10.1038/nrendo.2015.128

Clements KD (1997) Fermentation and gastrointestinal microorganisms in fishes. In: Mackie RI, White BA (eds) Gastrointestinal microbiology. Springer, Boston, MA, pp 156–198

Clarke SF, Murphy EF, O'Sullivan O, Lucey AJ, Humphreys M, Hogan A, Hayes P, O'Reilly M, Jeffery IB, Wood-Martin R, Kerins DM, Quigley E, Ross RP, O'Toole PW, Molloy MG, Falvey E, Shanahan F, Cotter PD (2014) Exercise and associated dietary extremes impact on gut microbial diversity. Gut 63(12):1913–1920

Del'duca A, Cesar DE, Diniz CG, Abreu PC (2013) Evaluation of the presence and efficiency of potential probiotic bacteria in the gut of tilapia (*Oreochromis niloticus*) using the fluorescent in situ hybridization technique. Aquaculture 388–391:115–121

D'Souza WN, Douangpanya J, Mu S, Jaeckel P, Zhang M, Maxwell JR et al (2017) Differing roles for short chain fatty acids and GPR43 agonism in the regulation of intestinal barrier function and immune responses. PLoS One 12(7):e0180190

De Schryver P, Vadstein O (2014) Ecological theory as a foundation to control pathogenic invasion in aquaculture. ISME J 8(12):2360–2368

de Bruijn I, Liu Y, Wiegertjes GF, Raaijmakers JM (2017) Exploring fish microbial communities to mitigate emerging diseases in aquaculture. FEMS Microbiol Ecol 94(1):fix161

de Souza FP, deLima ECS, Pandolfi VCF, Leite NG, Furlan-Murari PJ, SeinoLeal CN et al (2020) Effect of β-glucan in water on growth performance, blood status and intestinal microbiota in tilapia under hypoxia. Aquac Rep 17:100369. https://doi.org/10.1016/j.aqrep.2020.100369

Ekblom R, Galindo J (2011) Applications of next generation sequencing in molecular ecology of non-model organisms. Heredity (Edinb) 107(1):1–15

Falcinelli S, Picchietti S, Rodiles A, Cossignani L, Merrifield DL, Taddei AR, Maradonna F, Olivotto I, Gioacchini G, Carnevali O (2015) *Lactobacillus rhamnosus* lowers zebrafish lipid content by changing gut microbiota and host transcription of genes involved in lipid metabolism. Sci Rep 5:9336

Fan L, Chen J, Meng S, Song C, Qiu L, Hu G (2017) Characterization of microbial communities in intensive GIFT tilapia (*Oreochromis niloticus*) pond systems during the peak period of breeding. Aquac Res 48:459–472

FAO (2020) The state of world fisheries and aquaculture (SOFIA). Italy, Rome

Ferguson RMW, Merrifield DL, Harper GM, Rawling MD, Mustafa S, Picchietti S et al (2010) The effect of *Pediococcus acidilactici* on the gut microbiota and immune status of on-growing red tilapia (*Oreochromis niloticus*). J App Microbiol 109(3):815–862

Ferrer-Picón E, Dotti I, Corraliza AM, Mayorgas A, Esteller M, Perales JC et al (2019) Intestinal inflammation modulates the epithelial response to butyrate in patients with inflammatory bowel disease. Inflamm Bowel Dis 26:43. https://doi.org/10.1093/ibd/izz119

Fitzsimmons KM (2016) Global Tilapia Market update 2015. In: WAS 2016, Las Vegas

Fouhy F, Ross RP, Fitzgerald GF, Stanton C, Cotter PD (2012) Composition of the early intestinal microbiota: knowledge, knowledge gaps and the use of high-throughput sequencing to address these gaps. Gut Microbes 3(3):203–220. https://doi.org/10.4161/gmic.20169

Foysal MJ, Nguyen T, Chaklader MR, Siddik M, Tay CY, Fotedar R, Gupta SK (2019) Marked variations in gut microbiota and some innate immune responses of freshwater crayfish, marron (*Cherax cainii*, Austin 2002) fed dietary supplementation of *clostridium butyricum*. PeerJ 7: e7553

Furusawa Y, Yuuki Obata Y, Shinji Fukuda S, Takaho A, Endo TA, Nakato G, Takahashi D et al (2013) Commensal microbe-derived butyrate induces the differentiation of colonic regulatory T cells. Nature 504:446–450. https://doi.org/10.1038/nature12721

Fukuda S, Toh H, Hase K, Oshima K, Nakanishi Y, Yoshimura K, Tobe T, Clarke JM, Topping DL, Suzuki T, Taylor TD, Itoh K, Kikuchi J, Morita H, Hattori M, Ohno H (2012) Bifidobacteria can protect from enteropathogenic infection through production of acetate. Nature 469(5):543–547

Gatesoupe FJ (2007) Live yeasts in the gut: natural occurrence, dietary introduction, and their effects on fish health and development. Aquaculture 267(1–4):20–30

Galindo-Villegas J, García-Moreno D, de Oliveira S, Meseguer J, Mulero V (2012) Regulation of immunity and disease resistance by commensal microbes and chromatin modifications during zebrafish development. Proc Natl Acad Sci U S A 109(39):E2605–E2614

Giatsis C, Sipkema D, Smidt H, Heilig H, Benvenuti G, Verreth J, Verdegem M (2015) The impact of rearing environment on the development of gut microbiota in tilapia larvae. Sci Rep 5:18206

Gibson GR, Roberfroid MB (1995) Dietary modulation of the human colonic microbiota: introducing the concept of prebiotics. J Nutr 125(6):1401–1412. https://doi.org/10.1093/jn/125.6.1401

Giorgia G, Elia C, Andrea P, Cinzia C, Stefania S, Ana R (2018) Effects of Lactogen 13, a new probiotic preparation, on gut microbiota and endocrine signals controlling growth and appetite of *Oreochromis niloticus* juveniles. Microbiol Ecol 76:1063–1074

Banerjee G, Ray AK (2017) The advancement of probiotics research and its application in fish farming industries. Res Vet Sci 115:66–77. https://doi.org/10.1016/j.rvsc.2017.01.016

Guo TY, Zhao W, He JY, Liao SY (2019) Dietary DL-methionyl-DL-methionine supplementation increased growth performance, antioxidant ability, the content of essential amino acids and improved the diversity of intestinal microbiota in Nile tilapia (*Oreochromis niloticus*). Br J Nutr 14(1):72–83. https://doi.org/10.1017/S0007114519002289

Hallali E, Kokou F, Chourasia TK, Nitzan T, Con P, Harpaz S, Mizrahi I, Cnaani A (2018) Dietary salt levels affect digestibility, intestinal gene expression, and the microbiome, in Nile tilapia (Oreochromis niloticus). PLoS One 13(8):e0202351. https://doi.org/10.1371/journal.pone.0202351

Hassaan MS, Soltan MA, Ghonemy MMR (2014) Effect of synbiotics between bacillus licheniformis and yeast extract on growth, hematological and biochemical indices of the Nile tilapia (Oreochromis niloticus). Egypt J Aquat Res 40:199–208. https://doi.org/10.1016/j.ejar.2014.04.001

He S, Ran C, Qin C, Li S, Zhang H, de Vos WM, Ringø E, Zhou Z (2017) Anti-infective effect of adhesive probiotic lactobacillus in fish is correlated with their spatial distribution in the intestinal tissue. Sci Rep 7:13195. https://doi.org/10.1038/s41598-017-13466-1

He S, Zhang Y, Xu L, Yang Y, Marubshi T, Zhou Z, Yao B (2013) Effects of dietary *Bacillus subtilis* C-3102 on the production, intestinal cytokine expression and autochthonous bacteria of hybrid tilapia *Oreochromis niloticus* ♀ × *Oreochromis aureus* ♂. Aquaculture 412–413:125–130

He S, Zhou Z, Liu Y, Cao Y, Meng K, Shi P (2012) Do dietary betaine and the antibiotic florfenicol influence the intestinal autochthonous bacterial community in hybrid tilapia (*Oreochromis niloticus* ♀ × *O. aureus* ♂)? World J Microbiol Biotechnol 28:785–791

Ingerslev HC, Strube ML, von Jørgensen LG, Dalsgaard I, Boye M, Madsen L (2014) Diet type dictates the gut microbiota and the immune response against *Yersinia ruckeri* in rainbow trout (*Oncorhynchus mykiss*). Fish Shellfish Immunol 40:624–633. https://doi.org/10.1016/j.fsi.2014.08.021

Iwashita MKP, Nakandakare IB, Terhune JS, Wood T, Ranzani-Paiva MJT (2015) Dietary supplementation with *Bacillus subtilis, Saccharomyces cerevisiae* and *aspergillus oryzae* enhance immunity and disease resistance against *Aeromonas hydrophila* and *streptococcus iniae* infection in juvenile tilapia *Oreochromis niloticus*. Fish Shellfish Immunol 43:60–66. https://doi.org/10.1016/j.fsi.2014.12.008

Jatoba A, Viera FN, Buglione NCC et al (2011) Diet supplemented with probiotic for Nile tilapia in polyculture system with marine shrimp. Fish Physiol Biochem 37:725–732

Jha R, Berrocoso JD (2015) Review: dietary fiber utilization and its effects on physiological functions and gut health of swine. Anim: Int J Anim Biosci 9(9):1441–1452. https://doi.org/10.1017/S1751731115000919

Junning C, PingSun L, Yongju L, Xinhua Y, Yongming Y (2018) Improving the performance of tilapia farming under climate variation. FAO, FAO Fisheries and aquaculture technical paper. Rome, p 608

Kalia VC, Wood TK, Kumar P (2014) Evolution of resistance to quorum-sensing inhibitors. Microb Ecol 68(1):13–23. https://doi.org/10.1007/s00248-013-0316-y

Kamada N, Seo SU, Chen GY, Núñez G (2013) Role of the gut microbiota in immunity and inflammatory disease. Nat Rev Immunol 13(5):321–335

Lahti L, Salonen A, Kekkonen RA, Salojarvi J, Jalanka-Tuovinen J, Palva A, Oresic M, deVos WM (2013) Associations between the human intestinal microbiota, *lactobacillus rhamnosus* GG and serum lipids indicated by integrated analysis of high-throughput profiling data. PeerJ 1:e32. https://doi.org/10.7717/peerj.32

Layden BT, Angueira AR, Brodsky M, Durai V, Jr W LL (2012). Short chain fatty acids and their receptors: new metabolic targets. Transl Res 161:131–140

Leblanc JG, Chain F, Martín R, Humarán LGB, Courau S, Langella P (2017) Beneficial effects on host energy metabolism of short–chain fatty acids and vitamins produced by commensal and probiotic bacteria. Microb Cell Factories 1–10. https://doi.org/10.1186/s12934-017-0691-z

Levy-Pereira N, Yasui GS, Cardozo MV, Neto JD, Vaz Farias TH, Sakabe R, de Pádua SB (2018) Immunostimulation and increase of intestinal lactic acid bacteria with dietary mannan-oligosaccharide in Nile tilapia juveniles. R Bras Zootec 47:e20170006. https://doi.org/10.1590/rbz4720170006

Li H, Zhou Y, Ling H, Lou L, Qi D, Feng L (2019) The effect of dietary supplementation with clostridium butyricum on the growth performance, immunity, intestinal microbiota and disease resistance of tilapia (Oreochromis niloticus). PLoS One 14(12):e0223428. https://doi.org/10.1371/journal.pone.0223428

Limbu SM, Zhou L, Sun SX, Zhang ML, Du ZY (2018) Chronic exposure to low environmental concentrations and legal aquaculture doses of antibiotics cause systemic adverse effects in Nile tilapia and provoke differential human health risk. Environ Int 115:205–219

Liu W, Ren P, He S, Xu L, Yang Y, Gu Z, Zhou Z (2013) Comparison of adhesive gut bacteria composition, immunity, and disease resistance in juvenile hybrid tilapia fed two different *lactobacillus* strains. Fish Shellfish Immunol 35(1):54–62. https://doi.org/10.1016/j.fsi.2013.04.010

Lovell T, Limsuwan T (1982) Intestinal synthesis and dietary non essentiality of vitamin B12 for Tilapia nilotica. Trans of Amer Fish Soc 11:485–490

Molinari LM, De Oliveira SD, Pedroso RB, De Lucas Rodrigues Bittencourt N, Nakamura CV, Ueda-nakamura T, de Abreu Filho BA, Dias Filho BP, Oliveira SD, Lucas N, Ueda T, Carneiro-Filho BA, Scoaris DD (2003) Bacterial microflora in the gastrointestinal tract of Nile tilapia, *Oreochromis niloticus*, cultured in a semi-intensive system. Acta Sci Biol Sci 25(2):267–271

McFall-Ngai M, Hadfield MG, Bosch TC, Carey HV, Domazet-Lošo T, Douglas AE, Dubilier N, Eberl G, Fukami T, Gilbert SF, Hentschel U, King N, Kjelleberg S, Knoll AH, Kremer N, Mazmanian SK, Metcalf JL, Nealson K, Pierce NE, Rawls JF, Reid A, Ruby EG, Rumpho M, Sanders JG, Tautz D, Wernegreen JJ (2013) Animals in a bacterial world, a new imperative for the life sciences. Proc Natl Acad Sci U S A 110(9):3229–3236

Melo-Boliver JF, Ruiz Pardo RY, Hume ME, Nisbet DJ, Rodríguez-Villamizar F, Alzate JF, Junca H, Díaz LMV (2019) Establishment and characterization of a competitive exclusion bacterial culture derived Nile tilapia (*Oreocghromis niloticus*) gut microbiomes showing antibacterial activity against pathogenic *Streptococcus agalactiae*. PLoS One 14(5):e0215375

Merrifeld DL, Dimitroglou A, Foey A, Davies SJ, Baker RTM et al (2010) The current status and future focus of probiotic and prebiotic applications for salmonids. Aquaculture 302:1–18

Merrifield DL, Harper GM, Mustafa S et al (2011) Effect of dietary alginic acid on juvenile tilapia (*Oreochromis niloticus*) intestinal microbial balance, intestinal histology and growth performance. Cell Tissue Res 344:135–146

Nie L, Zhou QJ, Qiao Y, Chen J (2017) Interplay between the gut microbiota and immune responses of ayu (*Plecoglossus altivelis*) during *vibrio anguillarum* infection. Fish Shellfish Immunol 68: 479–487

Nayak SK (2010) Role of gastrointestinal microbiota in fish. Aquac Res 41:1553–1573

Nguyen NH (2015) Genetic improvement for important farmed aquaculture species with a reference to carp, tilapia and prawns in Asia: achievements, lessons and challenges. Fish Fish (Oxf) 17(2): 483–506

Opiyo MA, Jumbe J, Charles Ngugi C, Charo-Karisa H (2019) Dietary administration of probiotics modulates non-specific immunity and gut microbiota of Nile tilapia (*Oreochromis niloticus*) cultured in low input ponds. Int J Vet Sci Med 7(1):1–9

Oyarbide U, Iturria I, Rainieri S, Pardo MA (2015) Use of gnotobiotic zebrafish to study *vibrio anguillarum* pathogenicity. Zebrafish 12(1):71–80

Pascale A, Marchesi N, Marelli C, Coppola A, Luzi L, Govoni S et al (2018) Microbiota and metabolic diseases. Endocrine 61:357–371

Pakingking R, Palma P, Usero R (2015) Quantitative and qualitative analyses of the bacterial microbiota of tilapia (*Oreochromis niloticus*) cultured in earthen ponds in The Philippines. World J Microbiol Biotechnol 31:265–275

Pirarat N, Kobayashi T, Katagiri T, Maita M, Endo M (2006) Protective effects and mechanisms of a probiotic bacterium *lactobacillus rhamnosus* against experimental *Edwardsiella tarda* infection in tilapia (*Oreochromis niloticus*). Vet Immunol Immunopathol 113:339–347. https://doi.org/10.1016/j.vetimm.2006.06.003

Pirarat N, Pinpimai K, Rodkhum C, Chansue N, Ooi EL, Katagiri T, Maita M (2015) Viability and morphological evaluation of alginate-encapsulated *lactobacillus rhamnosus* GG under simulated tilapia gastrointestinal conditions and its effect on growth performance, intestinal morphology and protection against *Streptococcus agalactiae*. Anim Feed Sci Technol 207:93–103

Ponzoni RW, Nguyen HN, Khaw HL, Hamzah A, Abu Bakar KR et al (2011) Genetic improvement of Nile tilapia (Oreochromis niloticus) with special reference to the work conducted by the WorldFish center with the GIFT strain. Rev Aquac 3(1):27–41

Poolsawat L, Li X, He M, Ji D, Leng X (2019) *Clostridium butyricum* as probiotic for promoting growth performance, feed utilization, gut health and microbiota community of tilapia (*Oreochromis nilotics* ×*O. aureus*). Aquac Nutr 00:1–14

Posadas GA, Broadway PR, Thornton JA, Carroll JA, Lawrence A, Corley JR et al (2012) Yeast pro- and paraprobiotics have the capability to bind pathogenic bacteria associated with animal disease. Transl Anim Sci 1:60–68

Qin C, Zhang Y, Liu W, Xu L, Yang Y, Zhou Z (2014) Effects of chito-oligosaccharides supplementation on growth performance, intestinal cytokine expression, autochthonous gut bacteria and disease resistance in hybrid tilapia *Oreochromis niloticus* ♀ × *Oreochromis aureus* ♂. Fish Shellfish Immunol 40(1):267–274

Ramirez RF, Dixon BA (2003) Enzyme production by obligate intestinal anaerobic bacteria isolated from oscars (Astronotus ocellatus), angelfish (Pterophyllum scalare) and southern flounder (Paralichthys lethostigma). Aquaculture 227:417–426. https://doi.org/10.1016/S0044-8486(03)00520-9

Rawls JF, Samuel BS, Gordon JI (2004) Gnotobiotic zebrafish reveal evolutionarily conserved responses to the gut microbiota. Proc Natl Acad Sci U S A 101:4596–4601

Ray AK, Ghosh K, Ringø E (2012) Enzyme-producing bacteria isolated from fish gut: a review. Aquac Nutr 18:465–492

Ringø E, Zhou Z, Vecino JLG et al (2016) Effect of dietary components on the gut microbiota of aquatic animals. A never-ending story? Aquac Nutr 22:219–282

Roh SW, Abell GCJ, Kim K-H, Nam Y-D, Bae J-W (2010) Comparing microarrays and next-generation sequencing technologies for microbial ecology research. Trends Biotechnol 28:291–299

Romero J, Navarrete P (2006) 16S rDNA-based analysis of dominant bacterial populations associated with early life stages of coho salmon (*Oncorhynchus kisutch*). Microb Ecol 51:422–430

Round JL, Mazmanian SK (2009) The gut microbiota shapes intestinal immune responses during health and disease. Nat Rev Immunol 9(5):313–323. https://doi.org/10.1038/nri2515

Sayes C, Leyton Y, Riquelme C (2018) Probiotic bacteria as an healthy alternative for fish aquaculture. InTech. https://doi.org/10.5772/intechopen.71206

Shi F, Qiu X, Nie L, Hu L, Babu S, Lin Q et al (2020) Effects of oligochitosan on the growth, immune responses and gut microbes of tilapia (*Oreochromis niloticus)*. Fish Shellfish Immunol 106:563–573. https://doi.org/10.1016/j.fsi.2020.07.049

Smith PM, Howitt MR, Panikov N, Michaud M, Gallini CA, Bohlooly M, Glickman JN, Garrett WS (2013) The microbial metabolites, short-chain fatty acids, regulate colonic T_{reg} cell homeostasis. Science 341(6145):569–573

Smith P, Willemsen D, Popkes M, Metge F, Gandiwa E, Reichard M, Valenzano DR (2017) Regulation of life span by the gut microbiota in the short-lived African turquoise killifish. elife 6:e27014

Soto E, McGovern-Hopkins K, Klinger-Bowen R, Fox BK, Brock J, Antonio N, Waal ZV, Rushton S, Mill A, Tamaru CS (2013) Prevalence of Francisella noatunensis subsp. orientalis in cultured tilapia on the Island of Oahu, Hawaii. J Aquat Anim Health 25(2):104–109. https://doi.org/10.1080/08997659.2013.781554

Souza FP, Lima ECS, Urrea-Rojas AM, Suphoronski SA, Facimoto CT, Bezerra Júnior JS et al (2020) Effects of dietary supplementation with a microalga (*Schizochytrium* sp.) on the hemato-immunological, and intestinal histological parameters and gut microbiota of Nile tilapia in net cages. PLoS One 15(1):e0226977

Standen BT, Rodiles A, Peggs DL, Davies SJ, Santos GA, Merrifield DL (2015) Modulation of the intestinal microbiota and morphology of tilapia, *Oreochromis niloticus*, following the application of a multi-species probiotic. Appl Microbiol Biotechnol 99(20):8403–8417

Standen BT, Peggs DL, Rawling MD, Foey A, Davies SJ, Santos GA et al (2016) Dietary administration of a commercial mixed-species probiotic improves growth performance and modulates the intestinal immunity of tilapia, *Oreochromis niloticus*. Fish Shellfish Immunol 49:427–435

Sugita H, Miyajima C, Deguchi Y (1991) The vitamin B12-producing ability of intestinal bacteria isolated from tilapia and channel catfish. Nippon Suisan Gakkaishi 56:701

Sullam KE, Essinger SD, Lozupone CA, O'Connor MP, Rosen GL, Knight R, Kilham SS, Russell JA (2012) Environmental and ecological factors that shape the gut bacterial communities of fish: a meta-analysis. Mol Ecol 21(13):3363–3378. https://doi.org/10.1111/j.1365-294X.2012.05552.x

Sun M, Wu W, Chen L, Yang W, Huang X, Ma C et al (2018) Microbiota-derived short-chain fatty acids promote Th1 cell IL-10 production to maintain intestinal homeostasis. Nat Commun 9:3555

Tan HY, Chen SW, Hu SY (2019) Improvements in the growth performance, immunity, disease resistance, and gut microbiota by the probiotic *Rummeliibacillus stabekisii* in Nile tilapia (*Oreochromis niloticus*). Fish Shellfish Immunol 92:265–275

Tan J, Mckenzie C, Potamitis M, Thorburn AN, Mackay CR, Macia L (2014) The role of short-chain fatty acids in health and disease. Adv Immunol 121. https://doi.org/10.1016/B978-0-12-800100-4.00003-9

Taoka Y, Maeda H, Jo J, Kim S, Park S, Takeshi Yoshikawa T et al (2006) Use of live and dead probiotic cells in tilapia *Oreochromis niloticus*. Fish Sci 72:755–766

Taverniti V, Guglielmetti S (2011) The immunomodulatory properties of probiotic microorganisms beyond their viability (ghost probiotics: proposal of paraprobiotic concept). Genes Nutr 6:261–274. https://doi.org/10.1007/s12263-011-0218-x

Tsuchiya C, Sakata T, Sugita H (2008) Novel ecological niche of *Cetobacterium somerae*, an anaerobic bacterium in the intestinal tracts of freshwater fish. Lett Appl Microbiol 46:43–48

Villamil Diaz LM, Esguerra Rodriguez D (2017) Enterococcus, Myroidesy Exiguobacterium: Generos bacterianos con potencial probiotico para elcultivo de tilapia nilotica (Oreochromis niloticus). Acta Biologica Colomb 22:331–339

Watanabe T (2002) Strategies for further development of aquatic feeds. Fish Sci 68(2):242–252

Wang M, Lu M (2016) Tilapia polyculture: A global review. Aquac Res 47(8):2363–2374

Wang M, Liu G, Lu M, Ke X, Zhigang Liu Z, Fengying Gao F (2016) Effect of *Bacillus cereus* as a water or feed additive on the gut microbiota and immunological parameters of Nile tilapia. Aquac Res 48(6):3163–3173

Wu S, Ren Y, Peng C, Hao Y, Xiong F, Wang G, Li W, Zou H, Angert ER (2015) Metatranscriptomic discovery of plant biomass-degrading capacity from grass carp intestinal microbiomes. FEMS Microbiol Ecol 91:107

Wu X, Hao Q, Teame T, Dinge Q, Liu H, Ran C et al (2020) Gut microbiota induced by dietary GWF® contributes to growth promotion, immune regulation and disease resistance in hybrid sturgeon (*Acipenserbaerii x Acipenserschrenckii*): insights from a germ-free zebrafish model. Aquaculture 520:734966

Xia Y, Lu M, Chen G, Cao J, Gao F, Wang M, Liu Z, Zhang D, Zhu H, Yi M (2018a) Effects of dietary *lactobacillus rhamnosus* JCM1136 and *Lactococcus lactis* subsp. *lactis* JCM5805 on the growth, intestinal microbiota, morphology, immune response and disease resistance of juvenile Nile tilapia, *Oreochromis niloticus*. Fish Shellfish Immunol 76:368–379

Xia Y, Wang M, Gao F, Lu M, Chen G (2020) Effects of dietary probiotic supplementation on the growth, gut health and disease resistance of juvenile Nile tilapia (*Oreochromis niloticus*). Anim Nutr 6(1):69–79

Xia Y, Lu M, Chen G, Cao J, Gao F, Wang M et al (2018b) Effects of dietary *lactobacillus rhamnosus* JCM1136 and *Lactococcuslactis* subsp. *lactis* JCM5805 on the growth, intestinal microbiota, morphology, immune response and disease resistance of juvenile Nile tilapia, *Oreochromis niloticus*. Fish Shellfish Immunol 76:368–379

Xu L, Ran C, He S, Zhang J, Hu J, Yang Y et al (2015) Effects of dietary yeast nucleotides on growth, non-specific immunity, intestine growth and intestinal microbiota of juvenile hybrid tilapia Oreochromis niloticus ♀ × Oreochromis aureus ♂. Anim Nutr 1(3):244–251. https://doi.org/10.1016/j.aninu.2015.08.006

Yan Q, Li J, Yu Y, Wang J, He Z, Van Nostrand JD, Kempher ML, Wu L, Wang Y, Liao L, Li X, Wu S, Ni J, Wang C, Zhou J (2016) Environmental filtering decreases with fish development for the assembly of gut microbiota. Environ Microbiol 18(12):4739–4754

Yue GH, Lin HR, Li JL (2016) Tilapia is the fish for next–generation aquaculture. Int J Marine Sci Ocean Technol 3(1):11–13

Yusuf M, Hassan MA, Tag HM, Sarivistava K, Reddy PG, Hassan AM (2017) Influence of turmeric (Curcuma longa) on performance, histomorphology and microbiota of intestine in juvenile tilapia (Oreochromis niloticus). Int J Agric Sci Vet Med 5:7–16

Zarkasi KZ, Abell GC, Taylor RS, Neuman C, Hatje E, Tamplin ML, Katouli M, Bowman JP (2014) Pyrosequencing-based characterization of gastrointestinal bacteria of Atlantic salmon (*Salmo salar* L.) within a commercial mariculture system. J Appl Microbiol 117(1):18–27

Zhai Q, Yu L, Li T, Zhu J, Zhang C, Zhao J, Zhang H, Chen W (2017) Effect of dietary probiotic supplementation on intestinal microbiota and physiological conditions of Nile tilapia (*Oreochromis niloticus*) under waterborne cadmium exposure. Antonie Van Leeuwenhoek 110:501–513

Zheng Y, Wu W, Hu G, Qiu L, Meng S, Song C, Fan L, Zhao Z, Bing X, Chen J (2018) Gut microbiota analysis of juvenile genetically improved farmed tilapia (Oreochromis niloticus) by dietary supplementation of different resveratrol concentrations. Fish Shellfish Immunol 77:200–207. https://doi.org/10.1016/j.fsi.2018.03.040

Zhou L, Zhang J, Yan M, Tang S, Wang X, Qin J, Chen L, Li E (2020) Inulin alleviates hypersaline-stress induced oxidative stress and dysbiosis of gut microbiota in Nile tilapia (*Oreochromis niloticus*). Aquaculture 529:735681

Zhou Z, Liu Y, He S, Shi P, Gao X, Yao B, Ringø E (2009) Effects of dietary potassium diformate (KDF) on growth performance, feed conversion and intestinal bacterial community of hybrid tilapia (*Oreochromis niloticus* ♀ × *O. aureus* ♂). Aquaculture 291(1–2):89–94. https://doi.org/10.1016/j.aquaculture.2009.02.043

Zhu J, Dai W, Qiu Q, Dong C, Zhang J, Xiong J (2016) Contrasting ecological processes between intestinal bacterial community in healthy and diseased shrimp. Microb Ecol 72(4):975–985
Zhu HJ, Qiang J, Tao YF, Ngoepe TK, Bao JW, Chen DJ (2020) Physiological and gut microbiome changes associated with low dietary protein level in genetically improved farmed tilapia (GIFT, *Oreochromis niloticus*) determined by 16S rRNA sequence analysis. Microbiology Open 9(5): e1000

Chapter 5
Probiotics and their Application in Tilapia Culture

Hien Van Doan

Abstract Probiotics are live beneficial bacteria introduced into the gastrointestinal tract through food or water, promoting good health by enhancing the internal microbial balance. Probiotic microbes produce bacteriocins, siderophores, lysozymes, proteases, and hydrogen peroxides, inhibiting the growth of harmful pathogens. Such beneficial bacteria also produce many enzymes such as amylase enzyme by *Aeromonas* spp., *Bacillus subtilis,* Bacteridaceae*, Clostridium* spp., *Lactobacillus plantarum,* and *Staphylococcus* sp., and protease and cellulase enzymes by *B. subtilis*, *L. plantarum*, and *Staphylococcus* sp. In aquaculture, probiotics confer several benefits and play important roles in improving growth performances, disease resistance, immunity, health status, intestinal epithelial barrier integrity, gut microbiome, and water quality. In addition, the practical application of probiotics in aquaculture diets could minimize antibiotic side effects. Promoting these feed additives for fish would help to improve their product performance and feed utilization and, therefore, boost fish production and safeguard human health. This review provides updated information regarding definitions, sources of bacterial probiotics, probiotic use in fish diets against pathogenic bacteria, mechanisms of action, beneficial aspects, and potential applications of probiotics in fish. It is anticipated that these will be of significant value for nutritionists, agricultural engineers, researchers, pharmacists, scientists, pharmaceutical industries, and veterinarians.

5.1 Introduction

Aquaculture is one of the most important occupations, particularly in underdeveloped nations, as it not only helps to increase food security but also helps fish farmers to generate and enhance revenue (Dorji et al. 2022; El-Saadony et al. 2021). Moreover, the development of aquaculture has been necessitated by the exponential

H. Van Doan (✉)
Department of Animal and Aquatic Sciences, Faculty of Agriculture, Chiang Mai University, Chiang Mai, Thailand

© The Author(s), under exclusive license to Springer Nature Switzerland AG 2023
S. H. Hoseinifar, H. Van Doan (eds.), *Novel Approaches Toward Sustainable Tilapia Aquaculture*, Applied Environmental Science and Engineering for a Sustainable Future, https://doi.org/10.1007/978-3-031-38321-2_5

growth of the human population, the high demand for low-cost sources of proteins, and the decrease in fish capture from natural inland water bodies (Gebremedhin et al. 2021). As aquaculture is potentially the most promising activity for meeting the increasing global food demands, governments as well as development agencies have started to pay more attention to fish farming and aquaculture. The growth rate of aquaculture has been the fastest as compared to all other food production sectors, and thus it can play a vital role in combating malnutrition and increasing food security by supplying food to the ever-increasing population (FAO 2018, 2020).

The use of good husbandry techniques in aquaculture can decrease the stress faced by fish due to different reasons. However, this stress can never be eliminated and can eventually lead to an increased susceptibility of fish that are being produced to diseases and infections (Abdel-Latif et al. 2020; Eissa et al. 2018; Stevens et al. 2017). Vaccinations, biocontrol agents, chemicals as well as antibiotics are most commonly used to prevent or treat diseases in aquaculture (Assefa and Abunna 2018; Shao et al. 2021). However, in some countries, these treatments have either been restricted or prohibited, but many other countries are still widely using them (Shao et al. 2021; Guardone et al. 2022). As chemicals and antibiotics tend to bioaccumulate in the tissues of the fish, their use can pose a risk to the environment as well as consumers (Okocha et al. 2018). The potential development of antibiotic-resistant bacteria is another key concern related to the use of antibiotics in aquaculture (Pepi and Focardi 2021; Henriksson et al. 2018). Exposure to antibiotics at subtherapeutic doses for a long period can lead to the development of antimicrobial resistance in bacteria by allowing the plasmids or genes having antibiotic resistance to evolve (Zalewska et al. 2021; Amin et al. 2020). Moreover, these plasmids and genes can be horizontally passed on to other bacteria (Kent et al. 2020; Lerminiaux and Cameron 2019).

The transfer of antibiotic resistance genes and bacterial species from aquatic animals and environment to terrestrial animals and environment and consequently to humans and their environment, and vice versa, is very well known and has negative effects on the health of animals and humans as well as aquatic ecosystems (Santos and Ramos 2018; Koch et al. 2021). As aquaculture is at the crossroads of antimicrobial resistance, recent research has highlighted the importance of corrective and mitigative measures to protect the health of fish as well as humans (Reverter et al. 2020; Vaiyapuri et al. 2021). A large number of research studies have proved that there are many safe, economical, and ecologically appropriate supplements such as probiotics, prebiotics, and medicinal plants (Ghosh et al. 2021; Hernández-Contreras and Hernández 2020) as well as essential oils (Dawood et al. 2021) that can be added to feed and have significant benefits for the health of farmed fish (Butt et al. 2021; Rohani et al. 2021).

In this chapter, probiotics, their advantages for health, their sources, modes of action, safety concerns, categorization, applications, and side effects in the nutrition, production as well as the health of Nile tilapia are discussed.

5.2 Modes of Action

The most commonly known mechanisms of action of probiotics include (i) competitive exclusion by the production of inhibitory compounds, (ii) competition for energy, nutrients, or chemicals, (iii) competition for adhesion site, (iv) contribution to digestion, (v) contribution to macro- and micronutrients, (vi) augmentation of the immune response, and (vii) reduction of virulence by manipulation of quorum sensing (QS) (Balcázar et al. 2006; De Bidhan et al. 2014; Zorriehzahra et al. 2016; Noor et al. 2020; Dawood et al. 2020).

Probiotics being engaged in the competitive exclusion of pathogenic pathogens was one of the earliest theories to be put out (El-Saadony et al. 2021). It was thought that when beneficial microorganisms entered the host's digestive tract, they would either create inhibitory compounds or engage in competition with pathogens for adhesion sites, nutrients, chemicals, or energy sources, thus interfering with the growth or other activities of the pathogens (Verschuere et al. 2000; Decamp et al. 2008; Kuebutornye et al. 2020a, b, c). Incidentally, advantageous isolates from the intestines of both freshwater and marine species have shown antagonistic action against several fish and shellfish diseases. For instance, extracellular products of five potential probiotics isolated from common clownfish's *(Amphiprion percula)* stomach and intestine displayed possible inhibitory effects against *Carnobacterium piscicola, A. salmonicid, V. harveyi, A. hydrophila, V. anguillarum, V. damsela* and *V. alginolyticus* (Vine et al. 2004). *Saprolegnia* sp. was strongly inhibited by *Aeromonas* media obtained from rearing water used for eel culture (Lategan et al. 2004). Diffusible inhibitors against *E. tarda, V. parahaemolyticus, V. harveyi*, and *A. hydrophila* were discovered to be produced by *B. amyloliquefaciens* (Das et al. 2013).

Contrarily, substances that are known to have bactericidal or bacteriostatic effects such as hydrogen peroxide, lysozymes, bacteriocins, proteases, carbon dioxide, and siderophores, etc. Furthermore, the pH of the gut might change as a result of the generation of organic acids and volatile fatty acids (such as lactic, propionic, butyric, and acetic acids) (Tinh et al. 2008; Vine et al. 2006). By doing so, probiotics may prevent the growth of opportunistic pathogenic bacteria in vivo.

By competing for sites of attachment, some probiotic microbes can reduce the growth of pathogens on the surface of the gut (Dawood 2021). According to research by Chabrillón et al. (2005) lactic acid bacteria such as *L. lactis, L. plantarum*, and *L. fermentum*, respectively, reduced the adhesion of the fish pathogens *A. hydrophila, A. salmonicida, V. anguillarum*, and *Y. ruckeri* to the intestinal mucus of rainbow trout in vitro. The production of antimicrobial agents like antibiotics or siderophores may account for this anti-adhesion activity (Chabrillón et al. 2005).

Indeed, *Pseudomonas fluorescens*, a probiotic that produces siderophores, prevented the growth of *V. anguillarum* by competing for freely available iron (Gram et al. 1999; 2001). One of the most significant defenses against pathogen invasion by probiotics is their capacity to colonize the GIT by adhering to and proliferating in the mucus and epithelial cells (Amoah et al. 2019; Azad et al. 2019).

5.3 Probiotics in Tilapia

Probiotics work in a variety of ways to assist increases in aquaculture production, by encouraging improved growth (Silva et al. 2012), increasing feed consumption and nutrition (Zhou et al. 2010), lowering disease rates (Irianto and Austin 2002), and developing immunological responses (Nayak et al. 2007). Probiotics not only enhance the well-being of domesticated animals but also contribute to consumer safety (Rohani et al. 2021). The most commonly used probiotic bacterial strains include *Arthrobacter* sp., *Enterococcus* sp., *Bifidobacterium* sp., *Bacillus* sp., *Lactobacillus* sp., *Pseudomonas* sp., *Streptomyces* sp., *Phaeobacter* sp., *Streptococcus* sp., *Microbacterium* sp., *Lactococcus* sp., *Micrococcus* sp., etc., yeast probiotics include cell wall of yeast, *Debrayomyces hansenii, Saccharomyces cerevisiae,* etc., micro-algal probiotics include *Spirulina platensis*, etc., and *Tetrasehnis suecica,* Bacteriophages probiotics include *Bacteriophages* sp. Table 5.1 lists the potential impacts of adding probiotics to aquafeeds on fish development, immunity, feed consumption, and disease resistance.

5.3.1 Bacillus *Spp.*

Gram-positive, anaerobic *Bacillus* species are known to exhibit a variety of physiological traits (Jahangiri and Esteban 2018; England 2014). *Bacillus* species is one of the most commonly used probiotic organisms in aquaculture. They have also been discovered to be natural members of the intestinal microbiota of some fish species (Kuebutornye et al. 2020a, b, c). The addition of *Bacillus* species to the fish diet has increased growth performance, disease resistance, as well as resistance to infections from pathogenic bacteria in tilapia. According to Liu et al. (2017), Nile tilapia fed 10^8 cfu/g *B. subtilis* HAINUP40 which was isolated from aquatic ecosystems could improve disease resistance, growth performance, and immunological response. *B. subtilis* HAINUP40's capacity to release exoenzymes like protease and amylase may be responsible for the evaluation of these parameters. Additionally, *B. subtilis* HAINUP40 inclusion dramatically increased the activity of serum SOD and TAC, which may be related to its stronger microbial killing capacity of macrophages (Giri et al. 2013) and higher levels of antioxidant enzymes as well as antioxidants secreted in response to *Bacillus antigens* (Zhang et al. 2013). The addition of *B. subtilis* LT3–1 to the diet also dramatically improved weight gain and serum biochemical parameters. In addition, the hematocrit, total red and white blood cell counts, serum catalase, and lysozyme activity, were observed to be increased. While, the levels of serum complement 3, serum malondialdehyde, as well as immunoglobulin M, were

Table 5.1 Effects of probiotics on Nile tilapia

Bacillus subtilis HAINUP40	Aquatic environment	10^8 CFU/g	Nile tilapia (*Oreochromis niloticus*)	Final weight, WG, SGR FE ↑ Intestinal probiotic recovery, digestive enzyme activities ↑ Innate immunity and disease resistance ↑	Liu et al. (2017)
Bacillus pumilus AQAHBS01	Farmed tilapia	10^7, 10^8 and 10^9 CFU/kg 60 days	Nile tilapia (*Oreochromis niloticus*)	Final weight, WG, SGR FE ↑ FCR ↓ Resistance against *S. agalactiae* during the critical period of early	Srisapoome and Areechon (2017)
Bacillus cereus	Intestinal contents of tilapia	Experiment 1: 0, 1.0×10^4 and 1.0×10^5 CFU mL^{-1} Experiment 2: 0, 1.0×10^7 and 1.0×10^8 CFU g^{-1} (42 days)	Nile tilapia (*Oreochromis niloticus*)	Lysozyme and peroxidase activities ↑ Alkaline phosphatase and total superoxide dismutase activities ↑	Wang et al. (2017)
Bacillus licheniformis Dahb1		0; 10^5 and 10^7 CFU g^{-1} 4 weeks	Mozambique tilapia (*Oreochromis mossambicus*)	Final weight, WG, SGR FE ↑ FCR ↓ Total protein, alkaline phosphatase, myeloperoxidase, lysozyme, reactive oxygen species, reactive nitrogen species ↑	Gobi et al. (2018)
Bacillus paralicheniformis SO-1	Marine sediment samples	0, 5, 10 and 20 g/kg 50 days	Nile tilapia (*Oreochromis niloticus*)	Growth rates, digestive enzymes activities ↑ Immunological parameters ↑	Makled et al. (2019)
Bacillus licheniformis strain HGA8B	Commercial probiotic	10^6 and 10^8 CFU g^{-1} 60 days	Nile tilapia (*Oreochromis niloticus*)	Final weight, WG, SGR FE ↑ FCR ↓ Hepatosomatic index, modulated digestive and antioxidant ↑ Enzyme activity ↑	Midhun et al. (2019)

(continued)

Table 5.1 (continued)

Bacillus velezensis LF01	Nile tilapia (*Oreochromis niloticus*)	1,0 × 10^9 CFU/g 9 weeks	Nile tilapia (*Oreochromis niloticus*)	Lysozyme (LZY) and superoxide dismutase (SOD) activities ↑ Resistance against *Streptococcus agalactiae* ↑	Zhang et al. (2019b)
Bacillus subtilis LT3-1	Isolated from the gut of tilapia	0, 3.8 × 10^{10}, 7.6 × 10^{10}, 1.14 × 10^{11} and 1.52 × 10^{11} CFU kg^{-1} 6 weeks	Nile tilapia (*Oreochromis niloticus*)	FW, WG, and SGR ↑ Serum biochemical indices, haematocrit, the serum catalase and lysozyme activities, parameters for intestinal morphology ↑	Zhu et al. (2019)
Bacillus subtilis strains SB3086, SB3295, SB3615, or AP193	Commercial probiotic	4 × 10^7 CFU/g 21 days	Nile tilapia (*Oreochromis niloticus*)	Lysozyme activity ↑ Resistance against *Streptococcus iniae* ↑	Addo et al. (2017)
Bacillus strains	Commercial probiotic	(*Bacillus subtilis* 3.25 × 10^9 CFU/g, *Bacillus licheniformis* 3.50 × 10^9 CFU/g and *Bacillus pumilus* 3.25 × 10^9 CFU/g; Sanolife PRO-F, with a total number 1.0 × 10^{10} CFU/g)	Nile tilapia (*Oreochromis niloticus*)	FW, WG, and SGR ↑ FCR ↓ Intestinal lengths, anterior and terminal intestinal villi heights and anterior goblet cells ↑	Elsabagh et al. (2018)
Bacillus spp. (B. subtilis, B. megaterium and *B. licheniformis)*	Commercial probiotic	Add in water environment: 1 × 10^6 CFU g^{-1} 90 days	Nile tilapia (*Oreochromis niloticus*	FW, WG, and SGR ↑ Aspartate aminotransferase (AST) and alanine aminotransferase (ALT) ↓	Sutthi et al. (2018)
Bacillus velezensis TPS3N; *Bacillus subtilis* TPS4; *Bacillus amyloliquefaciens* TPS17	Gastrointestinal tract of Nile tilapia, *O. niloticus*	0, 1 × 10^8 CFU/ml (*B. velezensis* TPS3N), 1 × 10^8 CFU/ml (*B. subtilis* TPS4), 1 × 10^8 CFU/ml (*B. amyloliquefaciens* TPS17) and combination at a ratio of 1:1:1 with same concentration.	Nile tilapia (*Oreochromis niloticus*)	Mucosal immunity, intestinal health ↑ Proteobacteria ↑ Resistance of Nile tilapia against *A. hydrophila* ↑	Kuebutornye et al. (2020a, b, c)

Bacillus sp. KUAQ1 and *Bacillus sp.* KUAQ2,	Nile tilapia *Oreochromis niloticus*	0, 1×10^8, 3×10^8 and 5×10^8 CFU/g 8 weeks	Nile tilapia (*Oreochromis niloticus*)	Lysozyme, phagocytic, and respiratory burst activities ↑	Sookchaiyaporn et al. (2020)
Lactobacillus lactis D1813 (PowerLac™)	Commercial probiotic	For laboratory experiment: 0, 0.25, 0.5, 1.0 and 2.0 g kg^{-1} for 8 weeks For field experiments: 0.25, 0.5 and 1.0 g kg^{-1} for 22 weeks	Nile tilapia (*Oreochromis niloticus*)	FW, WG, and SGR ↑ FCR ↓ Resistance against *A. hydrophila* ↑	Suprayudi et al. (2017)
Lactobacillus delbrueckii	Commercial probiotic	0, 1×10^5, 1×10^6, 1×10^7 and 1×10^8 CFU g^{-1} 8 weeks	Nile tilapia (*Oreochromis niloticus*)	FW, WG, and SGR ↑ FCR ↓ Resistance against *A. hydrophila* ↑	Zhang et al. (2017)
Lactobacillus plantarum CCFM8661	Commercial probiotic	10^8 CFU g^{-1} 4 weeks	Nile tilapia (*Oreochromis niloticus*)	Growth performance ↑ Prevented the death of Pb-exposed fish. Pb accumulation in kidney, liver, gonad, brain, gills and muscle ↓	Zhai et al. (2017a)
Lactobacillus plantarum CCFM8610		10^8 CFU g^{-1} 4 weeks	Nile tilapia (*Oreochromis niloticus*	Growth performance ↑ Prevented the death of the Cd-exposed fish	Zhai et al. (2017a)
Lactococcus lactis subsp. *lactis* JCM5805	Commercial probiotic	0; 10^4 and 10^8 CFU ml^{-1} 15 days	Nile tilapia (*Oreochromis niloticus*)	Disease resistance Resistance against *Streptococcus agalactiae* ↑	Xia et al. (2019)
Bacillus subtilis and malic acid		1.1 1.1. $\times 10^5$ CFU/g 84 days	Nile tilapia (*Oreochromis niloticus*)	FBW, WG, SGR, PER, PPV and ER ↑ Haematocrit, haemoglobin, red blood cells, white blood cells, total protein, albumin and globulin ↑	Hassaan et al. (2018)

(continued)

Table 5.1 (continued)

Saccharomyces cerevisiae and *Bacillus subtilis*	Commercial probiotic	*S. cerevisiae* at 2; 4 and 6 g kg^{-1} *B. subtilis* at 5; 10 and 15 g kg^{-1} 7 months	Nile tilapia (*Oreochromis niloticus*)	Final weight, WG, SGR ↑ FCR ↓ Body composition ↑	Opiyo et al. (2019)
Rummeliibacillus stabekisii	Commercial probiotic	10^6 and 10^7 CFU/g 8 weeks	Nile tilapia (*Oreochromis niloticus*)	Final weight, WG, SGR FE ↑ FCR ↓ Innate immune response ↑ Resistance against *Streptococcus* and *Staphylococcus spp.*	Tan et al. (2019)
Paenibacillus ehimensis NPUST1	Water of local tilapia culture pools	0; 10^6 and 10^7 CFU/g	Nile tilapia (*Oreochromis niloticus*)	Final weight, WG, SGR FE ↑ FCR ↓ Resistance against *Streptococcus iniae* and *A. hydrophila* ↑	Chen et al. (2019)
Traditional Chinese medicine (TCM), *Bacillus subtilis* and *Bacillus lincheniformis*	Commercial probiotic	0, (TCM at 3 and BS at 7 g/Kg), (TCM at 5 and BS at 5 g/kg), and (TCM at 7 and BS at 3 g/kg) 4 weeks	Nile tilapia (*Oreochromis niloticus*)	Final weight, WG, SGR FE ↑ FCR ↓ Lysozyme, superoxide dismutase, catalase, protease and antiprotease activities ↑ Resistance against *Streptococcus agalactiae* ↑	Abarike et al. (2019a)
Lactobacillus plantarum N11 *Bacillus velezensis* H3.1	Nile tilapia (*Oreochromis niloticus*)	0, 10^7 $CFUg^{-1}$ (*B. velezensis* H3.1), 10^8 $CFUg^{-1}$ (*L. plantarum* N11), and 10^7 CFU g^{-1} (*B. velezensis* H3.1) + 10^8 CFU g^{-1} (*L.plantarum N11*).	Nile tilapia (*Oreochromis niloticus*)	Final weight, WG, SGR FE ↑ Skin mucus lysozyme and peroxidase activities ↑ Serum lysozyme activity ↑	Van Doan et al. (2018)
Lactobacillus rhamnosus (LR) JCM1136 và *Lactococcus lactis* subsp. *Lactis* (LL) JCM5805	Commercial probiotic	10^8 CFU/g 6 weeks	Nile tilapia (*Oreochromis niloticus*)	Final weight, WG, SGR FE ↑ FCR ↓ Gut microvilli length and microvilli density ↑ Resistance against *Streptococcus agalactiae* ↑	Xia et al. (2018)

Lactococcus garvieae	From raw cow mill	10^7 cells/g 10 days	Nile tilapia (*Oreochromis niloticus*)	Resistance against *Staphylococcus aureus* ↑	Abdelfatah et al. (2018)
Bacillus subtilis and *Lactococcus lactis*	Intestine of juvenile Japanese eel	*Bacillus subtilis* at 10^7 and at 10^8 CFU/g; *Lactococcus lactis* at 10^7 and at 10^8 CFU/g	Nile tilapia (*Oreochromis niloticus*)	Final weight, WG, SGR FE ↑ FCR ↓ Lysozyme activity ↑ Superoxide dismutase and myeloperoxidase activity ↑ Gene expression ↑	Won et al. (2020)
Lactobacillus plantarum and *Bacillus velezensis*	Isolated from tilapia gut	10^7 and 10^8 CFU/g 30 days	Nile tilapia (*Oreochromis niloticus*)	FW, WG, and SGR ↑ FCR ↓ Resistance against *S. agalactiae* ↑	Van Doan et al. (2018)
Clostridium butyricum	Commercial probiotic	0, 0.5, 1, 2, 4 and 8 g/kg commercial probiotic (containing *C. butyricum* 1.5×10^8 CFU/g) 8 weeks	Tilapia (Oreochromis niloticus × O. aureus)	WG ↑ FCR ↓ Protein retention, lipid retention and apparent digestibility coefficient ↑ Intestinal *E. coli* ↓ Resistance against *A. hydrophila* ↑	Poolsawat et al. (2019)
Micrococcus MCCB 104 and *Bacillus* MCCB 101		10^3 CFU animal^{-1} day^{-1} 28 days.	Mozambique tilapia (*Oreochromis mossambicus*)	Intestinal digestive enzymes, intestinal alpha-amylase and total alkaline proteases ↑	Sankar et al. (2017)
Bacillus sp., Pediococcus sp., Enterococcus sp., Lactobacillus sp.	Commercial probiotic	0; 1.0×10^6; 2.3×10^6 CFU/g 8 weeks	Nile tilapia (*Oreochromis niloticus*	FW, WG, and SGR ↑ Villi height and goblet cell counts ↑	Ramos et al. (2017)
B. subtilis and *B. licheniformis* with traditional Chinese medicine	Commercial probiotic	0, 3, 5, 7 and 10 gkg^{-1} 4 weeks	Nile tilapia (*Oreochromis niloticus*)	Final weight, WG, SGR FE ↑ FCR ↓ Lysozyme, protease, antiprotease, superoxide dismutase activities, and immunoglobulin M level ↑	Abarike et al. (2019b)

all observed to be decreased. Fish-fed diets enriched with *B. subtilis* appeared to have healthier intestines than fish fed the control diet, according to parameters for intestinal morphology. In tilapia-fed *B. subtilis*, the survival rate following *Streptococcus agalactiae* exposure increased (Zhu et al. 2019). In tilapia production, *Bacillus licheniformis* has been used as a functional feed supplement. According to Gobi et al. (2018), feeding tilapia (*O. mossambicus*) with *B. licheniformis* Dahb1 significantly improved growth rate, immunological parameters, feed conversion ratio, as well as antioxidant parameters in serum and mucus. *B. licheniformis* HGA8B can considerably improve growth parameters including feed conversion ratio, weight gain, specific growth rate, and hepato-somatic index when added to the diet, according to Midhun et al. (2019). The probiotic preparation adjusted the activity of the digestive and antioxidant enzymes. In the liver tissue, there was a significant modulation of growth-related genes such as the growth hormone receptor genes (GHR-1 and GHR-2) and the insulin-like growth factor genes (IGF-1 and IGF-2). IGF-1 gene expression was markedly elevated in the muscular tissue. Significant immune-related gene upregulation, as evidenced by an increase in TLR-2 and anti-inflammatory cytokines such as IL-10, was seen. *B. licheniformis* secreted digestive enzymes, which were found to be the cause of fish growth and health promotion. Furthermore, the presence of *B. licheniformis* can influence the expression of regional growth factors (IGF-1 and IGF-2) and the receptors (GHR-1 and GHR-2) that bind growth factors and hormones (growth hormone and steroid hormones) to promote greater growth. A new potential probiotic for Nile tilapia is *Bacillus pumilus* AQAHBS01. Srisapoome and Areechon (2017) have suggested that in both laboratory and field culture conditions, the addition of *B. pumilus* at 1×10^8 and 10^9 CFU/kg to diet could boost immune response as well as resistance against streptococcosis. An important topic that needs to be discussed is the mechanism that *Bacillus pumilus* uses to improve fish health and disease resistance. The possible mechanisms by which *B. pumilus* may shield fish from harmful bacterial infection include boosting superoxide anion generation, phagocytic activity, and the number of essential leukocytes. Theoretically, *B. pumilus* may possess pro-inflammatory tumor necrosis factor, bacteriocin-like activity, powerful anti-spore IgG titers, and complement activity against other Bacillus species (Duc et al. 2004; Sun et al. 2010), all of which may significantly increase the survival rate of experimental fish fed with *B. pumilus* (Aly et al. 2008a, b). Unknown are other *B. pumilus* mechanisms of activity, such as the colony and adhesive capacity inside and outside the fish body. *B. pumilus* has not been used much lately in Nile tilapia. Compared to various earlier findings using this probiotic species, there have been only minor improvements in growth properties. A study by Aly et al. (2008a, b) is the only one to demonstrate the potential value of *B. pumilus* as a probiotic for enhancing Nile tilapia growth. In a different study using *Bacillus cereus* isolated from tilapia's stomach, Wang et al. (2017) found that *B. cereus*, when added to water or feed, might improve tilapia's immunological status and have an impact on their gut microbiota. When used as a feed supplement as opposed to a water additive, *Bacillus cereus* was more successful at boosting tilapia's immune system. Similarly, Makled et al. (2019) observed that Nile tilapia were given *B. paralicheniformis* SO-1 isolated from marine habitats greatly increased growth rates, digestive enzyme activities, and immunological parameters with increasing supplementary SO-1 up to 10 g/kg. However, when the bacterial load was increased further to 20 g/kg, fish performance and immune response were significantly reduced. When Nile tilapia were fed SO-1-treated diets, the expression of the IL-1, IL-4, and IL-12 genes was considerably increased. Based on the results of this investigation, *B. paralicheniformis* SO-1 can be recommended as an effective probiotic growth promoter and immune system stimulator for farmed Nile tilapia. According to Zhang et al. (2019a, b), growth performances were observed to be significantly improved in tilapia-fed diets supplemented with LF01 (1.0×10^9 CFU/g) as compared to the control group. Lysozyme (LZY) as well as superoxide dismutase (SOD) activity was also considerably elevated in tilapia-fed diets supplemented with LF01. The gills, gut, and head kidney of tilapia from the continuous feeding group had higher expression levels of three immune-related genes (C3, lyzc, and MHC-II) as compared to the control group. After being infected with *S. agalactiae*, the survival rates of tilapia-fed diets supplemented with LF01 were noticeably higher, and an examination of the gut pathogens showed that the number of Edwardsiella as well as *Plesiomonas* had decreased considerably as compared to the control group. The effect of *Bacillus altitudinis* B61-34b (BAA) isolated from the intestine of Nile tilapia, on immunological response, growth, as well as disease resistance against *Streptococcus agalactiae* has recently been studied (Van Doan et al. 2021a, b, c). The results proved that BAA treatment increased skin mucus as well as peroxidase activity in comparison to the control, with maximal values found in BAA4 and BAA5. In comparison to the control, BAA1, BAA2, and BAA3 dietary groups, higher serum immunities, including serum lysozyme, serum peroxidase, complement, phagocytosis, and respiratory burst activities, were seen in the BAA4 and BAA5 dietary groups. The BAA4 and BAA5 groups also showed considerably greater relative percentages of survival, growth performance, and food conversion ratio (FCR). The authors hypothesized that nutritional supplementation with *B. altitudinis* at 108 CFU ml-1 can enhance Nile tilapia's disease resistance, growth performance, skin mucus as well as serum immunity. Numerous mechanisms have been suggested to achieve these improvements. Increased growth performance is a direct outcome of increased intake of feed and utilization of nutrients that have been digested (Dawood 2021). By improving feed digestion through the production of digestive enzymes, probiotics boost feed intake as well as utilization and make the nutrients that have been digested available for metabolic processes for the body (Wang et al. 2020). Beneficial bacterial cells increase the intestinal villi's surface area for absorption of digested nutrients, protect the villus from harmful microorganisms, and boost transcription of IGFI and inhibition of myostatin (Zaineldin et al. 2021). The starch-degrading enzymes known as α-Amylases, which are released by the bacteria *Bacillus altitudinis* play a critical role in digestion by catalyzing the hydrolysis of internal −1,4-o-glycosidic linkages in polysaccharides (Kumar et al. 2014). Furthermore, *B. altitudinis* can produce pectinase and xylanase enzymes simultaneously (Thite et al. 2020). Although the exact mechanism by which probiotics increase mucosal immunity is unknown, the improvement in mucosal immunity of fish that result is likely because of the impact of derivatives of beneficial bacteria on local intestine immunity (Guardiola et al. 2017). According to Caipang and Lazado (2015), probiotic cell derivatives are likely to penetrate via the local epithelial immune cells and into the stream of systemic lymphoid tissues, where they work to increase innate immunity. Probiotics' metabolic impact on the feed composition, which results in functional nutrients being available, is likely responsible for the increased systemic immunity that is responsible for the increased serum immunity (Terpou et al. 2019). Indirectly competing with pathogenic invaders, probiotics prevent their ability to decrease the immunity of intestine (Dawood and Koshio 2020). In this way, compounds which are derivatives of probiotics migrate via intestinal enterocytes, enter the blood circulation, and ultimately end up in the immunological organs (Kelly and Salinas 2017; Zhang et al. 2019a, b).

5.3.2 Lactobacillus *Spp.*

According to Zhai et al. (2017a, b), adding *Lactobacillus plantarum* CCFM8661 to feed improved the growth performance and kept fish exposed to Pb from dying. When Pb and CCFM8661 were combined in the treatment, Pb buildup in the kidney, liver, muscles, brain, gonads, and gills was found to be significantly reduced. Additionally, administering this probiotic reduced oxidative stress induced by Pb, recovered the activities of blood δ-aminolevulinic acid dehydratase as well as digestive enzymes, reversed changes to innate immune status, and reduced the occurrence of nuclear abnormalities in fish peripheral blood erythrocytes. These findings suggested that, at least in tilapia, CCFM8661 might be a potential dietary supplement against toxicity caused by Pb. The *L. plantarum* CCFM8661 probiotic used in this study has been shown to efficiently bind Pb in vitro, and its theoretical monolayer biosorption capacity ($Qmax$) has been calculated to be 86.96 mg g^{-1} dry biomass, outperforming some commercial lactobacilli like *L. casei* Shirota (70.40 mg g^{-1}) as well as *L. rhamnosus* GG (46.80 mg g^{-1}) (Yin et al. 2016). Because of this, the intestinal CCFM8661 may bind such released Pb before it gets reabsorbed from the intestines, increasing the excretion of Pb through fish feces and decreasing its intestinal absorption (Ringø et al. 2020). The significant reduction in the level of Pb in tissues of the Pb-plus-CCFM8661 group could be explained by these findings. A higher number of this probiotic colonizing in the gut of fish after a longer feeding duration, providing further beneficial effects against Pb toxicity, may also be the cause of the significant protection on improving growth performance, reducing tissue oxidative stress, and regulating innate immune responses in fish after supplementation for 2 months. Similar to this, Zhai et al. (2017a, b) found that adding *L. plantarum* CCFM8610 to the diet reversed the alterations in the composition of gut microbiota in the fish exposed to Cd and also decreased the abundance of *Flavobacterium* as well as *Pseudomonas*. The probiotic treatment greatly improved growth performance and stopped the death of the fish exposed to Cd as compared to the Cd-only group. Supplementing with *L. plantarum* CCFM8610 also reduced Cd accumulation, reduced oxidative stress in tissues, and reversed changes in hemato-biochemical parameters in fish blood. Suprayudi et al. (2017) used *Lactobacillus lactis* D1813 at three different levels, that is, 0.25, 0.5, 1.0, and 2.0 g kg^{-1} in feed as compared to a control group having 0 g kg^{-1}, in both laboratory and field conditions. They found that treatments with dietary probiotic supplements of 0.25 and 0.5 g kg^{-1} significantly enhanced the final body weight of fish, increased growth, and lowered food conversion ratio (FCR) compared to controls (0 g kg^{-1}) in laboratory conditions. In field conditions, higher growth, retention of proteins and lipids, lowered FCR, and reduction in death rate by *Aeromonas hydrophila* challenge was observed by dietary inclusion of 0.5 g kg^{-1}. The beneficial effects of *L. lactis* D1813 on Nile tilapia's growth, as well as feed conversion ratio, may be explained by the probiotics' well-established benefits for the targeted species, including improvements in feed digestibility (Dawood et al. 2015) and uptake of nutrients and their utilization through (1) the involvement of digestive enzymes (Yanbo and Zirong 2006; Suzer et al. 2008), (2) changes in gut microbiota (Yang et al. 2014), (3) contribution on the development of morphology of the digestive tract (an increase in microvilli) that works to increase the surface area of the intestine allowing for the uptake of nutrients (Frouël et al. 2008), and (4) increase in the activity of enzymes that are related to the utilization of nutrients in intestines, such as alkaline phosphatase and leucine aminopeptidase which are the enzymes involved in the absorption of nutrients in the intestinal brush border (Sáenz de Rodrigáñez et al. 2009; Panigrahi et al. 2010). Probiotics have been shown to promote antibacterial activity (Aly et al. 2008a, b), enhance phagocytic and respiratory burst activity (Pirarat et al. 2006), and stimulate complement activity (Wang et al. 2008) as part of their ability to stimulate innate as well as adaptive immune system responses. Different fish can produce more lymphocytes, erythrocytes, macrophages, and granulocytes as a result of some probiotics, and the levels of B-lymphocyte proliferation as well as immunoglobulin levels can also be actively stimulated (Nayak 2010). To improve immunological responses, probiotics interact with immune cells like natural killer (NK) cells, mononuclear phagocytic cells (monocytes, macrophages), and polymorphonuclear leucocytes (neutrophils) (Nayak 2010; Selim and Reda 2015). Similarly, supplementing the diet of *Cyprinus carpio* with a commercial probiotic (*L. delbrueckii*) considerably improved the growth performance, disease resistance against *A. hydrophila*, and immunity of the fish (Zhang et al. 2017).

According to recent research by Van Doan et al. (2021a, b, c) Nile tilapia given *Lactobacillus paracasei* l61-27b dramatically increased rate of growth, serum immunities, feed conversion ratio, disease resistance, and skin mucus. The immunostimulant factor associated with probiotics can be due to the influence of derivatives of beneficial bacteria which can reach mucosal tissues like the skin and systemic immune organs like the spleen and trigger their response (Lazado and Caipang 2014). Additionally, probiotics improve the digestion of functional nutrients like vitamins, which strengthen the local gut immunity and in turn strengthen the systemic immune system (Pérez-Sánchez et al. 2014). Another theory that probiotics have an immunostimulant effect on the body is that they reduce the impact of pathogenic microorganisms in fish intestines, which causes a pronounced effect of good probiotics on intestinal immune cells (Akhter et al. 2015).

5.3.3 Lactococcus *Sp.*

According to Xia et al. (2019), the probiotic strain JCM5805 from *Lactococcus lactis* subsp. *lactis* increased the expression of IFNα through the TLR7/TLR9-Myd88 pathway and improved larval disease resistance. Since JCM5805 was only transiently identified, it was excluded from the stable microbiota of the larva. Tilapia larvae's early exposure to microbes has an impact on the gut microbiome later in life. Further research is necessary to determine whether the upregulation of associated genes is connected to the presence of JCM5805 strain in the intestine. Abdelfatah and Mahboub (2018) reported that fish fed with a feed having dietary inclusion of *L. garvieae* isolated from raw cow milk exhibited no signs of sickness and showed a greater survival rate as compared to others. According to Xia et al. (2018), the dietary addition of *Lactobacillus rhamnosus* (LR) JCM1136 and *Lactococcus lactis* subsp. *lactis* (LL) JCM5805, as a feed additive at 108 CFU/g feed, may enhance the immune status and disease resistance, affect tilapia's gut microbiota, and improve its intestinal morphology. These additives may therefore be used as probiotics for juvenile Nile tilapia. When it came to improving growth, immunological function, and increasing disease resistance, JCM5805 was found to be more beneficial than JCM1136 or the two combined. The different effects of probiotics on the growth performance of tilapia can be attributed to variations in the antibiosis activities of particular strains of probiotics as well as the interactions between probiotics, diet, the host, conditions of research, as well as handling practices, among other factors. These different factors may have influenced the results, which in turn affected the success or failure of probiotics and their combinations to improve growth in each study. The growth performance and feed utilization of tilapia in this study were not impacted by feeding diets made by the combination of JCM1136 and JCM5805 or JCM1136 singly containing diets, in contrast to the effect of feeding JCM5805 singly.

5.3.4 Rummeliibacillus Stabekisii

Rummeliibacillus stabekisii was suggested as the type species for the genus *Rummeliibacillus*, and its entire genome sequence was revealed (Vaishampayan et al. 2009; Da Mota et al. 2016). The categorization of three *Rummeliibacillus* species, including *R. stabekisii, R. suwonensis,* and *R. pycnus*, has only been discussed in two publications, and no application studies have been published. In the current study, *Rummeliibacillus stabekisii* was isolated from the stomach of Nile tilapia, and its effects on the fish's growth, resistance to disease, immunity, and gut microbiota were examined. According to the findings, Nile tilapia's growth, immunity, disease resistance, and gut microflora were all enhanced by dietary supplementation with *R. stabekisii*. According to this study, *R. stabekisii* can be added to feed to help Nile tilapia grow and maintain good health. This is the first study to propose that the genus *Rummeliibacillus* may be a promising probiotic in animals (Tan et al. 2019). The improvement of the analyzed parameters in this study has been attributed to several different mechanisms. The mechanism of *R. stabekisii* growth improvement in fish may have been due to enhanced nutrient utilization through the secretion of protease- and xylanase enzymes and stimulation of intestinal digestive enzyme synthesis after *R. stabekisii* treatment. The mechanism by which *R. stabekisii* enhanced fish immunity may have been due to an increase in lysozyme, phagocytic, and respiratory burst activity as well as the release of cytokines (IL-1, TNF-α, and transforming growth factor (TGF)). Additionally, *R. stabekisii* can influence the intestinal microbiota of Nile tilapia, which helps with disease resistance, immunomodulation, and growth promotion.

5.3.5 Paenibacillus Ehimensis

Due to their shared morphological and physiological traits with the type species *Bacillus subtilis, Paenibacillus* species were first included under the *Bacillus* genus. *Paenibacillus* was given a new genus in 1993 based on phylogenetic research using 16S rRNA gene sequences. There are numerous habitats important to humans, animals, as well as plants where the genus *Paenibacillus* has been identified. Numerous *Paenbacillus* species produce bacteriocins, which are proteins or antimicrobial peptides that are used as pesticides in biocontrol or medicine (Abdel-Latif et al. 2020). In the current investigation, the impact of *P. ehimensis* NPUST1 supplementation on Nile tilapia (*Oreochromis niloticus*) growth and immunity against pathogenic infection was assessed (Chen et al. 2019).

5.3.6 Clostridium Butyricum

Clostridium butyricum is an endospore-forming, gram-positive, and obligate anaerobic bacterium (Liu et al. 2018; 2019). It can create short-chain fatty acids (SCFAs) like butyric acid to offer nutrients for the growth of microbes and sustain epithelial cells in the gut (Junghare et al. 2012). Additionally, *C. butyricum* can resist some antibiotics, thrive in low-acidity and high-temperature environments, and protect fish from infections (Gao et al. 2013; Zhang et al. 2014). In aquaculture, *C. butyricum* has been shown to improve immune responses in *Miichthys miiuy* (Pan et al. 2008), promote growth performance in silver pomfret (Gao et al. 2016), improve nutrient digestibility in hybrid grouper (He et al. 2017), and increase the activity of antioxidants in the intestine as well as the content of short-chain fatty acid in kuruma shrimp (Duan et al. 2017; 2018). In tilapia, Poolsawat et al. (2020) demonstrated that the C-2, C-3, and C-4 groups saw considerably higher weight gain and lower feed conversion ratios. Except for the C-1 group, probiotic supplementation groups showed a considerable increase in the retention of proteins, lipids, and apparent digestibility coefficient of dry matter (ADC); improvement was also observed in the ADC of proteins in the C-4 group. Probiotic supplementation greatly enhanced anterior intestine villus height and decreased intestinal *Escherichia coli* counts. High-throughput sequencing revealed that the top three phyla had higher levels than the NC group, including *Planctomycetes* in all groups containing probiotics, *Proteobacteria* in the C-1 and C-2 groups, and *Chloroflexi* in the C-3 group. After the challenge with *Aeromonas hydrophila*, dietary probiotics helped to lower the overall mortality. To conclude, *C. butyricum* can be added to feed at a rate of 1–2 g/kg to support tilapia's growth, intestinal health, feed utilization, and gut microbiota.

The beneficial effects of dietary *C. butyricum* on feed consumption and growth performance may be related to the production of short-chain fatty acids (SCFAs) in the gut, such as butyric acid, acetic acid, and propionic acid. The production of these acids, especially butyric acid, is stimulated by the presence of this bacteria in the gut, which is an essential source of energy for the proliferation of epithelial cells (Junghare et al. 2012; Van Doan et al. 2018). To lower the intracellular pH, butyric acid can release protons into the neutral cytoplasm. The gut microbiota may then use the leftover protons to retain cells that have a lower mortality rate during metabolism (Duan et al. 2017). Butyric acid may also improve feed utilization by achieving balance in the micro-ecological environment of the stomach (Liang et al. 2021). The immunological indices, as well as disease resistance, were improved because probiotics can boost antibiotic substances as SCFAs in the stomach to support the antioxidative enzyme activities and reduce reactive oxygen metabolites, which improves feed utilization, growth, and overall health of the host (Liao et al. 2015a, b). Probiotics can also boost mucus production, which inhibits the adherence of pathogens and increases the competitive exclusion of pathogens from the spaces as well as food present in the gut (Monteagudo-Mera et al. 2019). The increased intestinal morphology caused by probiotic dietary supplementation may be related to the germination as well as the growth of probiotic cells in the gut that can produce gastrointestinal peptides and raise intestinal epithelial barriers and villus height (Gāliņa et al. 2020). Increased intestinal absorption surface area can help the host's nutrient digestion and absorption, producing favorable impacts on growth performance and feed utilization (Reda and Selim 2015).

5.3.7 Combination of Probiotics

It has been shown that combining different Bacillus species improves tilapia's growth and health. According to Addo et al. (2017), Nile tilapia fed four different strains of *Bacillus subtilis* (SB3086, SB3295, SB3615, and AP193) and a mixture of

SB3086 and SB3615 did not show any differences in growth performance. However, results from serum bactericidal and lysozyme activities demonstrated a substantial difference between treatments and the control, except for the diet modified by SB3295. The challenge's results also revealed a sizable distinction between treatments and controls. Fish fed with strain SB3615 had the lowest death rate (44.0 7.2%), while fish fed the control diet had the highest. Since the fish were grown under ideal conditions in the experiment, there may not have been any discernible growth improvement in this study. Due to the essential and unique metabolic and trophic roles played by the intestinal microbiota, the results may have been affected by the colony of the gastrointestinal (GI) tract that was present (Denev et al. 2009). According to Gutowska et al. (2004), the bacterial flora in fishes' gastrointestinal tracts generally has a very significant and varied enzyme potential with the ability to produce lipolytic, amylolytic, proteolytic, cellulolytic, and chitinolytic enzymes. These enzymes are crucial for the digestion of lipids, proteins, cellulose, carbohydrates, and chitin to promote growth. Several mechanisms have been put forth to explain the impact of probiotics on disease resistance in fish. Competition for nutrition and energy sources, generation of secondary metabolites with bactericidal effects on other microbial populations, competition for adhesion sites on the epithelium of the intestine or other tissue surfaces, and improvement of the host immune response are a few of these. The ability to cling to intestinal wall surfaces and enteric mucus is necessary for probiotic bacteria to establish themselves in fish intestines (Lesel 1990; Westerdahl et al. 1991) because bacterial attachment to tissue surface is crucial during the early phases of pathogenic infection (La Ragione and Woodward 2003; La Ragione et al. 2004).

Montes (1993) postulated that competition with pathogens for adhesion receptors would be a fundamental probiotic trait. When given in sufficient doses, probiotics improve the host's health, according to FAO/WHO (2001). The probiotic diets used in this study likely provided some health benefits to the fish because the survival rate of Nile tilapia when challenged with *S. iniae* infection was found to be much higher than the control diet (Addo et al. 2017). Elsabagh et al. (2018) discovered that feeding farmed Nile tilapia a diet containing certain Bacillus strains (*Bacillus pumilus* 3.25×10^9 CFU/g; *Bacillus licheniformis* 3.50×10^9 CFU/g; *Bacillus subtilis* 3.25×10^9 CFU/g, and Sanolife PRO-F, INVE Aquaculture, Belgium, with a total number 1.0×10^{10} CFU/g) improved the stress responses, gut health and function, growth, immunity, and water quality of farmed Nile tilapia. *Bacilli* can invade the guts of fish, enhancing the generation of organic acids, activating digestive enzymes, and detoxifying hazardous feed ingredients. This all helps to maintain a healthy gut, which then improves nutritional digestibility and absorption (Adeoye et al. 2016; Silva et al. 2015). Additionally, it has been shown that *Bacillus* can remove harmful bacteria from the stomach, increasing disease resistance and enhancing fish performance (Addo et al. 2017; Hostins et al. 2017; Srisapoome and Areechon 2017) Sutthi et al. (2018) also reported that fish reared in water treated with probiotic *Bacillus* spp. and yeast (*S. cerevisiae*) showed a synergistic effect that significantly improved growth performances such as average daily growth gain (ADG), final body weight (FBW) and weight gain (WG) as compared to the control group having lower alanine aminotransferase (ALT) and aspartate aminotransferase (AST) levels. Malondialdehyde (MDA) levels did not alter significantly between any of the treatments. The findings indicated that adding *S. cerevisiae* or *Bacillus* spp. to water as a probiotic improved Nile tilapia growth performance. This strategy can be used to boost production of tilapia in the Thai aquaculture sector. A recent study by Kuebutornye et al. (2020a, b, c) suggested that adding host-associated probiotics to Nile tilapia's diet, such as *B. velezensis* TPS3N, *B. subtilis* TPS4, and *B. amyloliquefaciens* TPS17, either separately or in combination, could improve the fish's mucosal immunity, gut health and resistance against infection by *A. hydrophila.* According to the authors, the greater mucosal immune parameters seen in the CB group may be the result of the three *Bacillus* species' synergistic effects (Kuebutornye et al. 2020a, b, c). Additionally, *Bacillus* species added to the diet of Nile tilapia may improve the health of the fish by positively influencing its gut microbiota. Similarly, Sookchaiyaporn et al. (2020) found that the dietary inclusion of two potential probiotics, *Bacillus* sp. KUAQ1 and *Bacillus* sp. KUAQ2, isolated from the intestine of Nile tilapia *Oreochromis niloticus*, could be beneficial in terms of disease control as well as the stimulation of the immune response of cultured tilapia. In their study, Ramos et al. (2017) sought to determine the effects of a blend of probiotics (*Lactobacillus* sp., *Pediococcus* sp., *Bacillus* sp., *Enterococcus* sp.) given to juvenile Nile tilapia (*Oreochromis niloticus*) on growth performance, innate immunity, feed utilization, and oxidative stress. The findings showed that after a long time of supplementation, no benefit of probiotic feeding was seen. A beneficial nutritional strategy for sustainable tilapia aquaculture may be the dietary supplementation of mixed species probiotics. In a competitive and expanding industry like aquaculture, outcomes like improved immune responses and intestinal morphology are crucial for boosting growth performance, nutritional absorption, and disease resistance in fish. Similar to this, Nile tilapia have been shown to have improved growth performance, immunological responses, digestive enzymes, and disease resistance when given the dietary strains *Bacillus* MCCB 101 and *Micrococcus* MCCB 104 or *Bacillus velezensis* H3.1 and *Lactobacillus plantarum* N11 (Sankar et al. 2017; Van Doan et al. 2018). An eight-week feeding trial was carried out to assess the effects of a basal control diet (CON), *Lactococcus lactis* at 10^7 CFU/g (LL7) and at 10^8 CFU/g diet (LL8), *Bacillus subtilis* at 10^7 (BS7) and at 10^8 CFU/g diet (BS8), and oxytetracycline (OTC) at 4 g/kg diet on Nile tilapia, by Won et al. (2020). The findings demonstrated that fish given the BS8, LL8, and LL7 diets experienced considerably greater feed efficiency, weight increase, lysozyme activity, specific growth rate, and protein efficiency ratio, than fish fed the CON diet. Fish fed the BS8, LL8, BS7, LL7, and OTC diets had considerably higher superoxide dismutase and myeloperoxidase activities than fish fed the CON diet. Fish fed the BS8, LL8, and LL7 diets had much longer intestinal villi and thicker muscle layers than fish fed the CON and OTC diets. Additionally, fish fed the BS8 and LL8 diets had considerably higher levels of interferon-gamma (IFN-γ), heat shock protein 70 (HSP70), interleukin (IL-1 β), and tumor necrosis factor (TNF-α) gene expression than fish fed the CON diet. Following a 13-day challenge test, fish fed the BS8 and LL8 diets had a cumulative survival rate that was considerably greater than fish fed the CON, BS7, and OTC diets. These findings suggest that *B. subtilis* and *L. lactis* at 10^8 (CFU/g) could substitute antibiotics and have positive effects on Nile tilapia growth, histology, immunology, gene expression, and disease resistance. The increased intestinal histology and enzyme activity of Nile tilapia fed probiotic diets may be the cause of their superior growth performance and feed utilization. By enhancing intestinal villi length, trypsin activity, and muscle layer thickness, probiotics may have promoted efficient food absorption. The improvement of phagocytic activity and reactive oxygen metabolites by macrophages is one of the causes of increased immunity. Recent research by Van Doan et al. (2021a, b, c) demonstrated that a mixture of Bacillus species (*B. licheniformis* DF001, *Bacillus subtilis* TISTR001, and *B. megaterium* TISTR067) increased Nile tilapia growth, immunological response, and disease resistance.

References

Abdelfatah EN, Mahboub HH (2018) Studies on the effect of *Lactococcus garvieae* of dairy origin on both cheese and Nile tilapia (*O. niloticus*). Int. J Vet Sci 6(2):201–207

Abdel-Latif HM, Dawood MA, Menanteau-Ledouble S, El-Matbouli M (2020) The nature and consequences of co-infections in tilapia: a review. J Fish Dis 6:651–664

Addo S, Carrias AA, Williams MA, Liles MR, Terhune JS, Davis DA (2017) Effects of *Bacillus subtilis* strains on growth, immune parameters, and *streptococcus iniae* susceptibility in Nile tilapia, *Oreochromis niloticus*. J World Aquac Soc 48(2):257–267

Adeoye AA, Yomla R, Jaramillo-Torres A, Rodiles A, Merrifield DL, Davies SJ (2016) Combined effects of exogenous enzymes and probiotic on Nile tilapia (*Oreochromis niloticus*) growth, intestinal morphology and microbiome. Aquac 463:61–70

Akhter N, Wu B, Memon AM, Mohsin M (2015) Probiotics and prebiotics associated with aquaculture: a review. Fish Shellfish Immunol 45(2):733–741

Aly SM, Abd-El-Rahman AM, John G, Mohamed MF (2008a) Characterization of some bacteria isolated from *Oreochromis niloticus* and their potential use as probiotics. Aquaculture 277(1–2): 1–6

Aly SM, Ahmed YA, Ghareeb AA, Mohamed MF (2008b) Studies on *Bacillus subtilis* and *lactobacillus acidophilus*, as potential probiotics, on the immune response and resistance of Tilapia nilotica (*Oreochromis niloticus*) to challenge infections. Fish Shellfish Immunol 25(1–2):128–136

Amin A, Zahra T, Raja H, Amin M, Dilshad E, Naveed M, Ahmed I (2020) Major natural sinks for harboring microorganisms with altered antibiotic resistance versus major human contributing sources of antibiotic resistance: a detailed insight. In: Antibiotics and antimicrobial resistance genes in the environment. Elsevier, pp 70–98

Amoah K, Huang QC, Tan BP, Zhang S, Chi SY, Yang QH, Liu HY, Dong XH (2019) Dietary supplementation of probiotic *Bacillus coagulans* ATCC 7050, improves the growth performance, intestinal morphology, microflora, immune response, and disease confrontation of Pacific white shrimp, *Litopenaeus vannamei*. Fish Shellfish Immunol 87:796–808

Assefa A, Abunna F (2018) Maintenance of fish health in aquaculture: review of epidemiological approaches for prevention and control of infectious disease of fish. Vet Med Int 5432497:1

Azad MA, Islam SS, Sithi IN, Ghosh AK, Banu GR, Bir J, Huq KA (2019) Effect of probiotics on immune competence of giant freshwater prawn *Macrobrachium rosenbergii*. Aquac Res 50(2): 644–657

Balcázar JL, De Blas I, Ruiz-Zarzuela I, Cunningham D, Vendrell D, Múzquiz JL (2006) The role of probiotics in aquaculture. Vet Microbiol 114(3–4):173–186
De Bidhan C, Meena DK, Behera BK, Das P, Das Mohapatra PK, Sharma AP (2014) Probiotics in fish and shellfish culture: immunomodulatory and ecophysiological responses. Fish Physiol Biochem 3:921–971
Butt UD, Lin N, Akhter N, Siddiqui T, Li S, Wu B (2021) Overview of the latest developments in the role of probiotics, prebiotics and synbiotics in shrimp aquaculture. Fish Shellfish Immunol 114:263–281
Caipang CM, Lazado CC (2015) Nutritional impacts on fish mucosa: immunostimulants, pre- and probiotics. In: Mucosal health in aquaculture. Academic Press, pp 211–272
Chabrillón M, Rico RM, Arijo S, Diaz-Rosales P, Balebona MC, Moriñigo MA (2005) Interactions of microorganisms isolated from gilthead sea bream, *Sparus aurata* L., on *Vibrio harveyi,* a pathogen of farmed Senegalese sole, *Solea senegalensis* (Kaup). J Fish Dis 28(9):531–537
Chen SW, Liu CH, Hu SY (2019) Dietary administration of probiotic *Paenibacillus ehimensis* NPUST1 with bacteriocin-like activity improves growth performance and immunity against *Aeromonas hydrophila* and *streptococcus iniae* in Nile tilapia (*Oreochromis niloticus*). Fish Shellfish Immunol 84:695–703
da Mota FF, Vollú RE, Jurelevicius D, Seldin L (2016) Whole-genome sequence of *Rummeliibacillus stabekisii* strain PP9 isolated from Antarctic soil. Genome Announc 4(3): e00416–e00416
Das A, Nakhro K, Chowdhury S, Kamilya D (2013) Effects of potential probiotic bacillus amyloliquifaciens FPTB16 on systemic and cutaneous mucosal immune responses and disease resistance of catla (Catla catla). Fish Shellfish Immunol 35(5):1547–1553
Dawood MA (2021) Nutritional immunity of fish intestines: important insights for sustainable aquaculture. Rev Aquac 13(1):642–663
Dawood MA, Abo-Al-Ela HG, Hasan MT (2020) Modulation of transcriptomic profile in aquatic animals: probiotics, prebiotics and synbiotics scenarios. Fish Shellfish Immunol 97:268–282
Dawood MA, El Basuini MF, Zaineldin AI, Yilmaz S, Hasan MT, Ahmadifar E, El Asely AM, Abdel-Latif HM, Alagawany M, Abu-Elala NM, Van Doan H (2021) Antiparasitic and antibacterial functionality of essential oils: an alternative approach for sustainable aquaculture. Pathogens 10(2):185
Dawood MA, Koshio S (2020) Application of fermentation strategy in aquafeed for sustainable aquaculture. Rev Aquac 12(2):987–1002
Dawood MA, Koshio S, Ishikawa M, Yokoyama S (2015) Effects of partial substitution of fish meal by soybean meal with or without heat-killed *lactobacillus plantarum* (LP20) on growth performance, digestibility, and immune response of amberjack, *Seriola dumerili* juveniles. Biomed Res Int 2015:1
Decamp O, Moriarty DJ, Lavens P (2008) Probiotics for shrimp larviculture: review of field data from Asia and Latin America. Aquac Res 39(4):334–338
Denev S, Beev G, Staykov Y, Moutafchieva R (2009) Microbial ecology of the gastrointestinal tract of fish and the potential application of probiotics and prebiotics in finfish aquaculture. Int Aquat Res 1(1):1–29
Dorji N, Yamazaki S, Thinley P (2022) Productivity improvement to sustain small-scale fish production in developing countries: the case of Bhutan. Aquac 548:737612
Duan Y, Dong H, Wang Y, Zhang Y, Zhang J (2018) Effects of the dietary probiotic *clostridium butyricum* on intestine digestive and metabolic capacities, SCFA content and body composition in *Marsupenaeus japonicus*. J Ocean Univ China 17(3):690–696
Duan Y, Zhang Y, Dong H, Wang Y, Zhang J (2017) Effect of the dietary probiotic *clostridium butyricum* on growth, intestine antioxidant capacity and resistance to high temperature stress in kuruma shrimp *Marsupenaeus japonicus*. J Therm Biol 66:93–100
Duc LH, Hong HA, Barbosa TM, Henriques AO, Cutting SM (2004) Characterization of bacillus probiotics available for human use. Appl Environ Microbiol 70(4):2161–2171

Eissa N, Wang HP, Yao H, Abou-ElGheit E (2018) Mixed bacillus species enhance the innate immune response and stress tolerance in yellow perch subjected to hypoxia and air- exposure stress. Sci Rep 8(1):1–0

El-Saadony MT, Alagawany M, Patra AK, Kar I, Tiwari R, Dawood MA, Dhama K, Abdel- Latif HM (2021) The functionality of probiotics in aquaculture: an overview fish. Shellfish Immunol 117:36–52

Elsabagh M, Mohamed R, Moustafa EM, Hamza A, Farrag F, Decamp O, Dawood MA, Eltholth M (2018) Assessing the impact of bacillus strains mixture probiotic on water quality, growth performance, blood profile and intestinal morphology of Nile tilapia, *Oreochromis niloticus*. Aquac Nutr 6:1613–1622

England PH (2014) UK standards for microbiology investigations. Investigations of Faecal specimens for enteric pathogens. Public Health England, London

FAO (2018) The state of world fisheries and aquaculture meeting the sustainable development goals. Rome. Licence: CC BY-NC-SA 3.0 IGO

FAO (2020) The state of world fisheries and aquaculture (SOFIA), in: F.a.A.O.o.t.U. Nations (Ed.) FAO, Rome, Italy

FAO/WHO (2001) Health and nutritional properties of probiotics in food including powder milk with liver lactic acid bacteria, Food and Agriculture Organization and World Health Organization Joint report p. 34

Frouël S, Le Bihan E, Serpentini A, Lebel JM, Koueta N, Nicolas JL (2008) Preliminary study of the effects of commercial lactobacilli preparations on digestive metabolism of juvenile sea bass (*Dicentrarchus labrax*). Microb Physiol 14(1–3):100–106

Gāliņa D, Ansonska L, Valdovska A (2020) Effect of probiotics and herbal products on intestinal histomorphological and immunological development in piglets. Vet Med Int 24:2020

Gao Q, Xiao C, Min M, Zhang C, Peng S, Shi Z (2016) Effects of probiotics dietary supplementation on growth performance, innate immunity and digestive enzymes of silver pomfret, *Pampus argenteus*. Indian J Anim Res 50(6):936–941

Gao QX, Shi ZH, Peng SM (2013) Probiotics in aquaculture: recent progress and outlook. Mar Fish Rev 35(3):364–372

Gebremedhin S, Bruneel S, Getahun A, Anteneh W, Goethals P (2021) Scientific methods to understand fish population dynamics and support sustainable fisheries management. Water 13(4):574

Ghosh AK, Panda SK, Luyten W (2021) Anti-vibrio and immune-enhancing activity of medicinal plants in shrimp: a comprehensive review. Fish Shellfish Immunol 117:192–210

Giri SS, Sukumaran V, Oviya M (2013) Potential probiotic *lactobacillus plantarum* VSG3 improves the growth, immunity, and disease resistance of tropical freshwater fish, *Labeo rohita*. Fish Shellfish Immunol 34(2):660–666

Gobi N, Vaseeharan B, Chen JC, Rekha R, Vijayakumar S, Anjugam M, Iswarya A (2018) Dietary supplementation of probiotic *bacillus licheniformis* Dahb1 improves growth performance, mucus and serum immune parameters, antioxidant enzyme activity as well as resistance against *Aeromonas hydrophila* in tilapia *Oreochromis mossambicus*. Fish Shellfish Immunol 74:501–508

Gram L, Løvold T, Nielsen J, Melchiorsen J, Spanggaard B (2001) In vitro antagonism of the probiont *Pseudomonas fluorescens* strain AH2 against *Aeromonas salmonicida* does not confer protection of salmon against furunculosis. Aquaculture 199(1–2):1–1

Gram L, Melchiorsen J, Spanggaard B, Huber I, Nielsen TF (1999) Inhibition of *vibrio anguillarum* by *Pseudomonas fluorescens* AH2, a possible probiotic treatment of fish. AEM 65(3):969–973

Guardiola FA, Bahi A, Bakhrouf A, Esteban MA (2017) Effects of dietary supplementation with fenugreek seeds, alone or in combination with probiotics, on gilthead seabream (*Sparus aurata* L.) skin mucosal immunity. Fish Shellfish Immunol 65:169–178

Guardone L, Tinacci L, Armani A, Trevisani M (2022) Residues of veterinary drugs in fish and fish products: an analysis of RASFF data over the last 20 years. Food Control 1(135):108780

Gutowska MA, Drazen JC, Robison BH (2004) Digestive chitinolytic activity in marine fishes of Monterey Bay, California. Comparative biochemistry and physiology part a. Comp Biochem Physiol 139(3):351–358
He RP, Feng J, Tian XL, Dong SL, Wen B (2017) Effects of dietary supplementation of probiotics on the growth, activities of digestive and non-specific immune enzymes in hybrid grouper (*Epinephelus lanceolatus*♂× *Epinephelus fuscoguttatus*♀). Aquac Res 48(12):5782–5790
Henriksson PJ, Rico A, Troell M, Klinger DH, Buschmann AH, Saksida S, Chadag MV, Zhang W (2018) Unpacking factors influencing antimicrobial use in global aquaculture and their implication for management: a review from a systems perspective. Sustain Sci 13(4):1105–1120
Hernández-Contreras Á, Hernández MD (2020) Application of aromatic plants and their extracts in aquaculture. In: Feed additives. Academic Press, pp 239–259
Hostins B, Lara G, Decamp O, Cesar DE, Wasielesky W Jr (2017) Efficacy and variations in bacterial density in the gut of *Litopenaeus vannamei* reared in a BFT system and in clear water supplemented with a commercial probiotic mixture. Aquaculture 480:58–64
Irianto A, Austin B (2002) Probiotics in aquaculture. J Fish Dis 25(11):633–642
Jahangiri L, Esteban MÁ (2018) Administration of probiotics in the water in finfish aquaculture systems: a review. Aust Fish 3(3):33
Junghare M, Subudhi S, Lal B (2012) Improvement of hydrogen production under decreased partial pressure by newly isolated alkaline tolerant anaerobe, *clostridium butyricum* TM-9A: optimization of process parameters. Int J Hydrog Energy 37(4):3160–3168
Kelly C, Salinas I (2017) Under pressure: interactions between commensal microbiota and the teleost immune system. Front Immunol 8:559
Kent AG, Vill AC, Shi Q, Satlin MJ, Brito IL (2020) Widespread transfer of mobile antibiotic resistance genes within individual gut microbiomes revealed through bacterial hi-C. Nat Commun 11(1):1–9
Koch N, Islam NF, Sonowal S, Prasad R, Sarma H (2021) Environmental antibiotics and resistance genes as emerging contaminants: methods of detection and bioremediation. Curr Res Microb Sci 1(2):100027
Kuebutornye FK, Abarike ED, Lu Y, Hlordzi V, Sakyi ME, Afriyie G, Wang Z, Li Y, Xie CX (2020a) Mechanisms and the role of probiotic bacillus in mitigating fish pathogens in aquaculture. Fish Physiol Biochem 46(3):819–841
Kuebutornye FK, Lu Y, Abarike ED, Wang Z, Li Y, Sakyi ME (2020b) In vitro assessment of the probiotic characteristics of three bacillus species from the gut of Nile tilapia, *Oreochromis niloticus*. Probiotics Antimicrob 12(2):412–424
Kuebutornye FK, Wang Z, Lu Y, Abarike ED, Sakyi ME, Li Y, Xie CX, Hlordzi V (2020c) Effects of three host-associated bacillus species on mucosal immunity and gut health of Nile tilapia, *Oreochromis niloticus* and its resistance against *Aeromonas hydrophila* infection. Fish Shellfish Immunol 97:83–95
Kumar YA, Singh PK, Singh AK, Masih HA, Peter KJ, Benjamin JC, Rath S (2014) Production optimization of alpha amylase from *Bacillus altitudinis*. Int J Sci Eng Technol Res 3(4):654–673
La Ragione RM, Narbad A, Gasson MJ, Woodward MJ (2004) In vivo characterization of *lactobacillus johnsonii* FI9785 for use as a defined competitive exclusion agent against bacterial pathogens in poultry. Lett Appl Microbiol 38(3):197–205
La Ragione RM, Woodward MJ (2003) Competitive exclusion by *Bacillus subtilis* spores of *salmonella enterica* serotype Enteritidis and Clostridium perfringens in young chickens. Vet Microbio 94(3):245–256
Lategan MJ, Torpy FR, Gibson LF (2004) Control of saprolegniosis in the eel Anguilla australis Richardson, by Aeromonas media strain A199. Aquaculture 240(1–4):19–27
Lazado CC, Caipang CM (2014) Mucosal immunity and probiotics in fish. Fish Shellfish Immunol 39(1):78–89
Lerminiaux NA, Cameron AD (2019) Horizontal transfer of antibiotic resistance genes in clinical environments. Can J Microbiol 65(1):34–44

Lesel R (1990) Microbiology in poecilotherms. InInternational Symposium on Microbiology in Poecilotherms (1989: Paris, France) sole distributors for the USA and Canada, Elsevier Science Pub. Co

Liang J, Zhang M, Wang X, Ren Y, Yue T, Wang Z, Gao Z (2021) Edible fungal polysaccharides, the gut microbiota, and host health. Carbohydr Polym 1(273):118558

Liao X, Wu R, Ma G, Zhao L, Zheng Z, Zhang R (2015a) Effects of *clostridium butyricum* on antioxidant properties, meat quality and fatty acid composition of broiler birds. Lipids Health Dis 14(1):1–9

Liao XD, Ma G, Cai J, Fu Y, Yan XY, Wei XB, Zhang RJ (2015b) Effects of *clostridium butyricum* on growth performance, antioxidation, and immune function of broilers. Poult Sci 94(4): 662–667

Liu H, Wang S, Cai Y, Guo X, Cao Z, Zhang Y, Liu S, Yuan W, Zhu W, Zheng YU, Xie Z (2017) Dietary administration of Bacillus subtilis HAINUP40 enhances growth, digestive enzyme activities, innate immune responses and disease resistance of tilapia, *Oreochromis niloticus*. Fish Shellfish Immunol 60:326–333

Liu L, Zeng D, Yang M, Wen B, Lai J, Zhou Y, Sun H, Xiong L, Wang J, Lin Y, Pan K (2019) Probiotic *clostridium butyricum* improves the growth performance, immune function, and gut microbiota of weaning rex rabbits. Probiotics Antimicrob 11(4):1278–1292

Liu Y, Jia Y, Liu C, Ding L, Xia Z (2018) RNA-Seq transcriptome analysis of breast muscle in Pekin ducks supplemented with the dietary probiotic *clostridium butyricum*. BMC Genomics 19(1):1–4

Makled SO, Hamdan AM, El-Sayed AF (2019) Effects of dietary supplementation of a marine thermotolerant bacterium, *bacillus paralicheniformis* SO-1, on growth performance and immune responses of Nile tilapia, *Oreochromis niloticus*. Aquac Nutr 25(4):817–827

Midhun SJ, Neethu S, Arun D, Vysakh A, Divya L, Radhakrishnan EK, Jyothis M (2019) Dietary supplementation of *bacillus licheniformis* HGA8B improves growth parameters, enzymatic profile and gene expression of *Oreochromis niloticus*. Aquaculture 505:289–296

Monteagudo-Mera A, Rastall RA, Gibson GR, Charalampopoulos D, Chatzifragkou A (2019) Adhesion mechanisms mediated by probiotics and prebiotics and their potential impact on human health. Appl Microbiol Biotechnol 103(16):6463–6472

Montes AJ (1993) The use of probiotics in food-animal practice. Vet Med 88:282–289

Nayak SK (2010) Probiotics and immunity: a fish perspective. Fish Shellfish Immunol 29(1):2–14

Nayak SK, Swain P, Mukherjee SC (2007) Effect of dietary supplementation of probiotic and vitamin C on the immune response of Indian major carp, *Labeo rohita* (ham.). Fish Shellfish Immunol 23(4):892–896

Noor Z, Noor M, Khan I, Khan SA (2020) Evaluating the lucrative role of probiotics in the aquaculture using microscopic and biochemical techniques. Microsc Res Tech 83(3):310–317

Okocha RC, Olatoye IO, Adedeji OB (2018) Food safety impacts of antimicrobial use and their residues in aquaculture. Public Health Rev 39(1):1–22

Pan X, Wu T, Song Z, Tang H, Zhao Z (2008) Immune responses and enhanced disease resistance in Chinese drum, *Miichthys miiuy* (Basilewsky), after oral administration of live or dead cells of *clostridium butyrium* CB2. J Fish Dis 31(9):679–686

Panigrahi A, Kiron V, Satoh S, Watanabe T (2010) Probiotic bacteria *lactobacillus rhamnosus* influences the blood profile in rainbow trout *Oncorhynchus mykiss* (Walbaum). Fish Physiol Biochem 36(4):969–977

Pepi M, Focardi S (2021) Antibiotic-resistant bacteria in aquaculture and climate change: a challenge for health in the Mediterranean area. IJERPH 18(11):5723

Pérez-Sánchez T, Ruiz-Zarzuela I, de Blas I, Balcázar JL (2014) Probiotics in aquaculture: a current assessment. Rev Aquac 6(3):133–146

Pirarat N, Kobayashi T, Katagiri T, Maita M, Endo M (2006) Protective effects and mechanisms of a probiotic bacterium *lactobacillus rhamnosus* against experimental *Edwardsiella tarda* infection in tilapia (*Oreochromis niloticus*). Vet Immunol Immunopathol 113(3–4):339–347

Poolsawat L, Li X, He M, Ji D, Leng X (2020) *Clostridium butyricum* as probiotic for promoting growth performance, feed utilization, gut health and microbiota community of tilapia (*Oreochromis niloticus*× *O. aureus*). Aquac Nutr 26(3):657–670

Ramos MA, Batista S, Pires MA, Silva AP, Pereira LF, Saavedra MJ, Ozório RO, Rema P (2017) Dietary probiotic supplementation improves growth and the intestinal morphology of Nile tilapia. Animal 11(8):1259–1269

Reda RM, Selim KM (2015) Evaluation of *bacillus amyloliquefaciens* on the growth performance, intestinal morphology, hematology and body composition of Nile tilapia, *Oreochromis niloticus*. Aquac Int 23(1):203–217

Reverter M, Sarter S, Caruso D, Avarre JC, Combe M, Pepey E, Pouyaud L, Vega-Heredía S, De Verdal H, Gozlan RE (2020) Aquaculture at the crossroads of global warming and antimicrobial resistance. Nat Commun 11(1):1–8

Ringø E, Doan HV, Lee S, Song SK (2020) Lactic acid bacteria in shellfish: possibilities and challenges. Rev Fish Sci Aquac 28(2):139–169

Rohani MF, Islam SM, Hossain MK, Ferdous Z, Siddik MA, Nuruzzaman M, Padeniya U, Brown C, Shahjahan M (2021) Probiotics, prebiotics and synbiotics improved the functionality of aquafeed: upgrading growth, reproduction, immunity and disease resistance in fish fish shellfish Immunol

Sáenz de Rodrigáñez MA, Díaz-Rosales P, Chabrillón M, Smidt H, Arijo S, León-Rubio JM, Alarcón FJ, Balebona MC, Moriñigo MA, Cara JB, Moyano FJ (2009) Effect of dietary administration of probiotics on growth and intestine functionality of juvenile Senegalese sole (*Solea senegalensis*, Kaup 1858). Aquac Nutr 15(2):177–185

Sankar H, Philip B, Philip R, Singh IS (2017) Effect of probiotics on digestive enzyme activities and growth of cichlids, *Etroplus suratensis* (pearl spot) and *Oreochromis mossambicus* (tilapia). Aquac Nutr 23(4):852–864

Santos L, Ramos F (2018) Antimicrobial resistance in aquaculture: current knowledge and alternatives to tackle the problem. IJAA 52(2):135–143

Selim KM, Reda RM (2015) Improvement of immunity and disease resistance in the Nile tilapia, *Oreochromis niloticus*, by dietary supplementation with *bacillus amyloliquefaciens*. Fish Shellfish Immunol 44(2):496–503

Shao Y, Wang Y, Yuan Y, Xie Y (2021) A systematic review on antibiotics misuse in livestock and aquaculture and regulation implications in China. Sci Total Environ 1(798):149205

Silva EF, Soares MA, Calazans NF, Vogeley JL, de Valle BC, Soares R, Peixoto S (2012) Effect of probiotic (*bacillus* spp.) addition during larvae and postlarvae culture of the white shrimp *Litopenaeus vannamei*. Aquac Res 44(1):13–21

Silva TF, Petrillo TR, Yunis-Aguinaga J, Marcusso PF, da Silva CG, de Moraes FR, de Moraes JR (2015) Effects of the probiotic *bacillus amyloliquefaciens* on growth performance, hematology and intestinal morphometry in cage-reared Nile tilapia. Lat Am J Aquat Res 43(5):963–971

Sookchaiyaporn N, Srisapoome P, Unajak S, Areechon N (2020) Efficacy of bacillus spp. isolated from Nile tilapia *Oreochromis niloticus* Linn. On its growth and immunity, and control of pathogenic bacteria. Fish Sci 86(2):353–365

Srisapoome P, Areechon N (2017) Efficacy of viable *Bacillus pumilus* isolated from farmed fish on immune responses and increased disease resistance in Nile tilapia (*Oreochromis niloticus)*: laboratory and on-farm trials. Fish Shellfish Immunol 67:199–210

Stevens CH, Croft DP, Paull GC, Tyler CR (2017) Stress and welfare in ornamental fishes: what can be learned from aquaculture? J Fish Biol 2:409–428

Sun YZ, Yang HL, Ma RL, Lin WY (2010) Probiotic applications of two dominant gut bacillus strains with antagonistic activity improved the growth performance and immune responses of grouper *Epinephelus coioides*. Fish Shellfish Immunol 29(5):803–809

Suprayudi MA, Maeda M, Hidayatullah H, Widanarni W, Setiawati M, Ekasari J (2017) The positive contributions of PowerLac™ supplementation to the production performance, feed utilization and disease resistance of Nile tilapia *Oreochromis niloticus* (L.). Aquac Res 48(5): 2145–2156

Sutthi N, Thaimuangphol W, Rodmongkoldee M, Leelapatra W, Panase P (2018) Growth performances, survival rate, and biochemical parameters of Nile tilapia (*Oreochromis niloticus*) reared in water treated with probiotic. Comp Clin Path 27(3):597–603

Suzer C, Çoban D, Kamaci HO, Saka Ş, Firat K, Otgucuoğlu Ö, Küçüksari H (2008) *Lactobacillus* spp. bacteria as probiotics in gilthead sea bream (*Sparus aurata*, L.) larvae: effects on growth performance and digestive enzyme activities. Aquaculture 280(1–4):140–150

Tan HY, Chen SW, Hu SY (2019) Improvements in the growth performance, immunity, disease resistance, and gut microbiota by the probiotic *Rummeliibacillus stabekisii* in Nile tilapia (*Oreochromis niloticus*). Fish Shellfish Immunol 92:265–275

Terpou A, Papadaki A, Lappa IK, Kachrimanidou V, Bosnea LA, Kopsahelis N (2019) Probiotics in food systems: significance and emerging strategies towards improved viability and delivery of enhanced beneficial value. Nutrients 11(7):1591

Thite VS, Nerurkar AS, Baxi NN (2020) Optimization of concurrent production of xylanolytic and pectinolytic enzymes by *bacillus safensis* M35 and *Bacillus altitudinis* J208 using agro-industrial biomass through response surface methodology. Sci Rep 10(1):1–2

Tinh NT, Dierckens K, Sorgeloos P, Bossier P (2008) A review of the functionality of probiotics in the larviculture food chain mar. Biotechnol 10(1):1–2

Vaishampayan P, Miyashita M, Ohnishi A, Satomi M, Rooney A, La Duc MT, Venkateswaran K (2009) Description of *Rummeliibacillus stabekisii* gen. Nov., sp. nov. and reclassification of Bacillus pycnus Nakamura et al. 2002 as Rummeliibacillus pycnus comb. nov. Int J Syst Evol Microbiol 59(5):1094–1099

Vaiyapuri M, Pailla S, Rao Badireddy M, Pillai D, Chandragiri Nagarajarao R, Prasad Mothadaka M (2021) Antimicrobial resistance in Vibrios of shrimp aquaculture: incidence, identification schemes, drivers and mitigation measures. Aquac Res 52(7):2923–2941

Van Doan H, Hoseinifar SH, Khanongnuch C, Kanpiengjai A, Unban K, Srichaiyo S (2018) Host-associated probiotics boosted mucosal and serum immunity, disease resistance and growth performance of Nile tilapia (Oreochromis niloticus). Aquaculture 491:94–100

Van Doan H, Lumsangkul C, Jaturasitha S, Meidong R, Hoseinifar SH, Dawood MA (2021a) Modulation of growth, skin mucus and serum immunities, and disease resistance of Nile tilapia fed host-associated probiotic (*lactobacillus paracasei* l61-27b). Aquac Nutr 27:3–12

Van Doan H, Lumsangkul C, Ruangwong OU, Meidong R, Hoseinifar SH, Dawood MA, Azra MN, Jatursitha S, Carnevali O (2021b) Effects of host-associated probiotic *Bacillus altitudinis* B61-34b on growth performance, immune response and disease resistance of Nile tilapia (*Oreochromis niloticus*) raised under biofloc system. Aquac Nutr 27:61–72

Van Doan H, Wangkahart E, Thaimuangphol W, Panase P, Sutthi N (2021c) Effects of *bacillus* spp. mixture on growth, immune responses, expression of immune-related genes, and resistance of Nile tilapia against *Streptococcus agalactiae* infection. Probiotics Antimicrob 1:1–6

Verschuere L, Rombaut G, Sorgeloos P, Verstraete W (2000) Probiotic bacteria as biological control agents in aquaculture. MMBR 64(4):655–671

Vine NG, Leukes WD, Kaiser H (2004) In vitro growth characteristics of five candidate aquaculture probiotics and two fish pathogens grown in fish intestinal mucus. FEMS Microbiol Lett 231(1): 145–152

Vine NG, Leukes WD, Kaiser H (2006) Probiotics in marine larviculture. FEMS Microbiol Rev 30(3):404–427

Wang C, Chuprom J, Wang Y, Fu L (2020) Beneficial bacteria for aquaculture: nutrition, bacteriostasis and immunoregulation. J Appl Microbiol 128(1):28–40

Wang M, Liu G, Lu M, Ke X, Liu Z, Gao F, Cao J, Zhu H, Yi M, Yu D (2017) Effect of *Bacillus cereus* as a water or feed additive on the gut microbiota and immunological parameters of Nile tilapia. Aquac Res 48(6):3163–3173

Wang YB, Tian ZQ, Yao JT, Li WF (2008) Effect of probiotics, *Enteroccus faecium*, on tilapia (*Oreochromis niloticus*) growth performance and immune response. Aquaculture 277(3–4): 203–207

Westerdahl AL, Olsson JC, Kjelleberg ST, Conway PL (1991) Isolation and characterization of turbot (*Scophtalmus maximus*)-associated bacteria with inhibitory effects against *vibrio anguillarum*. Appl Environ Microbiol 57(8):2223–2228

Won S, Hamidoghli A, Choi W, Park Y, Jang WJ, Kong IS, Bai SC (2020) Effects of *Bacillus subtilis* WB60 and *Lactococcus lactis* on growth, immune responses, histology and gene expression in Nile tilapia, *Oreochromis niloticus*. Microorganisms 8(1)

Xia Y, Cao J, Wang M, Lu M, Chen G, Gao F, Liu Z, Zhang D, Ke X, Yi M (2019) Effects of *Lactococcus lactis* subsp. lactis JCM5805 on colonization dynamics of gut microbiota and regulation of immunity in early ontogenetic stages of tilapia. Fish Shellfish Immunol 86:53–63

Xia Y, Lu M, Chen G, Cao J, Gao F, Wang M, Liu Z, Zhang D, Zhu H, Yi M (2018) Effects of dietary *lactobacillus rhamnosus* JCM1136 and Lactococcus lactis subsp. lactis JCM5805 on the growth, intestinal microbiota, morphology, immune response and disease resistance of juvenile Nile tilapia, *Oreochromis niloticus*. Fish Shellfish Immunol 76:368–379

Yanbo W, Zirong X (2006) Effect of probiotics for common carp (*Cyprinus carpio*) based on growth performance and digestive enzyme activities. Anim Feed Sci Technol 127(3–4): 283–292

Yang HL, Xia HQ, Ye YD, Zou WC, Sun YZ (2014) Probiotic *Bacillus pumilus* SE5 shapes the intestinal microbiota and mucosal immunity in grouper *Epinephelus coioides*. Dis Aquat Org 111(2):119–127

Yin R, Zhai Q, Yu L, Xiao Y, Wang G, Yu R, Tian F, Chen W (2016) The binding characters study of lead removal by lactobacillus plantarum CCFM8661. Eur Food Res Technol 242(10): 1621–1629

Zaineldin AI, Hegazi S, Koshio S, Ishikawa M, Dawood MA, Dossou S, Yukun Z, Mzengereza K (2021) Singular effects of Bacillus subtilis C-3102 or Saccharomyces cerevisiae type 1 on the growth, gut morphology, immunity, and stress resistance of red sea bream (*Pagrus major*). Ann AnI Sci 21(2):589–608

Zalewska M, Błażejewska A, Czapko A, Popowska M (2021) Antibiotics and antibiotic resistance genes in animal manure–consequences of its application in agriculture. Front Microbiol 640

Zhai Q, Wang H, Tian F, Zhao J, Zhang H, Chen W (2017a) Dietary lactobacillus plantarum supplementation decreases tissue lead accumulation and alleviates lead toxicity in Nile tilapia (*Oreochromis niloticus*). Aquac Res 48(9):5094–5103

Zhai Q, Yu L, Li T, Zhu J, Zhang C, Zhao J, Zhang H, Chen W (2017b) Effect of dietary probiotic supplementation on intestinal microbiota and physiological conditions of Nile tilapia (*Oreochromis niloticus*) under waterborne cadmium exposure. Antonie Van Leeuwenhoek 110(4):501–513

Zhang CN, Li XF, Xu WN, Jiang GZ, Lu KL, Wang LN, Liu WB (2013) Combined effects of dietary fructooligosaccharide and bacillus licheniformis on innate immunity, antioxidant capability and disease resistance of triangular bream (*Megalobrama terminalis*). Fish Shellfish Immunol 35(5):1380–1386

Zhang CN, Zhang JL, Guan WC, Zhang XF, Guan SH, Zeng QH, Cheng GF, Cui W (2017) Effects of *lactobacillus delbrueckii* on immune response, disease resistance against *Aeromonas hydrophila*, antioxidant capability and growth performance of *Cyprinus carpio* Huanghe var. Fish Shellfish Immunol 68:84–91

Zhang CX, Wang HY, Chen TX (2019a) Interactions between intestinal microflora/probiotics and the immune system. Biomed Res Int 6764919:1

Zhang D, Gao Y, Ke X, Yi M, Liu Z, Han X, Shi C, Lu M (2019b) *Bacillus velezensis* LF01: in vitro antimicrobial activity against fish pathogens, growth performance enhancement, and disease resistance against streptococcosis in Nile tilapia (*Oreochromis niloticus*). Appl Microbiol Biotechnol 103(21):9023–9035

Zhang L, Cao GT, Zeng XF, Zhou L, Ferket PR, Xiao YP, Chen AG, Yang CM (2014) Effects of *clostridium butyricum* on growth performance, immune function, and cecal microflora in broiler chickens challenged with Escherichia coli K88. Poult Sci 93(1):46–53

Zhou X, Tian Z, Wang Y, Li W (2010) Effect of treatment with probiotics as water additives on tilapia (*Oreochromis niloticus*) growth performance and immune response. Fish Physiol Biochem 36(3):501–509

Zhu C, Yu L, Liu W, Jiang M, He S, Yi G, Wen H, Liang X (2019) Dietary supplementation with *Bacillus subtilis* LT3-1 enhance the growth, immunity and disease resistance against *Streptococcus agalactiae* infection in genetically improved farmed tilapia, *Oreochromis niloticus*. Aquac Nutr 25(6):1241–1249

Zorriehzahra MJ, Delshad ST, Adel M, Tiwari R, Karthik K, Dhama K, Lazado CC (2016) Probiotics as beneficial microbes in aquaculture: an update on their multiple modes of action: a review. Vet Q 36(4):228–241

Chapter 6
Prebiotics and Synbiotics in Nile Tilapia Culture

Hien Van Doan

Abstract With the increase in the human population, the need for protein-rich food is increasing. To cope with the increased demand for fish and fish-related products, aquaculture is shifted towards semi-intensive and intensive culture practices. These cultivation practices exacerbate disease outbreaks. Farmers use extensive antibiotics to manage disease outbreaks and lessen the economic impact due to fish mortality. Inappropriate and extensive use of antibiotics results in the development of resistant strains and the presence of antibiotic residues in fish muscles pose a serious threat to food safety for human consumption. Water and dietary supplements that can be effectively used as alternatives to antibiotics are an active area of research. Prebiotics and synbiotics are promising candidates to replace antibiotics and have been used for the treatment and control of pathogens in tilapia aquaculture. In this chapter, the use of prebiotics and synbiotics as growth promoters and as alternatives to antibiotics in tilapia aquaculture has been addressed, along with their possible mechanism of action. Moreover, the positive effects of pre and synbiotics on tilapia health are also highlighted with emphases on growth, immune modulation, and alteration in gut microflora.

Keywords Prebiotic · Gut · Microbiota · Oligosaccharide · Fermentation

6.1 Introduction

Aquaculture is one of the fastest food-producing sectors that can provide protein-rich food to the growing human population (Fiorella et al. 2021; Edwards et al. 2019; Butt et al. 2021). In 2020, fisheries and aquaculture production reached a record of 214 million tons, worth about USD 424 billion (FAO 2022). Tilapia, a member of the Cichlid family, is the second most commonly farmed fish worldwide (Prabu et al. 2019), with a production of 5.584.4 million tons (FAO 2022). Production of tilapia is

H. Van Doan (✉)
Department of Animal and Aquatic Sciences, Faculty of Agriculture, Chiang Mai University, Chiang Mai, Thailand

© The Author(s), under exclusive license to Springer Nature Switzerland AG 2023

S. H. Hoseinifar, H. Van Doan (eds.), *Novel Approaches Toward Sustainable Tilapia Aquaculture*, Applied Environmental Science and Engineering for a Sustainable Future, https://doi.org/10.1007/978-3-031-38321-2_6

projected to increase to 7.3 MT by 2030 (FAO 2022; Kobayashi et al. 2015). Currently, tilapia is grown and traded commercially in more than 140 nations worldwide (Munguti et al. 2022). Tilapia is a great choice for warm-water aquaculture because of its many advantages, such as high growth rate, well-adapted, tolerance to stress and diseases, high fertility, rapid rate of generation, acceptance of artificial feeds right away following yolk-sac absorption and capacity to feed at a low trophic level (Ng and Romano 2013). To enhance tilapia production and financial returns, there is currently a paradigm shift away from an extensive aquaculture approach toward super intensive system due to its socioeconomic significance (Henriksson et al. 2021). However, super-intensive aquacultural practices have led to over-reliance on commercial feed, environmental contamination from the release of nutrient-rich water into the environment, and disease outbreaks are all consequences of these aquaculture operations (Abdel-Latif et al. 2020a). One of the biggest issues facing aquaculture growers today is a drop in production caused by pathogens. The most prevalent bacterial infections in tilapia that significantly reduce its yield are *Aeromonas hydrophila*, *Streptococcus iniae*, and *Streptococcus agalactiae* (Eissa et al. 2021; Gewaily et al. 2021). As a result, throughout the past few decades, disease outbreaks have been managed by using antibiotics and chemotherapeutics (Rico et al. 2014). However, their unauthorized use has significantly altered the fish's microbiota, which may harm their immunity (Kokou et al. 2020; Payne et al. 2021). Additionally, the unrestricted use of antibiotics in aquaculture has resulted in the development of antibiotic-resistant bacteria and their metabolites, which have the potential to contaminate aquaculture products intended for human consumption (Henriksson et al. 2018; Miranda et al. 2018). Consuming such tainted aquaculture products could alter human gut microbiota, which could impact human health (Cabello and Godfrey 2016; Henriksson et al. 2018). Since the use of antibiotics in aquaculture has recently been prohibited, research on the use of prebiotics and synbiotics against pathogenic bacteria has come to the fore as one of the most effective environmentally friendly and sustainable alternatives to control bacterial outbreaks in aquaculture and to enhance fish growth and immunity (Wee et al. 2022; Yilmaz et al. 2022; Rohani et al. 2022). In light of the importance of tilapia production, this chapter provides an in-depth analysis of the use of prebiotics and synbiotics in tilapia aquaculture while highlighting the growth, immunological parameters, and disease resistance utilized in evaluating fish health following their application.

6.1.1 Prebiotics

Gibson and Roberfroid were the first to introduce the idea of prebiotics, which they described as "an indigestible fiber that can enhance the growth and activity of health-promoting bacteria in the intestine and beneficially affect the host" (Gibson and Roberfroid 1995). Similarly, Bindels et al. (2015a) described prebiotics as "non-digestible food ingredients that are broken down into simpler substances, enhancing the growth and activity of preferred microorganisms found in the gastrointestinal tract and benefiting the health of the host". Guerreiro et al. (2017) presented them as "useful food ingredients that have numerous favorable effects on the host". Due to the development of antibiotic-resistant strains, the appearance of drug residues in the host, changes in the microbial community of the gut and the culture system, and lowered immunity, the use of antibiotics in aquaculture have recently come under heavy fire (Watts et al. 2017; Ibrahim et al. 2020; Henriksson et al. 2018). The manufacture, preservation, and pelleting processes for the probiotic-based feed could prevent probiotics from surviving or growing. The efficiency of the feed may be decreased if it is processed in an improper or unfavorable environment (Lauzon et al. 2014a). Prebiotics has so recently gained popularity as a probiotic and antibiotic substitute. Numerous studies and remarks have been published highlighting the advantages of prebiotics in aquaculture (Guerreiro et al. 2017; Akhter et al. 2015; Cavalcante et al. 2020; Nawaz et al. 2018). Prebiotics that are widely used in tilapia farming includes β-glucan, inulin, fructooligosaccharides (FOS), short-chain fructooligosaccharides (scFOS), mannan oligosaccharides (MOS), galactooligosaccharides (GOS), xylooligosaccharides (XOS), pectin, raffinose, and chitosan. Depending on how much they are polymerized into monosaccharides, oligosaccharides, and polysaccharides, certain carbohydrates are also categorized as prebiotics (Ringø et al. 2010b).

Short-chain fatty acids (SCFAs) and lactate can be produced by the metabolizing of prebiotics by the helpful bacteria *Lactobacillus* and *Bifidobacterium* (Markowiak-Kopeć and Śliżewska 2020; Davani-Davari et al. 2019). While lactate serves as a stimulant for gluconeogenesis, SCFAs can also be absorbed through the digestive system to serve as an energy source (Denev et al. 2009). The colon's pH is decreased by the formation of SCFAs, which slows the spread of infections and promotes the growth of lactic acid-producing bacteria like *Bifidobacterium* (Markowiak-Kopeć and Śliżewska 2020; Parada Venegas et al. 2019; Tiwari et al. 2019; Nagpal et al. 2018). Prebiotics work similarly by competing with glycoconjugates on the epithelial lining to boost SCFAs, cytokines, and mucus production while decreasing the number of pathogenic bacteria (Lauzon et al. 2014a; Hoseinifar et al. 2017a). They have an advantageous impact on the host's growth characteristics, including ultimate weight, weight increase, daily weight gain, specific growth rate, dietary conversion ratio, feed efficiency ratio, and protein efficiency ratio (Mohammadian et al. 2021; Nazarudin et al. 2020; Butt et al. 2021; Hahor et al. 2019; Jami et al. 2019).

Prebiotics added to the diet can help boost the activity of digestive enzymes. Digestive enzyme levels were found to be higher in animals that gained weight and had better feed conversion efficiency (Nedaei et al. 2019; Abasubong et al. 2019; Li et al. 2019; El Basuini et al. 2020). However, no sustained improvements were seen because prebiotics was also identified in certain investigations to act ineffectively (Hoseinifar et al. 2015a; Eshaghzadeh et al. 2015; Dobšíková et al. 2013). By triggering the non-specific immune system and fostering healthy microorganisms, prebiotics can improve immunity even further (Hoseinifar et al. 2015a; Song et al. 2014). The favorable microbiota enhances immunity in the host, protects against dangerous bacteria in the colon, and maintains a healthy balance between the microbiota of the digestive and immune systems (Gomez and Balcazar 2008). To qualify as possible prebiotics, prebiotic candidates must meet the following requirements in both *in vivo* and *in vitro* studies: they must be able to (i) withstand stomach and abdominal acid, (ii) start a variety of enzymatic reactions and processes, (iii) improve intestinal digesting, (iv) encouragement of microbial fermentation, and (v) encourage the growth of beneficial microorganisms that can improve health status (Bindels et al. 2015b; Schrezenmeir and de Vrese 2001). Yet not all prebiotics meet these requirements (Lauzon et al. 2014a).

6.1.2 Synbiotics

The word "synbiotic" is a mixture of two Greek terms "συν" and "βίος", meaning "joint or together" and "life" respectively (Kolida and Gibson 2011). A dietary supplement known as a "synbiotic" beneficially combines probiotics and prebiotics. A substance that modifies the opportunistic bacterial community of the host species' digestive system to enable probiotics to thrive and survive there (Chauhan and Chorawala 2012; Das et al. 2017). Prebiotics are chosen for synbiotic formulation based on their capacity to subtly encourage the growth and survival of advantageous bacteria within their desired range (Davani-Davari et al. 2019). Probiotics are selected based on how they specifically benefit the host (Hill et al. 2014). The selection of probiotics, on the other hand, is based on their overall advantages for the host, whereas the selection of prebiotics is based on their potential to stimulate the growth of probiotic bacteria as their primary function (Kolida and Gibson 2011; Markowiak and Śliżewska 2017). The synergistic effects of synbiotics have been the subject of most recent investigations, which have been emphasized exclusively (Torrecillas et al. 2018; Cavalcante et al. 2020; Sewaka et al. 2019; Hasan et al. 2018; Kumar et al. 2018; Villumsen et al. 2020).

The prebiotics' mode of action in the host species mostly determines how synbiotics are composed (Zepeda-Hernández et al. 2021; Li et al. 2020). Prebiotics that are low in polymerization, however, is better suited for the generation of synbiotics than those that are high in polymerization (Grimoud et al. 2010), since they hydrolyze the developed beneficial strains and produce primary and secondary metabolites that may have favorable effects on the host (Mazzola et al. 2015). Additionally, the administration of synbiotics is more beneficial than probiotic use alone due to increased microbial survival when combined with prebiotics and vice versa (Merrifield et al. 2010). However, prebiotics offers greater tolerance to large variations in oxygen, temperature, and pH (Cunningham et al. 2021). As a beneficial substitute for Nile tilapia, the following research has demonstrated the usefulness of synbiotics. Further evidence of its efficiency in fish farming comes from the greater stimulation of host species' growth and survival. This happens through the production of pancreatic enzymes, which improve enzymatic digestion, modify intestinal microbiota, and lowers mortality rates by altering potential immune responses, and antagonistic species of harmful organisms.

6.2 Modes of Action

Prebiotics' mode of action has been suggested by several in recent years, but the paths by which they function in aquaculture are not fully understood. Prebiotics' application in tilapia farming has received a lot of recent attention from aquaculturists (Zhou et al. 2020; Poolsawat et al. 2021; Koch et al. 2021; Wan-Mohtar et al. 2021; Souza et al. 2020; Mohammadi et al. 2020). Prebiotics has been shown to boost tilapia's resistance to infections (Poolsawat et al. 2021; Koch et al. 2021; Pilarski et al. 2017), immune system activation and regulation (Koch et al. 2021; Mohammadi et al. 2020), intestinal microbiota augmentation, regulation, and modification (Poolsawat et al. 2021; Zhou et al. 2020; Souza et al. 2020), enhancement of the host species' capacity for development and survival (Pilarski et al. 2017; Poolsawat et al. 2021), antioxidant activity (Abdel-Latif et al. 2020b; Dawood et al. 2020c), improvements in enzyme activity (Poolsawat et al. 2021; Zhou et al. 2020), and opposed to toxicity (Abdelhamid et al. 2020).

The use of probiotics and prebiotics in tilapia farming has been explored independently, while the idea of synbiotics has only lately come into existence (Mugwanya et al. 2022). Since being used in tilapia, synbiotics have proven a variety of action mechanisms. According to Laice et al. (2021), synbiotics can increase production and improve the sustainability of Nile tilapia. *Nile tilapia* fed a combination of probiotic *DBA®, MOS, and chitosan*, led to the enhancement of growth and disease resistance against *Aeromonas hydrophila* (Cavalcante et al. 2020). They can also multiply immunological functions including lysozyme activity and boost resistance to bacterial infection (Sewaka et al. 2019). Additionally, it has been discovered that the tilapia gut has greater digestive enzyme activity, which has led to enhanced growth (Hassaan et al. 2020).

6.3 Application of Prebiotics in Tilapia

6.3.1 Prebiotics as Growth Promoters and Disease Protection

The aquaculture industry with the goal is to achieve maximum production with minimum cost. Prebiotics has been used as an effective tool to obtain this goal. In the past decades, many studies have been conducted to access the effects of different prebiotics on Nile tilapia growth and feed utilization.

6.3.1.1 β-glucan

β-glucan has been used as a growth promoter and disease prevention in fish. In Nile tilapia, Sherif and Mahfouz (2019) reported that dietary incorporation of β-glucans at 0.5 $g.kg^{-1}$ feed for 2 weeks significantly increased relative level protection against

Aeromonas hydrophila. Recently, Souza et al. (2020) indicated that feeding Nile tilapia with β-glucan led to a higher survival rate and disease resistance against the *Vibrionaceae* family. More recently, Nile tilapia supplemented with β-glucan at 0.1% significantly enhanced growth and disease resistance against both *Aeromonas sobria* and *Streptococcus agalactiae* strains (Koch et al. 2021). Similarly, significant enhancement of growth, feed conversion ratio, villus length, villus width, goblet cell count, and disease resistance were observed in Nile tilapia fed β-glucan (Chirapongsatonkul et al. 2019; Dawood et al. 2020e). The beneficial effects of β-glucan on the gut flora may provide a credible explanation for growth enhancement (Meena et al. 2013). β-glucans are non-digestible feed components that function as prebiotics, enhancing the health of the host by promoting the growth and/or activity of specific gut bacteria (Miest et al. 2016; Ghaedi et al. 2016).

6.3.1.2 Fructooligosaccharide (FOS)

FOS has been applied in aquaculture as a functional supplement to improve farmed fish growth and health status (Rohani et al. 2022). Supplementation of FOS could improve Nile tilapia's growth performance (Abd El-Gawad et al. 2016b). However, no significant differences in growth rates were observed in fish-fed scFOS (Liu et al. 2017b). In contrast, gut adherence microbiota and disease resistance was significantly increased in fish-fed scFOS (Liu et al. 2017b). The authors hypothesized that dietary scFOS would have a selective promotion on various Lactobacillus strains in tilapia because it enhanced the relative abundance of *L. brevies*, *L. plantarum*, and *L. rhamnosus*. Although the exact mode of action is unknown, FOS's benefits for growth performance may be related to better digestion, immunity, and intestinal homeostasis maintenance through manipulation of the intestinal microbiota (Hu et al. 2019; Markowiak and Śliżewska 2017).

6.3.1.3 Mannanoligosaccharide (MOS)

MOS is one of the popular prebiotics used in aquaculture. Two studies have been conducted on Nile tilapia. Yuji-Sado et al. (2015) revealed that MOS-supplemented diets significantly enhanced intestinal fold height; whereas no impacts on growth performance were observed. In contrast, fish given MOS showed higher growth and survival rates as well as a lower FCR than CPF-intoxicated fish (Dawood et al. 2020c). The influence of MOS on the GIT microbial community, which can cause beneficial bacteria to generate digestive enzymes, determines how effective feed utilization will be (Jami et al. 2019; Mohammadian et al. 2019). Digestive enzymes disassemble the nutrients so that they are readily available within the GIT, improving their rates of absorption (Dawood et al. 2019a). Since the rate of nutrient digestion and absorption increased, it is anticipated that fish will develop faster and be in better overall health (Dawood et al. 2020b).

6.3.1.4 Inulin

In terms of final weight, weight increase, specific growth rate, and feed conversion ratio, dietary inclusion of inulin at 5 g kg^{-1} improved growth responses and feed utilization efficiency (Tiengtam et al. 2015b). Supplementation of 5 g kg^{-1} inulin led to greater intestinal villi height, larger relative goblet cells, increased lactic acid bacteria, and *Bifidobacterium* spp., but decreased *Vibrio* spp. (Boonanuntanasarn et al. 2018). Likewise, the administration of inulin at 0.4% resulted in a better growth rate and feed conversion ratio (Zhou et al. 2020). Fish administered a higher dose of inulin (0.8%) did not exhibit a higher impact on their ability to grow. Similarly, fish fed with diets enriched with 3% inulin exhibited poorer weight gains than fish fed with basal, 1%, and 2% inulin (Reza et al. 2009). Arctic char *Salvelinus alpinus* suffered negative effects from high supplementation levels of inulin (15%) (Olsen et al. 2001). These adverse effects of higher inulin supplementation may be attributed to the intestinal microbiota's inability to ferment excessive amounts of inulin, and extra inulin that builds up in the gut may be harmful to the enterocytes (Olsen et al. 2001; Hunt et al. 2019), forcing Nile tilapia to expend energy on inulin detoxification rather than using it to benefit the host.

6.3.1.5 Xylooligosaccharides (XOS)

Doan et al. (2018) reported that dietary inclusion of corncob-derived XOS significantly improved the growth performance and survival rate of Nile tilapia against *Streptococcus agalactiae.* Similar results were observed in tilapia fed 1-2 g kg^{-1} feed, where weight gain, nutrient retention, feed conversion ratio, protease, amylase, villus height, muscular thickness, numbers of lactic acid bacteria, and *Bacillus* were significantly enhanced. Furthermore, dietary XOS dramatically decreased tilapia's overall mortality following an *Aeromonas hydrophila* injection (Poolsawat et al. 2021). Although the precise mechanism by which XOS dramatically enhances fish development performance is yet unknown, it may be because XOS serves as a prebiotic and immunostimulant. Additionally, studies on dietary XOS have shown positive effects on gut lactic acid bacteria (LAB) and an increase in the activity of the amylase and protease enzymes (Hoseinifar et al. 2014; Xu et al. 2009). The role of LAB in enhancing growth performance, controlling pathogens by the competitive exclusion of adhesion sites, and producing SCFA, hydrogen peroxide, antibiotics, bacteriocins, siderophores, and lysozyme are all well documented (Hoseinifar et al. 2015a; Xia et al. 2018; Li et al. 2018). LAB also affects the fish's physiological and immune responses (Hoseinifar et al. 2015a; Nayak 2010; Merrifield et al. 2014; Ringø et al. 2010a). On the other hand, the metabolites acetic, lactic, propionic, and butyric acid are increased when prebiotics are supplemented (Ríos-Covián et al. 2016; Hoseinifar et al. 2017b). Further research on tilapia and CDXOS is warranted since, like in vertebrates, these SCFAs may benefit the immune cells in the gut-associated lymphoid tissue. (Bach Knudsen et al. 2003; Hoseinifar et al.

2015a). SCFAs are also taken up by the host and used as an energy source, with butyrate serving as the primary fuel for the colonic wall (den Besten et al. 2013). The dietary XOS, on the other hand, was shown in certain experiments to have just a small impact on the growth performance of white shrimp (Wang et al. 2010; Sun et al. 2019), Caspian white fish (*Rutilus frisii kutum*) (Hoseinifar et al. 2014), white sea bream (Guerreiro et al. 2016), and seabream (*Sparidentex hasta*) juvenile (Morshedi et al. 2018). Intestinal microbiota, administration techniques, diet composition, dosage, and species may all have an impact on how effective XOS is at promoting growth (Hoseinifar et al. 2016a).

6.3.1.6 Other Prebiotics

According to Abu-Elala et al. (2015), dietary regimens containing 1% chitosan have significantly boosted resistance to *Aeromonas hydrophila* and enhanced water quality. This could be a result of chitosan's activating impact on phagocytic cells, which causes them to discharge their mediators, including lysosomal enzymes, cationic peptides, complement components, and reactive oxygen species, all of which have antimicrobial effects against fish infections (Sajid et al. 2010; Niu et al. 2011). At a dosage of 0.1%, chitosan exhibited bactericidal effects that were generally more potent against gram-positive bacteria than gram-negative bacteria (No et al. 2002). According to some theories, its capacity to improve the permeability of the outer membrane of Gram-negative bacteria is what causes it to have an antibacterial effect (Chirkov 2002; Bégin and Van Calsteren 1999). By presumably competing with Ca^{++} for electronegative positions on the membrane without providing dimensional stability, the positive charges of chitosan disrupt the negatively charged residues of macromolecules at the surface of bacterial cells, making the membrane permeable (Bégin and Van Calsteren 1999). According to Meng et al. (2017) study employing chitooligosaccharides (COS), supplementing with 0.4 and 0.8 g kg^{-1} COS considerably improved growth performance, with the best value seen in fish fed 0.4 g kg^{-1} diet. COS could positively enhance growth rate because of its ability to improve feed intake (Tang et al. 2005), nutrient digestibility (Chen et al. 2009; Wang et al. 2009), growth hormone, and IGF-I concentrations (Tang et al. 2005).

Doan et al. (2018) conducted a study using pectin obtained from orange peels and found that Nile tilapia had greater growth metrics and were more resistant to disease from *Streptococcus agalactiae*. This improvement may be the outcome of improved digestive function brought on by the prevalence of beneficial bacteria in the gut (Wang An et al. 2017; Zhigang et al. 2013). Ho et al. (2017) demonstrated that pectin can be fermented by gut bacteria and highlighted its potential for application as a novel prebiotic. Additionally, positive effects on the gut microbiota and an increase in the activity of digestive enzymes including amylase, lipase, and protease (Dawood and Koshio 2016; Eshaghzadeh et al. 2015; Hoseinifar et al. 2016b; Kühlwein et al. 2014) or liver enzyme activities (Zhang et al. 2013; Hoseinifar et al. 2015a), along with improved appetite, increased vitamins, dissection of indigestible elements, and optimizing gut morphology (Irianto and Austin 2002; Hoseinifar et al. 2015c) have been recorded after consuming prebiotics. The fish's innate immune system was activated after being treated with pectin, which explains why resistance was significantly increased. Pectin may be a potential component for preventing diseases in both humans and animals, according to some research (Lattimer and Haub 2010; Wen et al. 2022; Wang et al. 2022).

Dietary inclusion of raffinose and pistachio hull-derived polysaccharide (PHDP) also led to better growth rates and disease resistance to *Aeromonas hydrophila* (Abdel-Latif et al. 2020b; Mohammadi et al. 2020). Raffinose greatly increased the growth rates of tilapia, which is likely due to the induced feed efficiency, as seen by the fish's low FCR. Raffinose causes the intestinal microbiota to secrete digesting enzymes, which improves fish's GIT's ability to absorb nutrients (Berrocoso et al. 2017). Because PHDP serves as a prebiotic and promotes the formation of healthy probiotic flora, it has positive benefits on growth rate and disease resistance (Akbari-Alavijeh et al. 2018; Li et al. 2019), or powerful antibacterial activities of polyphenols, which are implicated in bacterial cell breakdown through their lipophilic characteristics (Stratev et al. 2018).

6.3.1.7 Combination of Prebiotics

A combination of prebiotics with each other or with other bioactive compounds could improve the growth performance and disease resistance of Nile tilapia. Dietary inclusion of β-glucan with MOS led to better growth performance and disease resistance to *L. gravieae*, *A. hydrophila*, and *S. iniae* (Abu-Elala et al. 2018; Chen et al. 2019). The manipulation of MOS on intestinal microbiota may be a factor in the underpinning mechanisms of the current growth results. Increased villus integrity, improved digestion and absorption (Fernandez et al. 2002; Dimitroglou et al. 2009), and the production of energy and building protein from β-glucan degraded by glucanases in the digestive gland were all outcomes of MOS (Misra et al. 2006). The Nile tilapia's growth performance significantly improved when fed β-glucan and *Chlorella vulgaris* (Abdelhamid et al. 2020). This outcome could be attributable to CV high-quality protein, which increases fish weight gain and final weight (Kang et al. 2013). Additionally, CV contains growth-promoting ingredients including S-nucleotide adenosyl peptide complex, which could alter an animal's ability to grow and digest the food (Han et al. 2002). The distinct, varied macro- and micronutrients found in CV include proteins, polysaccharides, -carotene, -carotene, minerals, vitamins, pro-vitamins, chlorophyll, and lutein (Panahi et al. 2016). Additionally, our findings demonstrated that β-glucan enhanced growth efficiency following DZN toxicity. This increase in β-glucan's growth efficiency may be caused by the glucanase's likely breakdown of the glucan, which encourages the transfer of additional proteins for growth. Furthermore, β-glucan enhanced the release of digestive enzymes, which may have contributed to the improvement of growth (Guzmán-Villanueva et al. 2014). Similar to this, combining β-glucan with either propylene glycol or *Aspergillus oryzae* dramatically improved Nile tilapia growth performance and disease resistance (Abd El Tawab et al. 2016; Dawood et al. 2020a).

6.3.2 *Prebiotics as Immunomodulatory*

6.3.2.1 β-glucan

In teleost fish, β-glucans' ability to modulate immunity has been extensively explored (Petit et al. 2019). β-glucan treatment frequently has an immune-stimulating impact and can lead to an increase in resistance to both viral and bacterial illnesses, regardless of the administration route or fish species (Dalmo and Bogwald 2008; Meena et al. 2013; Petit and Wiegertjes 2016). There have been documented distinct mechanisms for β-glucan identification and/or downstream signaling in both mammalian vertebrates (Brown and Williams 2009; Legentil et al. 2015) and invertebrates (Cerenius et al. 2009; Soltanian et al. 2009), specific pathways for β-glucan identification and/or downstream signaling have been identified. Although β-glucans are frequently used in aquaculture, the precise processes underlying the produced effects for teleost fish are poorly understood (Petit et al. 2019).

Sado et al. (2016) indicated that Nile tilapia fed β-glucan for 15 days significantly enhanced serum lysozyme, leukocyte count, and neutrophil number. In another study, Sherif and Mahfouz (2019) reported that dietary inclusion of β-glucan at 0.5 g kg^{-1} feed for 2 weeks significantly improved the fish immune status both innate and adapted immune response. Similarly, dietary incorporation of β-glucan from mushroom and yeast at 10 μg per fish significantly up-regulated relative immune gene expressions (Chirapongsatonkul et al. 2019). Dawood et al. (2020e) also revealed that dietary inclusion of β-glucan significantly increased lysozyme and phagocytic activities, and up-regulated INF-γ, TNF-α, IL-1β, and HSP70 gene expressions. In a recent study, Koch et al. (2021) also reported that a β-glucan incorporated diet resulted in higher lysozyme, respiratory burst, and myeloperoxidase activities of Nile tilapia. However, no effects of β-glucan on serum lysozyme activity of Nile tilapia were found (Souza et al. 2020). β-glucan is well known for its capacity to stimulate the complement system, reactive oxygen metabolite formation, and phagocytic cells' ability to create antimicrobial compounds like lysosomal enzymes (Ji et al. 2017; Goodridge et al. 2009). Additionally, Engstad (1993) noted that salmon possessed BG receptors on its phagocyte membrane and that these receptors could be influenced by BG supplementation. The interaction between helpful microorganism cells and intestinal epithelial cells, which is thought to improve mucosal immunity in the host gut and tighten the epithelial junction, produce antimicrobial peptides and mucosal immunoglobulin, control inflammatory response, and subsequently stimulate the immune response, is responsible for BG's immunomodulating activity (Raa 2015).

6.3.2.2 Fructooligosaccharide (FOS)

Numerous edible plants, including onions and edible burdock, contain fructooligosaccharides (FOS), which are created commercially when fructosyltransferase reacts with sucrose (Mitsuoka et al. 1987). FOS is an example of a prebiotic, which are foods that pass undigested by pancreatic and small intestine enzymes in the human gut and instead reaches the colon, where they have advantageous effects. According to reports, dietary FOS affects the host by promoting the proliferation and activity of beneficial bacteria in the colon (Cummings et al. 2001), which in turn affects many aspects of intestinal function (Grisdale-Helland et al. 2008). These bacteria produce organic acids that are advantageous to the host (Gibson 1998; Roberfroid 2000). Inhibiting the growth of dangerous germs, enhancing the absorption of critical nutrients, synthesizing specific vitamins, and enhancing immunological function are additional examples of potential health-promoting advantages (Li et al. 2007). According to reports, FOS could increase production through improved growth, stimulated immune response, and higher resistance to fish infections, overcoming the drawbacks and side effects of antibiotics and other medications (Akrami et al. 2013; Guerreiro et al. 2014; Zhang et al. 2013; Soleimani et al. 2012). The dietary addition of FOS at 0, 1, 2, and 3% for 6 weeks significantly increased serum IgM and lysozyme activity in Nile tilapia (Abd El-Gawad et al. 2016b). The enhanced lysozyme activity was likely caused by the increased leukocyte production brought on by dietary FOS (Zhang et al. 2013) because neutrophils and macrophages are the primary producers of fish lysozyme (Fischer et al. 2006). FOS's ability to stimulate the growth of helpful bacteria like lactobacilli and bifidobacteria, which produce lipopolysaccharides with immunostimulatory qualities, may be responsible for its ability to stimulate the immune system (Passos and Park 2003). Additionally, the FOS fermentation's final products, acetate, propionate, and lactic acid, are essential in influencing the immune system (Passos and Park 2003). Additionally, FOS was able to interact with the toll-like receptor 2 (TLR2) found on macrophages (Vogt et al. 2013) and upregulate the expression of antimicrobial peptides, which play a crucial role in fish disease resistance due to their importance in innate immune defense (Zhang et al. 2014).

6.3.2.3 Mannanoligosaccharide (MOS)

One well-known prebiotic that has lately been used in aquaculture is mannanoligosaccharide (MOS). MOS is made from the cell walls of the yeast *Saccharomyces cerevisiae* (Gelibolu et al. 2018). By preventing pathogenic bacteria from adhering to the digestive tract, regulating the intestinal microbiota, and maintaining intestinal integrity, MOS aided in the improvement of fish growth performance and health conditions (Ringø et al. 2010b; Fernández-Montero et al. 2019). The dietary inclusion of MOS at 1% considerably increased lysozyme and phagocytic activities in Nile tilapia, according to Dawood et al. (2020c). Prebiotics can maintain a neutral pH in the GI tract, creating an ideal environment for good bacteria to boost phagocytic activity and upregulate the expression of the CAT, GPX, GPX, IL-1, IL-8, and IFN-γ genes (Dawood et al. 2020b). *However,* Yuji-Sado et al. (2015) reported that MOS had no impact on Nile tilapia's immunological parameters. It is well recognized that the effects of MOS largely depend on the biotic factors of farmed fish, such as species, rearing circumstances, length of supplementation, age, and size of fish (Sweetman et al. 2010; Song et al. 2014).

6.3.2.4 Inulin

Inulin is a non-digestible carbohydrate found in a variety of fruits, vegetables, and grains. It is frequently used as an ingredient in functional foods and is commercially made from the chicory root (Madrigal and Sangronis 2007; de Almeida Gualtieri et al. 2013). Inulin has lately been used as an immunostimulant in aquaculture. Tiengtam et al. (2015b) found that inulin-containing diets at 0, 2.5, and 5 g kg^{-1} for 8 weeks increased lysozyme activity in Nile tilapia. Inulin predominantly stimulates the synthesis of IL-10 and NK cell cytotoxicity in Peyer's patches of immunological cells in animal studies. IgA signaling in the ileum and caecum, splenic NK cell cytotoxicity, and splenocyte cytokine levels are further immunological processes that inulin modulates (de Almeida Gualtieri et al. 2013). According to Watzl et al. (2005) finding, inulin predominantly modifies immunological GALT parameters.

6.3.2.5 Xylooligosaccharides (XOS)

Xylooligosaccharides (XOS), which include xylobiose, xylotriose, xylotetrose, and others, are polymers of sugar oligomers made up of xylose units through -(1, 4)-xylosidic connections (Kumar and Satyanarayana 2011). It is possible to generate xylooligosaccharides using chemical (Samanta et al. 2012), enzymatic (Jayapal et al. 2013), or a mix of chemical and enzymatic processes (Samanta et al. 2015). Apart from specifically promoting the development of healthy gut flora (Manisseri and Gudipati 2012), XOS was also shown to stimulate the immune system in hybrid catfish (Hahor et al. 2019), lessen the activity of pro-carcinogenic enzymes in the gastrointestinal tract of Sprague-Dawley rats treated with 1, 2-dimethyl hydrazine (Hsu et al. 2004), and improve intestinal mineral absorption and immune stimulation in albino rats (Mumtaz et al. 2009). Doan et al. (2018) revealed that consuming XOS produced from maize cobs dramatically improved skin mucus and serum immunity in Nile tilapia. Similar outcomes were observed in Nile tilapia fed XOS for 8 weeks at doses of 0 (CON), 0.5 (X-1), 1 (X-2), 2 (X-3), and 4 (X-4) g kg^{-1} (Poolsawat et al. 2021). The short-chain fatty acids that the gut microbiota produces during prebiotic fermentation, which can serve as an antibiotic substrate and a messenger for enhancing immune function, may be connected to the enhanced immunological and antioxidant characteristics (Bermudez-Brito et al. 2012; Song et al. 2014). According to Yu et al. (2015) adding xylooligosaccharides generated from corncob to the diet had an impact on the rising levels of gut microbiota and short-chain fatty acids (CDXOS).

6.3.2.6 Other Prebiotics

The second-most prevalent biopolymer in nature is chitin, while chitosan is a polycationic linear polysaccharide that is produced when chitin is partially deacetylated (Crini 2022). The pharmaceutical, biotech, and agricultural industries have recently used chitin and chitosan extensively (Wang and Zhuang 2022). In recent years, the aquaculture sector has employed chitin and chitosan as feed supplements in the diets of fish and shellfish (Kamilya and Khan 2020). In Nile tilapia, Abu-Elala et al. (2015) indicated that dietary inclusion of chitosan obtained from shrimp shell at 0, 0.5%, 1%, and 2% significantly increased phagocytic activity/index, NBT, lysozyme activity, and ACH50. In another study with low molecular weight chitooligosaccharide (LMV-COS), Meng et al. (2017) reported that inclusion levels of 0, 0.1, 0.2, 0.4, and 0.8 g kg^{-1} resulted in higher phagocytic and lysozyme activities. Similar results were noticed in Nile tilapia-fed orange-derived pectin, raffinose, and pistachio hulls-derived polysaccharide (Abdel-Latif et al. 2020b; Mohammadi et al. 2020).

6.3.2.7 Combination of Prebiotics

A combination of two prebiotics or prebiotics with other active components has been widely applied in aquaculture. Abd El Tawab et al. (2016) dietary inclusion of β-glucan + propylene glycol at 0; 0.5; 1.0 and 1.5 mL resulted in higher lysozyme and nitroblue-tetrazolium activities of Nile tilapia. Similar results were noticed in Nile tilapia fed β-glucan (βG) and mannan-oligosaccharides (MOS), where they reported significantly enhanced phagocytic and lysozyme activities, as well as immune-related genes expressions (Abu-Elala et al. 2018; Chen et al. 2019). Combination of β-glucan with *Aspergillus oryzae* or *Chlorella vulgaris* significantly improved non-specific immunity and up-regulated TNF-α and IL-10 gene expressions (Dawood et al. 2019b; Abdelhamid et al. 2020).

6.4 Synbiotics in Tilapia

6.4.1 Synbiotics as Growth Promoters and Disease Protection

Due to their positive effects on immune response and, consequently, disease resistance, synbiotics have received more attention in aquaculture recently (Carbone and Faggio 2016; Huynh et al. 2018). Liu et al. (2017b) reported that synbiotics between scFOS and *Lactobacilli* resulted in better growth performance, feed conversion ratio, disease resistance, and gut microbiota. These could result from partial dietary protein or amino acid consumption by allochthonous bacteria (Dai et al. 2012; Dai et al. 2010). Similarly, the dietary combination of lactic acid bacteria with Jerusalem artichoke, low molecular weight sodium alginate, pectin, xylooligosaccharide, and β-glucan significantly enhanced growth performance and disease resistance of tilapia (Sewaka et al. 2019; Van Doan et al. 2016a; Van Doan et al. 2019b; Dawood et al. 2020f; Van Doan et al. 2019c). A combination of kefir and LMWSA, MOS and chitosan with DBA®, and MOS with probiotics also increased growth performance and disease resistance (Van Doan et al. 2017b; Cavalcante et al. 2020; Laice et al. 2021). This may be caused by the addition of an unsuitable prebiotic as the substrate in a synbiotic mixture, which prevented a chosen probiotic from fermenting to form short-chain fatty acids (Hoseinifar et al. 2015b). When fish are fed a synbiotic diet, their growth performance is significantly improved, which could be attributed to the probiotic's increased survival and colonization or the prebiotic's improved digestibility (Rodriguez-Estrada et al. 2009; Ai et al. 2011; Ye et al. 2011; Liew et al. 2015). It is generally known that consuming dietary probiotics and prebiotics can cause the fermentation of bioactive microbial metabolites like organic acids, fatty acids, bioactive peptides, and vitamins (Munir et al. 2016; Stanton et al. 2005). These bioactive substances will speed up digestion in the intestine, which will boost growth rates. Additionally, the presence of advantageous bacteria in the stomach may boost microbial balance, which in turn improves nutrient absorption and utilization (Gatesoupe 1999; Lara-Flores et al. 2003).

6.4.2 Synbiotics as Immunomodulatory

Synbiotic supplementation in aquaculture has been proven to boost innate and specific immune responses (Amenyogbe et al. 2020). Van Doan et al. (2017b) indicated that dietary inclusion of kefir and LMWSA significantly boosted skin mucus and serum immunities of Nile tilapia. Although the precise processes through which kefir affects fish's innate immune system have not been shown, they may be related to kefir's potential probiotic effects (Ali et al. 2022). Previous research has shown that advantageous bacteria can increase host resistance, non-specific humoral responses, and the exclusion of prospective pathogens (Nikoskelainen et al. 2003; Rengpipat et al. 2000; Newaj-Fyzul and Austin 2015; Akhter et al. 2015; Lazado et al. 2015; De et al. 2014; Llewellyn et al. 2014; Nayak 2010). Probiotics have also been shown to modulate the host's immune response through interactions with epithelial cells and by raising the secretion of anti-inflammatory cytokines, which may result in decreasing inflammation (Denev et al. 2009). The activity and number of lymphocytes, intraepithelial cells, and acidophilic granulocytes were all affected at the cellular level by the addition of probiotics (Picchietti et al. 2007). Additionally, fish can have their systemic immune responses triggered by probiotics (Lazado et al. 2015).

Improvements to the Nile tilapia's skin mucus immune response, serum immunities, and gene expression have been documented when various prebiotics and lactic acid bacteria are combined (Van Doan et al. 2016a, 2019b, c, Dawood et al. 2020f). Prebiotics and lactic acid bacteria may have an impact on the fish immunological response, as evidenced by the remarkable improvement that was seen in this study. Prebiotics and LAB may work together to enhance the number of lactobacilli and bifidobacteria in mouse feces while decreasing the viability of Enterococcus, Enterobacter, and *Clostridia* species (Markowiak and Śliżewska 2017; Vieco-Saiz et al. 2019). Corn cob-derived xylooligosaccharide (CDXOS), which was fermented with *L. plantarum*, was found to have strong superoxide anion radical-scavenging abilities, according to an in vitro antioxidant assay (Yu et al. 2015).

6.5 Conclusion

Using prebiotics and synbiotics in tilapia aquaculture has advantages for sustainable food supply to fulfill the rising worldwide need for animal protein. For the host organism to perform better and develop immunity, knowledge of the dosage and duration of administration is crucial. Under both normal and environmental stress circumstances, longer administration times appear to have a more favorable impact on fish performance and immune response. Prebiotics and synbiotics also appear to have effects that are dose-dependent and reliant on the fish's culture circumstances. Although numerous studies have been carried out to clarify their mode of action, further research is required to fully comprehend how they affect the activity of digestive enzymes, and the utilization of carbohydrates, fatty acids, and minerals. Similarly, further work needs to be done to understand how these feed additives affect tilapia's metabolic profile, gut microbiome, and gene expression using omics technology. Additionally, insufficient research has been done, as far as we are aware, to determine how these feed supplements affect tilapia when they are exposed to pesticide, heavy metal, and fungal toxicity, as well as virus and fungus infections. We, therefore, urge future research on tilapia to concentrate on these areas to protect the sustainable production of tilapia under various ecological systems.

References

Abasubong KP, Liu W-B, Adjoumani YJJ, Xia S-L, Xu C, Li X-F (2019) Xylooligosaccharides benefits the growth, digestive functions and TOR signaling in Megalobrama amblycephala fed diets with fish meal replaced by rice protein concentrate. Aquaculture 500:417–428. https://doi.org/10.1016/j.aquaculture.2018.10.048

Abd El Tawab AA, El-Hofy FA-A, Ammar AM, Saleh OAE-R, Tolba HAJBVMJ (2016) Effect of prebiotic on the immune status of Oreochromis niloticus. Benha Vet Med J 30(1):20–28

Abd El-Gawad EA, Abd El-latif AM, Shourbela RMJJARD (2016b) Enhancement of antioxidant activity, non-specific immunity and growth performance of Nile tilapia, Oreochromis niloticus by dietary fructooligosaccharide. J Aquac Res Dev 7(427):2

Abdelhamid FM, Elshopakey GE, Aziza AE (2020) Ameliorative effects of dietary Chlorella vulgaris and β-glucan against diazinon-induced toxicity in Nile tilapia (Oreochromis niloticus). Fish Shellfish Immunol 96:213–222. https://doi.org/10.1016/j.fsi.2019.12.009

Abdel-Latif HMR, Dawood MAO, Menanteau-Ledouble S, El-Matbouli M (2020a) The nature and consequences of co-infections in tilapia: a review. J Fish Dis 43(6):651–664. https://doi.org/10.1111/jfd.13164

Abdel-Latif HMR, Soliman AA, Sewilam H, Almeer R, Van Doan H, Alagawany M, Dawood MAO (2020b) The influence of raffinose on the growth performance, oxidative status, and immunity in Nile tilapia (Oreochromis niloticus). Aquac Rep 18:100457. https://doi.org/10.1016/j.aqrep.2020.100457

Abu-Elala NM, Mohamed SH, Zaki MM, Eissa AE (2015) Assessment of the immune-modulatory and antimicrobial effects of dietary chitosan on Nile tilapia (Oreochrmis niloticus) with special emphasis to its bio-remediating impacts. Fish Shellfish Immunol 46(2):678–685. https://doi.org/10.1016/j.fsi.2015.08.004

Abu-Elala NM, Younis NA, AbuBakr HO, Ragaa NM, Borges LL, Bonato MA (2018) Efficacy of dietary yeast cell wall supplementation on the nutrition and immune response of Nile tilapia. Egypt J Aquat Res 44(4):333–341. https://doi.org/10.1016/j.ejar.2018.11.001

Ai Q, Xu H, Mai K, Xu W, Wang J, Zhang W (2011) Effects of dietary supplementation of *Bacillus subtilis* and fructooligosaccharide on growth performance, survival, non-specific immune response and disease resistance of juvenile large yellow croaker, *Larimichthys crocea*. Aquaculture 317(1–4):155–161. https://doi.org/10.1016/j.aquaculture.2011.04.036

Akbari-Alavijeh S, Soleimanian-Zad S, Sheikh-Zeinoddin M, Hashmi S (2018) Pistachio hull water-soluble polysaccharides as a novel prebiotic agent. Int J Biol Macromol 107:808–816

Akhter N, Wu B, Memon AM, Mohsin M (2015) Probiotics and prebiotics associated with aquaculture: a review. Fish Shellfish Immunol 45(2):733–741

Akrami R, Iri Y, Khoshbavar Rostami H, Razeghi Mansour M (2013) Effect of dietary supplementation of fructooligosaccharide (FOS) on growth performance, survival, lactobacillus bacterial population and hemato-immunological parameters of stellate sturgeon (*Acipenser stellatus*) juvenile. Fish Shellfish Immunol 35(4):1235–1239. https://doi.org/10.1016/j.fsi.2013.07.039

Ali M, Soltanian S, Mirghaed AT, Akhlaghi M, Hossein Hoseinifar S (2022) The potential benefits of Lactobacillus farraginis and Enterococcus durans, isolated from kefir, as probiotic candidates on innate immune responses, expression of some immune genes, and resistance to Lactococcosis disease in juvenile rainbow trout. Aquac Res 53(13):4588–4604. https://doi.org/10.1111/are.15419

de Almeida Gualtieri K, Losi Guembarovski R, Oda JMM, Fiori-Lopes L, Ketelut Carneiro N, de Castro VD, Soni Neto J, Watanabe MAE (2013) Inulin: therapeutic potential, prebiotic properties and immunological aspects. Food Agric Immunol 24(1):21–31. https://doi.org/10.1080/09540105.2011.640993

Amenyogbe E, Chen G, Wang Z, Huang J, Huang B, Li H (2020) The exploitation of probiotics, prebiotics and synbiotics in aquaculture: present study, limitations and future directions: a review. Aquac Int 28(3):1017–1041. https://doi.org/10.1007/s10499-020-00509-0

Bach Knudsen KE, Serena A, Canibe N, Juntunen KS (2003) New insight into butyrate metabolism. Proc Nutr Soc 62(1):81–86. https://doi.org/10.1079/pns2002212

Bégin A, Van Calsteren M-R (1999) Antimicrobial films produced from chitosan. Int J Biol Macromol 26(1):63–67. https://doi.org/10.1016/S0141-8130(99)00064-1

Bermudez-Brito M, Plaza-Díaz J, Muñoz-Quezada S, Gómez-Llorente C, Gil A (2012) Probiotic mechanisms of action. Ann Nutr Metab 61(2):160–174

Berrocoso J, Kida R, Singh A, Kim Y, Jha R (2017) Effect of in ovo injection of raffinose on growth performance and gut health parameters of broiler chicken. Poult Sci 96(6):1573–1580 den Besten G, van Eunen K, Groen AK, Venema K, Reijngoud D-J, Bakker BM (2013) The role of short-chain fatty acids in the interplay between diet, gut microbiota, and host energy metabolism. J Lipid Res 54(9):2325–2340. https://doi.org/10.1194/jlr.R036012

Bindels LB, Delzenne NM, Cani PD, Walter J (2015a) Opinion: towards a more comprehensive concept for prebiotics. Nat Rev Gastroenterol Hepatol 12(5):303–310. https://doi.org/10.1038/nrgastro.2015.47

Bindels LB, Delzenne NM, Cani PD, Walter J (2015b) Towards a more comprehensive concept for prebiotics. Nat Rev Gastroenterol Hepatol 12(5):303

Boonanuntanasarn S, Tiengtam N, Pitaksong T, Piromyou P, Teaumroong N (2018) Effects of dietary inulin and Jerusalem artichoke (Helianthus tuberosus) on intestinal microbiota community and morphology of Nile tilapia (Oreochromis niloticus) fingerlings. Aquac Nutr 24(2): 712–722. https://doi.org/10.1111/anu.12600

Brown GD, Williams DL (2009) (1, 3)-β-Glucans in innate immunity: mammalian systems. In: Chemistry, biochemistry, and biology of 1-3 beta glucans and related polysaccharides. Elsevier, pp 579–619

Butt UD, Lin N, Akhter N, Siddiqui T, Li S, Wu B (2021) Overview of the latest developments in the role of probiotics, prebiotics and synbiotics in shrimp aquaculture. Fish Shellfish Immunol 114:263–281. https://doi.org/10.1016/j.fsi.2021.05.003

Cabello FC, Godfrey HP (2016) Even therapeutic antimicrobial use in animal husbandry may generate environmental hazards to human health. Environ Microbiol 18(2):311–313. https://doi.org/10.1111/1462-2920.13247

Carbone D, Faggio C (2016) Importance of prebiotics in aquaculture as immunostimulants. Effects on immune system of *Sparus aurata* and *Dicentrarchus labrax*. Fish Shellfish Immunol 54:172–178

Cavalcante RB, Telli GS, Tachibana L, Dias DC, Oshiro E, Natori MM, Silva WF, Ranzani-Paiva MJ (2020) Probiotics, prebiotics and synbiotics for Nile tilapia: growth performance and protection against Aeromonas hydrophila infection. Aquac Rep 17:100343. https://doi.org/10.1016/j.aqrep.2020.100343

Cerenius L, Kawabata S-I, Söderhäll K (2009) Biological and immunological aspects of innate defence mechanisms activated by (1, 3)-β-glucans and related polysaccharides in invertebrates. In: Chemistry, biochemistry, and biology of 1-3 beta glucans and related polysaccharides. Elsevier, pp 563-577

Chauhan SV, Chorawala MR (2012) Probiotics, prebiotics and synbiotics. Int J Pharm Sci Res 3(3): 711

Chen X-Q, Zhao W, Xie S-W, Xie J-J, Zhang Z-H, Tian L-X, Liu Y-J, Niu J (2019) Effects of dietary hydrolyzed yeast (Rhodotorula mucilaginosa) on growth performance, immune response, antioxidant capacity and histomorphology of juvenile Nile tilapia (Oreochromis niloticus). Fish Shellfish Immunol 90:30–39. https://doi.org/10.1016/j.fsi.2019.03.068

Chen YJ, Kim IH, Cho JH, Yoo JS, Wang Y, Huang Y, Kim HJ, Shin SO (2009) Effects of chitooligosaccharide supplementation on growth performance, nutrient digestibility, blood characteristics and immune responses after lipopolysaccharide challenge in weanling pigs. Livest Sci 124(1):255–260. https://doi.org/10.1016/j.livsci.2009.02.006

Chirapongsatonkul N, Mueangkan N, Wattitum S, U-taynapun K (2019) Comparative evaluation of the immune responses and disease resistance of Nile tilapia (*Oreochromis niloticus*) induced by yeast β-glucan and crude glucan derived from mycelium in the spent mushroom substrate of *Schizophyllum commune*. Aquac Rep 15:100205. https://doi.org/10.1016/j.aqrep.2019.100205

Chirkov SN (2002) The antiviral activity of Chitosan (review). Appl Biochem Microbiol 38(1):1–8. https://doi.org/10.1023/A:1013206517442

Crini G (2022) Chapter seven—chitin and chitosan: production, properties, and applications. In: Crini G (ed) Chitin and chitosan. Academic Press, pp 149–207. https://doi.org/10.1016/B978-0-323-96119-6.00004-9

Cummings JH, Macfarlane GT, Englyst HN (2001) Prebiotic digestion and fermentation. Am J Clin Nutr 73(2):415s–420s

Cunningham M, Vinderola G, Charalampopoulos D, Lebeer S, Sanders ME, Grimaldi R (2021) Applying probiotics and prebiotics in new delivery formats—is the clinical evidence transferable? Trends Food Sci Technol 112:495–506. https://doi.org/10.1016/j.tifs.2021.04.009

Dai Z-L, Li X-L, Xi P-B, Zhang J, Wu G, Zhu W-Y (2012) Metabolism of select amino acids in bacteria from the pig small intestine. Amino Acids 42(5):1597–1608

Dai Z-L, Zhang J, Wu G, Zhu W-Y (2010) Utilization of amino acids by bacteria from the pig small intestine. Amino Acids 39(5):1201–1215

Dalmo RA, Bogwald J (2008) Beta-glucans as conductors of immune symphonies. Fish Shellfish Immunol 25(4):384–396. https://doi.org/10.1016/j.fsi.2008.04.008

Das S, Mondal K, Haque S (2017) A review on application of probiotic, prebiotic and synbiotic for sustainable development of aquaculture. Growth 14:15

Davani-Davari D, Negahdaripour M, Karimzadeh I, Seifan M, Mohkam M, Masoumi SJ, Berenjian A, Ghasemi Y (2019) Prebiotics: definition, types, sources, mechanisms, and clinical applications. Foods (Basel, Switzerland) 8(3):92. https://doi.org/10.3390/foods8030092

Dawood MA, Eweedah NM, Moustafa EM, Shahin MG (2020a) Synbiotic effects of *Aspergillus oryzae* and β-glucan on growth and oxidative and immune responses of nile Tilapia, *Oreochromis niloticus*. Probiotics Antimicrob Proteins 12(1):172–183. https://doi.org/10.1007/s12602-018-9513-9

Dawood MA, Koshio S (2016) Recent advances in the role of probiotics and prebiotics in carp aquaculture: a review. Aquaculture 454:243–251. https://doi.org/10.1016/j.aquaculture.2015.12.033

Dawood MA, Koshio S, Fadl SE, Ahmed HA, El Asely A, Abdel-Daim MM, Alkahtani S (2020b) The modulatory effect of mannanoligosaccharide on oxidative status, selected immune parameters and tolerance against low salinity stress in red sea bream (Pagrus major). Aquac Rep 16:100278

Dawood MA, Mohsen M, El-dakar A, Abdelraouf E, Moustafa E, Ahmed HA (2019a) Effectiveness of exogenous digestive enzymes supplementation on the performance of rabbitfish (Siganus rivulatus). Slov Vet Res 56(22-Suppl):409–419

Dawood MAO, El-Shamaa IS, Abdel-Razik NI, Elkomy AH, Gewaily MS, Abdo SE, Soliman AA, Paray BA, Abdelkhalek N (2020c) The effect of mannanoligosaccharide on the growth performance, histopathology, and the expression of immune and antioxidative related genes in Nile tilapia reared under chlorpyrifos ambient toxicity. Fish Shellfish Immunol 103:421–429. https://doi.org/10.1016/j.fsi.2020.05.061

Dawood MAO, Eweedah NM, Moustafa EM, Shahin MG (2019b) Synbiotic effects of Aspergillus oryzae and β-glucan on growth and oxidative and immune responses of Nile Tilapia, Oreochromis niloticus. Probiotics Antimicrob Proteins. https://doi.org/10.1007/s12602-018-9513-9

Dawood MAO, Metwally AE-S, El-Sharawy ME, Atta AM, Elbialy ZI, Abdel-Latif HMR, Paray BA (2020e) The role of β-glucan in the growth, intestinal morphometry, and immune-related gene and heat shock protein expressions of Nile tilapia (Oreochromis niloticus) under different stocking densities. Aquaculture 523:735205. https://doi.org/10.1016/j.aquaculture.2020.735205

Dawood MAO, Moustafa EM, Elbialy ZI, Farrag F, Lolo EEE, Abdel-Daim HA, Abdel-Daim MM, Van Doan H (2020f) Lactobacillus plantarum L-137 and/or β-glucan impacted the histopathological, antioxidant, immune-related genes and resistance of Nile tilapia (Oreochromis niloticus) against Aeromonas hydrophila. Res Vet Sci 130:212–221. https://doi.org/10.1016/j.rvsc.2020.03.019

De BC, Meena D, Behera B, Das P, Mohapatra PD, Sharma A (2014) Probiotics in fish and shellfish culture: immunomodulatory and ecophysiological responses. Fish Physiol Biochem 40(3):921–971

Denev S, Staykov Y, Moutafchieva R, Beev G (2009) Microbial ecology of the gastrointestinal tract of fish and the potential application of probiotics and prebiotics in finfish aquaculture. Int Aquat Res 1:1–29

Dimitroglou A, Merrifield D, Moate R, Davies S, Spring P, Sweetman J, Bradley G (2009) Dietary mannan oligosaccharide supplementation modulates intestinal microbial ecology and improves gut morphology of rainbow trout, *Oncorhynchus mykiss* (Walbaum). J Anim Sci 87(10): 3226–3234

Doan HV, Hoseinifar SH, Elumalai P, Tongsiri S, Chitmanat C, Jaturasitha S, Doolgindachbaporn S (2018) Effects of orange peels derived pectin on innate immune response, disease resistance and growth performance of Nile tilapia (*Oreochromis niloticus*) cultured under indoor biofloc system. Fish Shellfish Immunol 80:56–62. https://doi.org/10.1016/j.fsi.2018.05.049

Dobšíková R, Blahová J, Mikulíková I, Modrá H, Prášková E, Svobodová Z, Škorič M, Jarkovský J, Siwicki A-K (2013) The effect of oyster mushroom β-1.3/1.6-D-glucan and oxytetracycline antibiotic on biometrical, haematological, biochemical, and immunological indices, and histopathological changes in common carp (Cyprinus carpio L.). Fish Shellfish Immunol 35(6):1813–1823

Edwards P, Zhang W, Belton B, Little DC (2019) Misunderstandings, myths and mantras in aquaculture: its contribution to world food supplies has been systematically over reported. Mar Policy 106:103547. https://doi.org/10.1016/j.marpol.2019.103547

Eissa AE, Attia MM, Elgendy MY, Ismail GA, Sabry NM, Prince A, Mahmoud MA, El-Demerdash GO, Abdelsalam M, Derwa HIM (2021) Streptococcus, Centrocestus formosanus and Myxobolus tilapiae concurrent infections in farmed Nile tilapia (Oreochromis niloticus). Microb Pathog 158:105084. https://doi.org/10.1016/j.micpath.2021.105084

El Basuini MF, Teiba II, Zaki MAA, Alabssawy AN, El-Hais AM, Gabr AA, Dawood MAO, Zaineldin AI, Mzengereza K, Shadrack RS, Dossou S (2020) Assessing the effectiveness of CoQ10 dietary supplementation on growth performance, digestive enzymes, blood health, immune response, and oxidative-related genes expression of Nile tilapia (Oreochromis niloticus). Fish Shellfish Immunol 98:420–428. https://doi.org/10.1016/j.fsi.2020.01.052

Engstad RE (1993) Recognition of yeast cell wall glucan by Atlantic salmon (Salmo salar L.) macrophages. Dev Comp Immunol 17(4):319–330

Eshaghzadeh H, Hoseinifar SH, Vahabzadeh H, Ringø E (2015) The effects of dietary inulin on growth performances, survival and digestive enzyme activities of common carp (*Cyprinus carpio*) fry. Aquac Nutr 21(2):242–247. https://doi.org/10.1111/anu.12155

FAO (2022) The state of world fisheries and aquaculture

Fernandez F, Hinton M, Gils BV (2002) Dietary mannan-oligosaccharides and their effect on chicken caecal microflora in relation to Salmonella Enteritidis colonization. Avian Pathol 31(1):49–58

Fernández-Montero Á, Torrecillas S, Izquierdo M, Caballero MJ, Milne DJ, Secombes CJ, Sweetman J, Da Silva P, Acosta F, Montero D (2019) Increased parasite resistance of greater amberjack (Seriola dumerili Risso 1810) juveniles fed a cMOS supplemented diet is associated with upregulation of a discrete set of immune genes in mucosal tissues. Fish Shellfish Immunol 86:35–45

Fiorella KJ, Okronipa H, Baker K, Heilpern S (2021) Contemporary aquaculture: implications for human nutrition. Curr Opin Biotechnol 70:83–90. https://doi.org/10.1016/j.copbio.2020.11.014

Fischer U, Utke K, Somamoto T, Köllner B, Ototake M, Nakanishi T (2006) Cytotoxic activities of fish leucocytes. Fish Shellfish Immunol 20(2):209–226

Gatesoupe FJ (1999) The use of probiotics in aquaculture. Aquaculture 180(1–2):147–165. https://doi.org/10.1016/S0044-8486(99)00187-8

Gelibolu S, Yanar Y, Genç MA, Genç E (2018) The effect of mannan-oligosaccharide (MOS) as a feed supplement on growth and some blood parameters of Gilthead Sea Bream (Sparus aurata). Turkish J Fisheries Aqua Sci 18:817–823

Gewaily MS, Shukry M, Abdel-Kader MF, Alkafafy M, Farrag FA, Moustafa EM, Doan HV, Abd-Elghany MF, Abdelhamid AF, Eltanahy A, Dawood MAO (2021) Dietary Lactobacillus plantarum Relieves Nile Tilapia (Oreochromis niloticus) juvenile from oxidative stress, immunosuppression, and inflammation induced by Deltamethrin and Aeromonas hydrophila. Front Mar Sci 8. https://doi.org/10.3389/fmars.2021.621558

Ghaedi G, Keyvanshokooh S, Azarm HM, Akhlaghi M (2016) Proteomic analysis of muscle tissue from rainbow trout (Oncorhynchus mykiss) fed dietary β-glucan. Iran J Vet Res 17(3):184
Gibson GR (1998) Dietary modulation of the human gut microflora using prebiotics. Br J Nutr 80 (S2):S209–S212
Gibson GR, Roberfroid MB (1995) Dietary modulation of the human colonic microbiota: introducing the concept of prebiotics. J Nutr 125(6):1401–1412. https://doi.org/10.1093/jn/125.6.1401
Gomez GD, Balcazar JL (2008) A review on the interactions between gut microbiota and innate immunity of fish. FEMS Immunol Med Microbiol 52(2):145–154. https://doi.org/10.1111/j.1574-695X.2007.00343.x
Goodridge HS, Wolf AJ, Underhill DM (2009) Beta-glucan recognition by the innate immune system. Immunol Rev 230(1):38–50. https://doi.org/10.1111/j.1600-065X.2009.00793.x
Grimoud J, Durand H, Courtin C, Monsan P, Ouarné F, Theodorou V, Roques C (2010) In vitro screening of probiotic lactic acid bacteria and prebiotic glucooligosaccharides to select effective synbiotics. Anaerobe 16(5):493–500. https://doi.org/10.1016/j.anaerobe.2010.07.005
Grisdale-Helland B, Helland SJ, Gatlin Iii DM (2008) The effects of dietary supplementation with mannanoligosaccharide, fructooligosaccharide or galactooligosaccharide on the growth and feed utilization of Atlantic salmon (*Salmo salar*). Aquaculture 283(1–4):163–167. https://doi.org/10.1016/j.aquaculture.2008.07.012
Guerreiro I, Couto A, Machado M, Castro C, Pousão-Ferreira P, Oliva-Teles A, Enes P (2016) Prebiotics effect on immune and hepatic oxidative status and gut morphology of white sea bream (*Diplodus sargus*). Fish Shellfish Immunol 50(Supplement C):168–174. https://doi.org/10.1016/j.fsi.2016.01.023
Guerreiro I, Oliva-Teles A, Enes P (2017) Prebiotics as functional ingredients: focus on Mediterranean fish aquaculture. Rev Aquac. https://doi.org/10.1111/raq.12201
Guerreiro I, Pérez-Jiménez A, Costas B, Oliva-Teles A (2014) Effect of temperature and short chain fructooligosaccharides supplementation on the hepatic oxidative status and immune response of turbot (Scophthalmus maximus). Fish Shellfish Immunol 40(2):570–576
Guzmán-Villanueva LT, Ascencio-Valle F, Macías-Rodríguez ME, Tovar-Ramírez D (2014) Effects of dietary β-1,3/1,6-glucan on the antioxidant and digestive enzyme activities of Pacific red snapper (Lutjanus peru) after exposure to lipopolysaccharides. Fish Physiol Biochem 40(3): 827–837. https://doi.org/10.1007/s10695-013-9889-0
Hahor W, Thongprajukaew K, Suanyuk N (2019) Effects of dietary supplementation of oligosaccharides on growth performance, gut health and immune response of hybrid catfish (Pangasianodon gigas × Pangasianodon hypophthalmus). Aquaculture 507:97–107. https://doi.org/10.1016/j.aquaculture.2019.04.010
Han J, Kang G, Kim J, Kim S (2002) The present status and future of Chlorella. Food Sci Industry 6: 64–69
Hasan MT, Jang WJ, Kim H, Lee B-J, Kim KW, Hur SW, Lim SG, Bai SC, Kong I-S (2018) Synergistic effects of dietary *Bacillus* sp. SJ-10 plus β-glucooligosaccharides as a synbiotic on growth performance, innate immunity and streptococcosis resistance in olive flounder (*Paralichthys olivaceus*). Fish Shellfish Immunol 82:544–553. https://doi.org/10.1016/j.fsi.2018.09.002
Hassaan MS, Mohammady EY, Soaudy MR, Palma J, Shawer EE, El-Haroun E (2020) The effect of dietary sericite on growth performance, digestive enzymes activity, gut microbiota and haematological parameters of Nile tilapia, Oreochromis niloticus (L.) fingerlings. Anim Feed Sci Technol 262:114400. https://doi.org/10.1016/j.anifeedsci.2020.114400
Henriksson PJG, Rico A, Troell M, Klinger DH, Buschmann AH, Saksida S, Chadag MV, Zhang W (2018) Unpacking factors influencing antimicrobial use in global aquaculture and their implication for management: a review from a systems perspective. Sustain Sci 13(4): 1105–1120. https://doi.org/10.1007/s11625-017-0511-8
Henriksson PJG, Troell M, Banks LK, Belton B, Beveridge MCM, Klinger DH, Pelletier N, Phillips MJ, Tran N (2021) Interventions for improving the productivity and environmental performance

of global aquaculture for future food security. One Earth 4(9):1220–1232. https://doi.org/10.1016/j.oneear.2021.08.009

Hill C, Guarner F, Reid G, Gibson GR, Merenstein DJ, Pot B, Morelli L, Canani RB, Flint HJ, Salminen S, Calder PC, Sanders ME (2014) Expert consensus document: The International Scientific Association for Probiotics and Prebiotics consensus statement on the scope and appropriate use of the term probiotic. Nat Rev Gastroenterol Hepatol 11(8):506–514. https://doi.org/10.1038/nrgastro.2014.66

Ho Y-Y, Lin C-M, Wu M-C (2017) Evaluation of the prebiotic effects of citrus pectin hydrolysate. J Food Drug Anal 25(3):550–558. https://doi.org/10.1016/j.jfda.2016.11.014

Hoseinifar S, Khalili M, Sun YZ (2016a) Intestinal histomorphology, autochthonous microbiota and growth performance of the oscar (Astronotus ocellatus Agassiz, 1831) following dietary administration of xylooligosaccharide. J Appl Ichthyol 32(6):1137–1141

Hoseinifar SH, Ahmadi A, Raeisi M, Hoseini SM, Khalili M, Behnampour N (2017a) Comparative study on immunomodulatory and growth enhancing effects of three prebiotics (galactooligosaccharide, fructooligosaccharide and inulin) in common carp (*Cyprinus carpio*). Aquac Res 48(7):3298–3307. https://doi.org/10.1111/are.13156

Hoseinifar SH, Eshaghzadeh H, Vahabzadeh H, Mana NP (2016b) Modulation of growth performances, survival, digestive enzyme activities and intestinal microbiota in common carp (*Cyprinus carpio*) larvae using short chain fructooligosaccharide. Aquac Res 47(10):3246–3253. https://doi.org/10.1111/are.12777

Hoseinifar SH, Esteban MÁ, Cuesta A, Sun Y-Z (2015a) Prebiotics and fish immune response: a review of current knowledge and future perspectives. Rev Fish Sci Aquac 23(4):315–328. https://doi.org/10.1080/23308249.2015.1052365

Hoseinifar SH, Mirvaghefi A, Amoozegar MA, Sharifian M, Esteban MÁ (2015b) Modulation of innate immune response, mucosal parameters and disease resistance in rainbow trout (Oncorhynchus mykiss) upon synbiotic feeding. Fish Shellfish Immunol 45(1):27–32. https://doi.org/10.1016/j.fsi.2015.03.029

Hoseinifar SH, Roosta Z, Hajimoradloo A, Vakili F (2015c) The effects of *Lactobacillus acidophilus* as feed supplement on skin mucosal immune parameters, intestinal microbiota, stress resistance and growth performance of black swordtail (*Xiphophorus helleri*). Fish Shellfish Immunol 42(2):533–538. https://doi.org/10.1016/j.fsi.2014.12.003

Hoseinifar SH, Sharifian M, Vesaghi MJ, Khalili M, Esteban MÁ (2014) The effects of dietary xylooligosaccharide on mucosal parameters, intestinal microbiota and morphology and growth performance of Caspian white fish (*Rutilus frisii kutum*) fry. Fish Shellfish Immunol 39(2):231–236. https://doi.org/10.1016/j.fsi.2014.05.009

Hoseinifar SH, Sun Y-Z, Caipang CM (2017b) Short chain fatty acids as feed supplements for sustainable aquaculture: an updated view. Aquac Res 48(4):1380–1391. https://doi.org/10.1111/are.13239

Hsu C-K, Liao J-W, Chung Y-C, Hsieh C-P, Chan Y-C (2004) Xylooligosaccharides and fructooligosaccharides affect the intestinal microbiota and precancerous colonic lesion development in rats. J Nutr 134(6):1523–1528

Hu X, Yang HL, Yan YY, Zhang CX, Ye Jd LKL, Hu LH, Zhang JJ, Ruan L, Sun YZ (2019) Effects of fructooligosaccharide on growth, immunity and intestinal microbiota of shrimp (Litopenaeus vannamei) fed diets with fish meal partially replaced by soybean meal. Aquac Nutr 25(1):194–204

Hunt AÖ, Çetinkaya M, Yılmaz FÖ, Yıldırım M, Berkoz M, Yalın S (2019) Effect of dietary supplementation of inulin on growth performance, digestion enzyme activities and antioxidant status of rainbow trout (Oncorhynchus mykiss). Turkish J Agric Food Sci Technol 7(9):1344–1353

Huynh T-G, Cheng A-C, Chi C-C, Chiu K-H, Liu C-H (2018) A synbiotic improves the immunity of white shrimp, *Litopenaeus vannamei*: metabolomic analysis reveal compelling evidence. Fish Shellfish Immunol 79:284–293

Ibrahim M, Ahmad F, Yaqub B, Ramzan A, Imran A, Afzaal M, Mirza SA, Mazhar I, Younus M, Akram Q, Ali Taseer MS, Ahmad A, Ahmed S (2020) Chapter 4—Current trends of antimicrobials used in food animals and aquaculture. In: Hashmi MZ (ed) Antibiotics and antimicrobial resistance genes in the environment, vol 1. Elsevier, pp 39–69. https://doi.org/10.1016/B978-0-12-818882-8.00004-8

Irianto A, Austin B (2002) Probiotics in aquaculture. J Fish Dis 25:11

Jami MJ, Abedian Kenari A, Paknejad H, Mohseni M (2019) Effects of dietary b-glucan, mannan oligosaccharide, *Lactobacillus plantarum* and their combinations on growth performance, immunity and immune related gene expression of Caspian trout, *Salmo trutta caspius* (Kessler, 1877). Fish Shellfish Immunol 91:202–208. https://doi.org/10.1016/j.fsi.2019.05.024

Jayapal N, Samanta A, Kolte AP, Senani S, Sridhar M, Suresh K, Sampath K (2013) Value addition to sugarcane bagasse: xylan extraction and its process optimization for xylooligosaccharides production. Ind Crops Prod 42:14–24

Ji L, Sun G, Li J, Wang Y, Du Y, Li X, Liu Y (2017) Effect of dietary β-glucan on growth, survival and regulation of immune processes in rainbow trout (Oncorhynchus mykiss) infected by Aeromonas salmonicida. Fish Shellfish Immunol 64:56–67

Kamilya D, Khan MIR (2020) Chapter 24—Chitin and chitosan as promising immunostimulant for aquaculture. In: Gopi S, Thomas S, Pius A (eds) Handbook of chitin and chitosan. Elsevier, pp 761–771. https://doi.org/10.1016/B978-0-12-817966-6.00024-8

Kang H, Salim H, Akter N, Kim D, Kim J, Bang H, Kim M, Na J, Hwangbo J, Choi H (2013) Effect of various forms of dietary Chlorella supplementation on growth performance, immune characteristics, and intestinal microflora population of broiler chickens. J Appl Poult Res 22(1): 100–108

Kobayashi M, Msangi S, Batka M, Vannuccini S, Dey MM, Anderson JL (2015) Fish to 2030: the role and opportunity for aquaculture. Aquac Econ Manage 19(3):282–300. https://doi.org/10.1080/13657305.2015.994240

Koch JFA, de Oliveira CAF, Zanuzzo FS (2021) Dietary β-glucan (MacroGard®) improves innate immune responses and disease resistance in Nile tilapia regardless of the administration period. Fish Shellfish Immunol 112:56–63. https://doi.org/10.1016/j.fsi.2021.02.014

Kokou F, Sasson G, Mizrahi I, Cnaani A (2020) Antibiotic effect and microbiome persistence vary along the European seabass gut. Sci Rep 10(1):10003. https://doi.org/10.1038/s41598-020-66622-5

Kolida S, Gibson GR (2011) Synbiotics in health and disease. Ann Rev Food Sci Technol 2:373–393

Kühlwein H, Merrifield DL, Rawling MD, Foey AD, Davies SJ (2014) Effects of dietary β-(1,3)(1,6)-D-glucan supplementation on growth performance, intestinal morphology and haemato-immunological profile of mirror carp (*Cyprinus carpio* L.). J Anim Physiol Anim Nutr 98(2): 279–289. https://doi.org/10.1111/jpn.12078

Kumar P, Jain KK, Sardar P (2018) Effects of dietary synbiotic on innate immunity, antioxidant activity and disease resistance of Cirrhinus mrigala juveniles. Fish Shellfish Immunol 80:124–132. https://doi.org/10.1016/j.fsi.2018.05.045

Kumar V, Satyanarayana T (2011) Applicability of thermo-alkali-stable and cellulase-free xylanase from a novel thermo-halo-alkaliphilic Bacillus halodurans in producing xylooligosaccharides. Biotechnol Lett 33(11):2279–2285

Laice LM, Corrêa Filho RAC, Ventura AS, Farias KNN, Silva ALDN, Fernandes CE, Silva ACF, Barbosa PTL, de Souza AI, Emerenciano MGC, Povh JA (2021) Use of symbiotics in biofloc (BFT)-based Nile tilapia culture: Production performance, intestinal morphometry and hematological parameters. Aquaculture 530:735715. https://doi.org/10.1016/j.aquaculture.2020.735715

Lara-Flores M, Olvera-Novoa MA, Guzmán-Méndez BZE, López-Madrid W (2003) Use of the bacteria Streptococcus faecium and Lactobacillus acidophilus, and the yeast Saccharomyces cerevisiae as growth promoters in Nile tilapia (Oreochromis niloticus). Aquaculture 216(1–4): 193–201. https://doi.org/10.1016/S0044-8486(02)00277-6

Lattimer JM, Haub MD (2010) Effects of dietary fiber and its components on metabolic health. Nutrients 2(12):1266–1289. https://doi.org/10.3390/nu2121266

Lauzon HL, Dimitroglou A, Merrifield DL, Ringø E, Davies SJ (2014a) Probiotics and prebiotics: concepts, definitions and history. In: Aquaculture nutrition: gut health, probiotics and prebiotics, pp 169–184. https://doi.org/10.1002/9781118897263.ch7

Lazado CC, Caipang CMA, Estante EG (2015) Prospects of host-associated microorganisms in fish and penaeids as probiotics with immunomodulatory functions. Fish Shellfish Immunol 45(1): 2–12. https://doi.org/10.1016/j.fsi.2015.02.023

Legentil L, Paris F, Ballet C, Trouvelot S, Daire X, Vetvicka V, Ferrières V (2015) Molecular interactions of β-(1→ 3)-glucans with their receptors. Molecules (Basel, Switzerland) 20(6): 9745–9766

Li C, Niu Z, Zou M, Liu S, Wang M, Gu X, Lu H, Tian H, Jha R (2020) Probiotics, prebiotics, and synbiotics regulate the intestinal microbiota differentially and restore the relative abundance of specific gut microorganisms. J Dairy Sci 103(7):5816–5829. https://doi.org/10.3168/jds.2019-18003

Li C, Ren Y, Jiang S, Zhou S, Zhao J, Wang R, Li Y (2018) Effects of dietary supplementation of four strains of lactic acid bacteria on growth, immune-related response and genes expression of the juvenile sea cucumber *Apostichopus japonicus* Selenka. Fish Shellfish Immunol 74:69–75. https://doi.org/10.1016/j.fsi.2017.12.037

Li P, Burr GS, Gatlin DM 3rd, Hume ME, Patnaik S, Castille FL, Lawrence AL (2007) Dietary supplementation of short-chain fructooligosaccharides influences gastrointestinal microbiota composition and immunity characteristics of Pacific white shrimp, Litopenaeus vannamei, cultured in a recirculating system. J Nutr 137(12):2763–2768

Li Z, Tran NT, Ji P, Sun Z, Wen X, Li S (2019) Effects of prebiotic mixtures on growth performance, intestinal microbiota and immune response in juvenile chu's croaker, *Nibea coibor*. Fish Shellfish Immunol 89:564–573. https://doi.org/10.1016/j.fsi.2019.04.025

Liew HJ, Fazio A, Faggio C, Blust R, De Boeck G (2015) Cortisol affects metabolic and ionoregulatory responses to a different extent depending on feeding ration in common carp, *Cyprinus carpio*. Comp Biochem Physiol Part A Mol Integr Physiol 189:45–57

Liu W, Wang W, Ran C, He S, Yang Y, Zhou Z (2017b) Effects of dietary scFOS and lactobacilli on survival, growth, and disease resistance of hybrid tilapia. Aquaculture 470:50–55. https://doi.org/10.1016/j.aquaculture.2016.12.013

Llewellyn M, Boutin S, Hoseinifar SH, Derome N (2014) Teleost microbiomes: progress towards their characterisation, manipulation and applications in aquaculture and fisheries. Front Microbiol 5:1–17. https://doi.org/10.3389/fmicb.2014.00207

Madrigal L, Sangronis E (2007) Inulin and derivates as key ingredients in functional foods. Arch Latinoam Nutr 57(4):387–396

Manisseri C, Gudipati M (2012) Prebiotic activity of purified xylobiose obtained from Ragi (Eleusine coracana, Indaf-15) Bran. Indian J Microbiol 52(2):251–257

Markowiak P, Śliżewska K (2017) Effects of probiotics, prebiotics, and synbiotics on human health. Nutrients 9(9):1021. https://doi.org/10.3390/nu9091021

Markowiak-Kopeć P, Śliżewska K (2020) The effect of probiotics on the production of short-chain fatty acids by human intestinal microbiome. Nutrients 12(4):1107. https://doi.org/10.3390/nu12041107

Mazzola G, Aloisio I, Biavati B, Di Gioia D (2015) Development of a synbiotic product for newborns and infants. LWT Food Sci Technol 64(2):727–734

Meena DK, Das P, Kumar S, Mandal SC, Prusty AK, Singh SK, Akhtar MS, Behera BK, Kumar K, Pal AK, Mukherjee SC (2013) Beta-glucan: an ideal immunostimulant in aquaculture (a review). Fish Physiol Biochem 39(3):431–457. https://doi.org/10.1007/s10695-012-9710-5

Meng X, Wang J, Wan W, Xu M, Wang T (2017) Influence of low molecular weight chitooligosaccharides on growth performance and non-specific immune response in Nile tilapia *Oreochromis niloticus*. Aquac Int 25(3):1265–1277

Merrifield DL, Balcázar JL, Daniels C, Zhou Z, Carnevali O, Sun Y, Hoseinifar SH, Ringø E (2014) Indigenous lactic acid bacteria in fish and crustaceans. In: Aquaculture nutrition. https://doi.org/10.1002/9781118897263.ch6

Merrifield DL, Dimitroglou A, Foey A, Davies SJ, Baker RT, Bøgwald J, Castex M, Ringø E (2010) The current status and future focus of probiotic and prebiotic applications for salmonids. Aquaculture 302(1-2):1–18. https://doi.org/10.1016/j.aquaculture.2010.02.007

Miest JJ, Arndt C, Adamek M, Steinhagen D, Reusch TB (2016) Dietary β-glucan (MacroGard®) enhances survival of first feeding turbot (Scophthalmus maximus) larvae by altering immunity, metabolism and microbiota. Fish Shellfish Immunol 48:94–104

Miranda CD, Godoy FA, Lee MR (2018) Current status of the use of antibiotics and the antimicrobial resistance in the chilean salmon farms. Front Microbiol 9:1284

Misra CK, Das BK, Mukherjee SC, Pattnaik P (2006) Effect of long term administration of dietary β-glucan on immunity, growth and survival of Labeo rohita fingerlings. Aquaculture 255(1–4): 82–94. https://doi.org/10.1016/j.aquaculture.2005.12.009

Mitsuoka T, Hidaka H, Eida T (1987) Effect of fructo-oligosaccharides on intestinal microflora. Food/Nahrung 31(5-6):427–436

Mohammadi G, Rafiee G, El Basuini MF, Abdel-Latif HMR, Dawood MAO (2020) The growth performance, antioxidant capacity, immunological responses, and the resistance against Aeromonas hydrophila in Nile tilapia (Oreochromis niloticus) fed Pistacia vera hulls derived polysaccharide. Fish Shellfish Immunol 106:36–43. https://doi.org/10.1016/j.fsi.2020.07.064

Mohammadian T, Ghanei-Motlagh R, Molayemraftar T, Mesbah M, Zarea M, Mohtashamipour H, Jangaran Nejad A (2021) Modulation of growth performance, gut microflora, non-specific immunity and gene expression of proinflammatory cytokines in shabout (Tor grypus) upon dietary prebiotic supplementation. Fish Shellfish Immunol 112:38–45. https://doi.org/10.1016/j.fsi.2021.02.012

Mohammadian T, Nasirpour M, Tabandeh MR, Mesbah M (2019) Synbiotic effects of β-glucan, mannan oligosaccharide and *Lactobacillus casei* on growth performance, intestine enzymes activities, immune-hematological parameters and immune-related gene expression in common carp, *Cyprinus carpio*: an experimental infection with *Aeromonas hydrophila*. Aquaculture 511: 634197. https://doi.org/10.1016/j.aquaculture.2019.06.011

Morshedi V, Agh N, Noori F, Jafari F, Tukmechi A, Marammazi J, Pagheh E (2018) Effects of dietary xylooligosaccharide on growth and feeding performance, body composition and physiological responses of sobaity seabream (Sparidentex hasta) juvenile. Aquac Nutr 24(6): 1796–1803

Mugwanya M, Dawood MAO, Kimera F, Sewilam H (2022) Updating the role of probiotics, prebiotics, and synbiotics for tilapia aquaculture as leading candidates for food sustainability: a review. Probiotics Antimicrob Proteins 14(1):130–157. https://doi.org/10.1007/s12602-021-09852-x

Mumtaz S, Rehman S, Huma N, Jamil A (2009) Effect of xylooligosaccharide enriched yogurt on serum profile in albino rats. Pak J Nutr 8(11):1756–1759

Munguti JM, Nairuti R, Iteba JO, Obiero KO, Kyule D, Opiyo MA, Abwao J, Kirimi JG, Outa N, Muthoka M, Githukia CM, Ogello EO (2022) Nile tilapia (Oreochromis niloticus Linnaeus, 1758) culture in Kenya: Emerging production technologies and socio-economic impacts on local livelihoods. Aquac Fish Fish. https://doi.org/10.1002/aff2.58

Munir MB, Hashim R, Chai YH, Marsh TL, Nor SAM (2016) Dietary prebiotics and probiotics influence growth performance, nutrient digestibility and the expression of immune regulatory genes in snakehead (Channa striata) fingerlings. Aquaculture 460:59–68. https://doi.org/10.1016/j.aquaculture.2016.03.041

Nagpal R, Wang S, Ahmadi S, Hayes J, Gagliano J, Subashchandrabose S, Kitzman DW, Becton T, Read R, Yadav H (2018) Human-origin probiotic cocktail increases short-chain fatty acid production via modulation of mice and human gut microbiome. Sci Rep 8(1):12649. https://doi.org/10.1038/s41598-018-30114-4

Nawaz A, Bakhsh Javaid A, Irshad S, Hoseinifar SH, Xiong H (2018) The functionality of prebiotics as immunostimulant: evidences from trials on terrestrial and aquatic animals. Fish Shellfish Immunol 76:272–278. https://doi.org/10.1016/j.fsi.2018.03.004

Nayak SK (2010) Probiotics and immunity: a fish perspective. Fish Shellfish Immunol 29(1):2–14

Nazarudin MF, Yusoff F, Idrus ES, Aliyu-Paiko M (2020) Brown seaweed Sargassum polycystum as dietary supplement exhibits prebiotic potentials in Asian sea bass Lates calcarifer fingerlings. Aquac Rep 18:100488. https://doi.org/10.1016/j.aqrep.2020.100488

Nedaei S, Noori A, Valipour A, Khanipour AA, Hoseinifar SH (2019) Effects of dietary galactooligosaccharide enriched commercial prebiotic on growth performance, innate immune response, stress resistance, intestinal microbiota and digestive enzyme activity in Narrow clawed crayfish (*Astacus leptodactylus* Eschscholtz, 1823). Aquaculture 499:80–89. https://doi.org/10.1016/j.aquaculture.2018.08.076

Newaj-Fyzul A, Austin B (2015) Probiotics, immunostimulants, plant products and oral vaccines, and their role as feed supplements in the control of bacterial fish diseases. J Fish Dis 38(11): 937–955. https://doi.org/10.1111/jfd.12313

Ng W-K, Romano N (2013) A review of the nutrition and feeding management of farmed tilapia throughout the culture cycle. Rev Aquac 5(4):220–254. https://doi.org/10.1111/raq.12014

Nikoskelainen S, Ouwehand AC, Bylund G, Salminen S, Lilius E-M (2003) Immune enhancement in rainbow trout (*Oncorhynchus mykiss*) by potential probiotic bacteria (*Lactobacillus rhamnosus*). Fish Shellfish Immunol 15(5):443–452. https://doi.org/10.1016/S1050-4648(03)00023-8

Niu J, Liu YJ, Lin HZ, Mai KS, Yang HJ, Liang GY, Tian LX (2011) Effects of dietary chitosan on growth, survival and stress tolerance of postlarval shrimp, Litopenaeus vannamei. Aquac Nutr 17(2):e406–e412. https://doi.org/10.1111/j.1365-2095.2010.00775.x

No HK, Young Park N, Ho Lee S, Meyers SP (2002) Antibacterial activity of chitosans and chitosan oligomers with different molecular weights. Int J Food Microbiol 74(1):65–72. https://doi.org/10.1016/S0168-1605(01)00717-6

Olsen RE, Myklebust R, Kryvi H, Mayhew TM, Ringø E (2001) Damaging effect of dietary inulin on intestinal enterocytes in Arctic charr (*Salvelinus alpinus* L.). Aquac Res 32(11):931–934

Panahi Y, Darvishi B, Jowzi N, Beiraghdar F, Sahebkar A (2016) Chlorella vulgaris: a multifunctional dietary supplement with diverse medicinal properties. Curr Pharm Des 22(2): 164–173

Parada Venegas D, De la Fuente MK, Landskron G, González MJ, Quera R, Dijkstra G, Harmsen HJM, Faber KN, Hermoso MA (2019) Short chain fatty acids (SCFAs)-mediated gut epithelial and immune regulation and its relevance for inflammatory bowel diseases. Front Immunol 10: 277–277. https://doi.org/10.3389/fimmu.2019.00277

Passos LML, Park YK (2003) Fructooligosaccharides: implications in human health being and use in foods. Ciência Rural 33(2):385–390

Payne CJ, Turnbull JF, MacKenzie S, Crumlish M (2021) Investigating the effect of an oxytetracycline treatment on the gut microbiome and antimicrobial resistance gene dynamics in Nile Tilapia (Oreochromis niloticus). Antibiotics 10(10):1213

Petit J, Bailey EC, Wheeler RT, de Oliveira CAF, Forlenza M, Wiegertjes GF (2019) Studies into β-Glucan recognition in fish suggests a key role for the C-Type Lectin pathway. Front Immunol 10:280. https://doi.org/10.3389/fimmu.2019.00280

Petit J, Wiegertjes GF (2016) Long-lived effects of administering β-glucans: indications for trained immunity in fish. Dev Comp Immunol 64:93–102

Picchietti S, Mazzini M, Taddei AR, Renna R, Fausto AM, Mulero V, Carnevali O, Cresci A, Abelli L (2007) Effects of administration of probiotic strains on GALT of larval gilthead seabream: immunohistochemical and ultrastructural studies. Fish Shellfish Immunol 22(1-2):57–67. https://doi.org/10.1016/j.fsi.2006.03.009

Pilarski F, Ferreira de Oliveira CA, Darpossolo de Souza FPB, Zanuzzo FS (2017) Different β-glucans improve the growth performance and bacterial resistance in Nile tilapia. Fish Shellfish Immunol 70:25–29. https://doi.org/10.1016/j.fsi.2017.06.059

Poolsawat L, Li X, Xu X, Rahman MM, Boonpeng N, Leng X (2021) Dietary xylooligosaccharide improved growth, nutrient utilization, gut microbiota and disease resistance of tilapia (Oreochromis niloticus x O. aureus). Anim Feed Sci Technol 275:114872. https://doi.org/10.1016/j.anifeedsci.2021.114872

Prabu E, Rajagopalsamy C, Ahilan B, Jeevagan IJMA, Renuhadevi M (2019) Tilapia–an excellent candidate species for world aquaculture: a review. Annu Res Rev Biol:1–14

Raa J (2015) Immune modulation by non-digestible and non-absorbable beta-1, 3/1, 6-glucan. Microb Ecol Health Dis 26(1):27824

Rengpipat S, Rukpratanporn S, Piyatiratitivorakul S, Menasaveta P (2000) Immunity enhancement in black tiger shrimp (*Penaeus monodon*) by a probiont bacterium (*Bacillus* S11). Aquaculture 191(4):271–288. https://doi.org/10.1016/S0044-8486(00)00440-3

Reza A, Abdolmajid H, Abbas M, Abdolmohammad AK (2009) Effect of dietary prebiotic inulin on growth performance, intestinal microflora, body composition and hematological parameters of juvenile Beluga, Huso huso (Linnaeus, 1758). J World Aquac Soc 40(6):771–779. https://doi.org/10.1111/j.1749-7345.2009.00297.x

Rico A, Oliveira R, McDonough S, Matser A, Khatikarn J, Satapornvanit K, Nogueira AJA, Soares AMVM, Domingues I, Van den Brink PJ (2014) Use, fate and ecological risks of antibiotics applied in tilapia cage farming in Thailand. Environ Pollut 191:8–16. https://doi.org/10.1016/j.envpol.2014.04.002

Ringø E, Løvmo L, Kristiansen M, Bakken Y, Salinas I, Myklebust R, Olsen RE, Mayhew TM (2010a) Lactic acid bacteria vs. pathogens in the gastrointestinal tract of fish: a review. Aquac Res 41(4):451–467. https://doi.org/10.1111/j.1365-2109.2009.02339.x

Ringø E, Olsen RE, Gifstad TØ, Dalmo RA, Amlund H, Hemre GI, Bakke AM (2010b) Prebiotics in aquaculture: a review. Aquac Nutr 16(2):117–136. https://doi.org/10.1111/j.1365-2095.2009.00731.x

Ríos-Covián D, Ruas-Madiedo P, Margolles A, Gueimonde M, de los Reyes-Gavilán CG, Salazar N (2016) Intestinal short chain fatty acids and their link with diet and human health. Front Microbiol 7:185. https://doi.org/10.3389/fmicb.2016.00185

Roberfroid MB (2000) Prebiotics and probiotics: are they functional foods? Am J Clin Nutr 71(6): 1682S–1687S

Rodriguez-Estrada U, Satoh S, Haga Y, Fushimi H, Sweetman J (2009) Effects of single and combined supplementation of *Enterococcus faecalis*, mannanoligosaccharide and polyhydrobutyric acid on growth performance and immune response of rainbow trout *Oncorhynchus mykiss*. Aquac Sci 57:609–617

Rohani MF, Islam SMM, Hossain MK, Ferdous Z, Siddik MAB, Nuruzzaman M, Padeniya U, Brown C, Shahjahan M (2022) Probiotics, prebiotics and synbiotics improved the functionality of aquafeed: Upgrading growth, reproduction, immunity and disease resistance in fish. Fish Shellfish Immunol 120:569–589. https://doi.org/10.1016/j.fsi.2021.12.037

Sado RY, Gimbo RY, Salles FB (2016) Routes of Β-glucan administration affect hematological and immune responses of Oreochromis niloticus. Arch Zootec 65(252):519–524

Sajid M, Prabjeet S, Samoon MH, Balange A (2010) Effect of dietary chitosan on non-specific immune response and growth of *Cyprinus carpio* challenged with *Aeromonas hydrophila*. Int Aquat Res 2(2):77–85

Samanta A, Senani S, Kolte AP, Sridhar M, Sampath K, Jayapal N, Devi A (2012) Production and in vitro evaluation of xylooligosaccharides generated from corn cobs. Food Bioprod Process 90(3):466–474

Samanta AK, Jayapal N, Jayaram C, Roy S, Kolte AP, Senani S, Sridhar M (2015) Xylooligosaccharides as prebiotics from agricultural by-products: production and applications. Bioact Carbohydrat Diet Fibre 5(1):62–71. https://doi.org/10.1016/j.bcdf.2014.12.003

Schrezenmeir J, de Vrese M (2001) Probiotics, prebiotics, and synbiotics—approaching a definition. Am J Clin Nutr 73(2):361s–364s

Sewaka M, Trullas C, Chotiko A, Rodkhum C, Chansue N, Boonanuntanasarn S, Pirarat N (2019) Efficacy of synbiotic Jerusalem artichoke and *Lactobacillus rhamnosus* GG-supplemented diets on growth performance, serum biochemical parameters, intestinal morphology, immune parameters and protection against Aeromonas veronii in juvenile red tilapia (*Oreochromis* spp.). Fish Shellfish Immunol 86:260–268. https://doi.org/10.1016/j.fsi.2018.11.026

Sherif AH, Mahfouz ME (2019) Immune status of Oreochromis niloticus experimentally infected with Aeromonas hydrophila following feeding with 1, 3 β-glucan and levamisole immunostimulants. Aquaculture 509:40–46. https://doi.org/10.1016/j.aquaculture.2019.05.016

Soleimani N, Hoseinifar SH, Merrifield DL, Barati M, Abadi ZH (2012) Dietary supplementation of fructooligosaccharide (FOS) improves the innate immune response, stress resistance, digestive enzyme activities and growth performance of Caspian roach (*Rutilus rutilus*) fry. Fish Shellfish Immunol 32(2):316–321. https://doi.org/10.1016/j.fsi.2011.11.023

Soltanian S, Stuyven E, Cox E, Sorgeloos P, Bossier P (2009) Beta-glucans as immunostimulant in vertebrates and invertebrates. Crit Rev Microbiol 35(2):109–138

Song SK, Beck BR, Kim D, Park J, Kim J, Kim HD, Ringø E (2014) Prebiotics as immunostimulants in aquaculture: a review. Fish Shellfish Immunol 40(1):40–48. https://doi.org/10.1016/j.fsi.2014.06.016

Souza FP, Lima ECSD, Pandolfi VCF, Leite NG, Furlan-Murari PJ, Leal CNS, Mainardi RM, Suphoronski SA, Favero LM, Koch JFA, Pereira UDP, Lopera-Barrero NM (2020) Effect of β-glucan in water on growth performance, blood status and intestinal microbiota in tilapia under hypoxia. Aquac Rep 17:100369. https://doi.org/10.1016/j.aqrep.2020.100369

Stanton C, Ross RP, Fitzgerald GF, Van Sinderen D (2005) Fermented functional foods based on probiotics and their biogenic metabolites. Curr Opin Biotechnol 16(2):198–203. https://doi.org/10.1016/j.copbio.2005.02.008

Stratev D, Zhelyazkov G, Noundou XS, Krause RW (2018) Beneficial effects of medicinal plants in fish diseases. Aquac Int 26(1):289–308

Sun Y, Wang G, Peng K, Huang Y, Cao J, Huang W, Chen B, Hu J (2019) Effects of dietary xylooligosaccharides on growth performance, immunity and Vibrio alginolyticus resistance of juvenile Litopenaeus vannamei. Aquac Res 50(1):358–365

Sweetman JW, Torrecillas S, Dimitroglou A, Rider S, Davies SJ, Izquierdo MS (2010) Enhancing the natural defences and barrier protection of aquaculture species. Aquac Res 41(3):345–355

Tang Z-R, Yin Y-L, Nyachoti CM, Huang R-L, Li T-J, Yang C, Yang X-j, Gong J, Peng J, Qi D-S, Xing J-J, Sun Z-H, Fan MZ (2005) Effect of dietary supplementation of chitosan and galacto-mannan-oligosaccharide on serum parameters and the insulin-like growth factor-I mRNA expression in early-weaned piglets. Domest Anim Endocrinol 28(4):430–441. https://doi.org/10.1016/j.domaniend.2005.02.003

Tiengtam N, Khempaka S, Paengkum P, Booanuntanasarn S (2015b) Effects of inulin and Jerusalem artichoke (*Helianthus tuberosus*) as prebiotic ingredients in the diet of juvenile Nile tilapia (*Oreochromis niloticus*). Anim Feed Sci Technol 207:10. https://doi.org/10.1016/j.anifeedsci.2015.05.008

Tiwari UP, Singh AK, Jha R (2019) Fermentation characteristics of resistant starch, arabinoxylan, and β-glucan and their effects on the gut microbial ecology of pigs: a review. Anim Nutr 5(3): 217–226. https://doi.org/10.1016/j.aninu.2019.04.003

Torrecillas S, Rivero-Ramírez F, Izquierdo MS, Caballero MJ, Makol A, Suarez-Bregua P, Fernández-Montero A, Rotllant J, Montero D (2018) Feeding European sea bass (Dicentrarchus labrax) juveniles with a functional synbiotic additive (mannan oligosaccharides and Pediococcus acidilactici): an effective tool to reduce low fishmeal and fish oil gut health effects? Fish Shellfish Immunol 81:10–20. https://doi.org/10.1016/j.fsi.2018.07.007

Trushenski JT, Boesenberg J, Kohler CC (2009) Influence of grow-out feed fatty acid composition on finishing success in Nile tilapia. N Am J Aquac 71(3):242–251. https://doi.org/10.1577/A08-051.1

Tubin JSB, Paiano D, Hashimoto GSO, Furtado WE, Martins ML, Durigon E, Emerenciano MGC (2020) Tenebrio molitor meal in diets for Nile tilapia juveniles reared in biofloc system. Aquaculture 519:734763. https://doi.org/10.1016/j.aquaculture.2019.734763

Van Doan H, Hoseinifar SH, Naraballobh W, Jaturasitha S, Tongsiri S, Chitmanat C, Ringø E (2019b) Dietary inclusion of Orange peels derived pectin and Lactobacillus plantarum for Nile tilapia (Oreochromis niloticus) cultured under indoor biofloc systems. Aquaculture 508:98–105

Van Doan H, Hoseinifar SH, Tapingkae W, Khamtavee P (2017b) The effects of dietary kefir and low molecular weight sodium alginate on serum immune parameters, resistance against *Streptococcus agalactiae* and growth performance in Nile tilapia (*Oreochromis niloticus*). Fish Shellfish Immunol 62:139–146. https://doi.org/10.1016/j.fsi.2017.01.014

Van Doan H, Hoseinifar SH, Tapingkae W, Seel-audom M, Jaturasitha S, Dawood MA, Wongmaneeprateep S, Thu TTN, Esteban MÁ (2019c) Boosted growth performance, mucosal and serum immunity, and disease resistance Nile Tilapia (Oreochromis niloticus) Fingerlings Using Corncob-Derived Xylooligosaccharide and Lactobacillus plantarum CR1T5. Probiotics and antimicrobial proteins:1-12

Van Doan H, Hoseinifar SH, Tapingkae W, Tongsiri S, Khamtavee P (2016a) Combined administration of low molecular weight sodium alginate boosted immunomodulatory, disease resistance and growth enhancing effects of *Lactobacillus plantarum* in Nile tilapia (*Oreochromis niloticus*). Fish Shellfish Immunol 58:678–685. https://doi.org/10.1016/j.fsi.2016.10.013

Vieco-Saiz N, Belguesmia Y, Raspoet R, Auclair E, Gancel F, Kempf I, Drider D (2019) Benefits and inputs from lactic acid bacteria and their bacteriocins as alternatives to antibiotic growth promoters during food-animal production. Front Microbiol 10:57. https://doi.org/10.3389/fmicb.2019.00057

Villumsen KR, Ohtani M, Forberg T, Aasum E, Tinsley J, Bojesen AM (2020) Synbiotic feed supplementation significantly improves lipid utilization and shows discrete effects on disease resistance in rainbow trout (Oncorhynchus mykiss). Sci Rep 10(1):1–12

Vogt L, Ramasamy U, Meyer D, Pullens G, Venema K, Faas MM, Schols HA, de Vos P (2013) Immune modulation by different types of β2→ 1-fructans is toll-like receptor dependent. PloS one 8(7):e68367

Wang An R, Ran C, Ringø E, Zhou Zhi G (2017) Progress in fish gastrointestinal microbiota research. Rev Aquac. https://doi.org/10.1111/raq.12191

Wang C, Li G, Karmakar B, AlSalem HS, Shati AA, El-kott AF, Elsaid FG, Bani-Fwaz MZ, Alsayegh AA, Salem Alkhayyat S, El-Saber Batiha G (2022) Pectin mediated green synthesis of Fe3O4/Pectin nanoparticles under ultrasound condition as an anti-human colorectal carcinoma bionanocomposite. Arab J Chem 15(6):103867. https://doi.org/10.1016/j.arabjc.2022.103867

Wang G-x, Zhou Y, Huang W-q, Huang Y-h, Liu X-h, Dong S-z (2010) Effects of xylooligosaccharide on grow, body composition and non-specific immunity in Litopenaeus vannamei. Freshwater Fish 40:55–58

Wang J, Zhuang S (2022) Chitosan-based materials: preparation, modification and application. J Clean Prod 355:131825. https://doi.org/10.1016/j.jclepro.2022.131825

Wang JP, Yoo JS, Kim HJ, Lee JH, Kim IH (2009) Nutrient digestibility, blood profiles and fecal microbiota are influenced by chitooligosaccharide supplementation of growing pigs. Livest Sci 125(2):298–303. https://doi.org/10.1016/j.livsci.2009.05.011

Wan-Mohtar WAAQI, Taufek NM, Thiran JP, Rahman JFP, Yerima G, Subramaniam K, Rowan N (2021) Investigations on the use of exopolysaccharide derived from mycelial extract of Ganoderma lucidum as functional feed ingredient for aquaculture-farmed red hybrid Tilapia (Oreochromis sp.). Future Foods 3:100018. https://doi.org/10.1016/j.fufo.2021.100018

Watts JEM, Schreier HJ, Lanska L, Hale MS (2017) The rising tide of antimicrobial resistance in aquaculture: sources, sinks and solutions. Mar Drugs 15(6):158. https://doi.org/10.3390/md15060158

Watzl B, Girrbach S, Roller M (2005) Inulin, oligofructose and immunomodulation. Br J Nutr 93 (S1):S49–S55

Wee W, Abdul Hamid NK, Mat K, Khalif RIAR, Rusli ND, Rahman MM, Kabir MA, Wei LS (2022) The effects of mixed prebiotics in aquaculture: a review. Aquac Fish. https://doi.org/10.1016/j.aaf.2022.02.005

Wen X, Zhong R, Dang G, Xia B, Wu W, Tang S, Tang L, Liu L, Liu Z, Chen L, Zhang H (2022) Pectin supplementation ameliorates intestinal epithelial barrier function damage by modulating intestinal microbiota in lipopolysaccharide-challenged piglets. J Nutr Biochem 109:109107. https://doi.org/10.1016/j.jnutbio.2022.109107

Xia Y, Lu M, Chen G, Cao J, Gao F, Wang M, Liu Z, Zhang D, Zhu H, Yi M (2018) Effects of dietary *Lactobacillus rhamnosus* JCM1136 and *Lactococcus lactis subsp. lactis* JCM5805 on the growth, intestinal microbiota, morphology, immune response and disease resistance of juvenile Nile tilapia, *Oreochromis niloticus*. Fish Shellfish Immunol 76:368–379. https://doi.org/10.1016/j.fsi.2018.03.020

Xu B, Wang Y, Li J, Lin Q (2009) Effect of prebiotic xylooligosaccharides on growth performances and digestive enzyme activities of allogynogenetic crucian carp (*Carassius auratus gibelio*). Fish Physiol Biochem 35(3):351–357. https://doi.org/10.1007/s10695-008-9248-8

Ye JD, Wang K, Li FD, Sun YZ (2011) Single or combined effects of fructo- and mannan oligosaccharide supplements and *Bacillus clausii* on the growth, feed utilization, body composition, digestive enzyme activity, innate immune response and lipid metabolism of the Japanese flounder *Paralichthys olivaceus*. Aquac Nutr 17(4):e902–e911. https://doi.org/10.1111/j.1365-2095.2011.00863.x

Yilmaz S, Yilmaz E, Dawood MAO, Ringø E, Ahmadifar E, Abdel-Latif HMR (2022) Probiotics, prebiotics, and synbiotics used to control vibriosis in fish: a review. Aquaculture 547:737514. https://doi.org/10.1016/j.aquaculture.2021.737514

Yu X, Yin J, Li L, Luan C, Zhang J, Zhao C, Li S (2015) Prebiotic potential of Xylooligosaccharides derived from corn cobs and their in vitro antioxidant activity when combined with *Lactobacillus*. J Microbiol Biotechnol 25(7):1084–1092. https://doi.org/10.4014/jmb.1501.01022

Yuji-Sado R, Raulino-Domanski F, de Freitas PF, Baioco-Sales F (2015) Growth, immune status and intestinal morphology of Nile tilapia fed dietary prebiotics (mannan oligosaccharides-MOS). Lat Am J Aquat Res 43(5):944–952. https://doi.org/10.3856/vol43-issue5-fulltext-14

Zepeda-Hernández A, Garcia-Amezquita LE, Requena T, García-Cayuela T (2021) Probiotics, prebiotics, and synbiotics added to dairy products: Uses and applications to manage type 2 diabetes. Food Res Int 142:110208. https://doi.org/10.1016/j.foodres.2021.110208

Zhang C-N, Li X-F, Jiang G-Z, Zhang D-D, Tian H-Y, Li J-Y, Liu W-B (2014) Effects of dietary fructooligosaccharide levels and feeding modes on growth, immune responses, antioxidant capability and disease resistance of blunt snout bream (*Megalobrama amblycephala*). Fish Shellfish Immunol 41(2):560–569. https://doi.org/10.1016/j.fsi.2014.10.005

Zhang C-N, Li X-F, Xu W-N, Jiang G-Z, Lu K-L, Wang L-N, Liu W-B (2013) Combined effects of dietary fructooligosaccharide and *Bacillus licheniformis* on innate immunity, antioxidant capability and disease resistance of triangular bream (*Megalobrama terminalis*). Fish Shellfish Immunol 35(5):1380–1386

Zhigang Z, Ørjan K, Suxu H, Erik OR, Bin Y, Einar R (2013) The effect of dietary chitin on the autochthonous gut bacteria of Atlantic cod (Gadus morhua L.). Aquac Res 44(12):1889–1900. https://doi.org/10.1111/j.1365-2109.2012.03194.x

Zhou L, Zhang J, Yan M, Tang S, Wang X, Qin JG, Chen L, Li E (2020) Inulin alleviates hypersaline-stress induced oxidative stress and dysbiosis of gut microbiota in Nile tilapia (Oreochromis niloticus). Aquaculture 529:735681. https://doi.org/10.1016/j.aquaculture.2020.735681

Chapter 7
Medicinal Plants in Tilapia Aquaculture

Mehwish Faheem, Seyed Hossein Hoseinifar, and Farid Firouzbakhsh

Abstract Tilapia are a group of fish belonging to the family Cichlidae. Because of their mild flavor and tasty flesh, tilapia has become the second most important aquaculture species after carp. Two highly cultured species of tilapia are *Oreochromis mossambicus* and *Oreochromis niloticus.* Tilapia aquaculture is expanding enormously and up to 6.6 million tons of tilapia is expected by 2030 annually. With the intensification of tilapia culture and the environmental manipulation of culture systems, there is an increase in infectious diseases and economic losses due to disease and stress. A lot of consideration has to be given to health management and the proper use of prophylactics. Various methods are used for disease prevention and control in tilapia aquaculture. Chemotherapeutic agents like antibiotics, biotherapeutic agents like probiotics, prebiotics symbiotics, and medicinal plants are often used. The use of medicinal plants is gaining much attention as they are eco-friendly, sustainable, and have the least environmental impact. Medicinal plants improve growth, appetite, and disease resistance and also boost the immune system of aquatic organisms. Powder of different parts of medicinal plants and extracts are used while routes of administration include oral, immersion, and in form of injections. Oral administration as feed additives is the most commonly used method. This chapter will focus on the use of different medicinal plants in tilapia

---

Mehwish Faheem and Seyed Hossein Hoseinifar contributed equally with all other contributors.

M. Faheem (✉)
Department of Zoology, Government College University, Lahore, Pakistan
e-mail: mehwishfaheem@gcu.edu.pk

S. H. Hoseinifar (✉)
Department of Fisheries, Faculty of Fisheries and Environmental Sciences, Gorgan University of Agricultural Sciences and Natural Resources, Gorgan, Iran
e-mail: hoseinifar@gau.ac.ir

F. Firouzbakhsh
Department of Fisheries, Faculty of Animal Science and Fisheries, Sari Agricultural Sciences and Natural Resources University, Sari, Iran

© The Author(s), under exclusive license to Springer Nature Switzerland AG 2023
S. H. Hoseinifar, H. Van Doan (eds.), *Novel Approaches Toward Sustainable Tilapia Aquaculture*, Applied Environmental Science and Engineering for a Sustainable Future, https://doi.org/10.1007/978-3-031-38321-2_7

aquaculture, their routes of administration, and their effects on fish physiology. Moreover, various modes of action of these medicinal plants are also discussed.

Keywords Herbs · Immune parameters · Disease resistance · Antioxidant defense · Bioactive compounds

7.1 Introduction

The human population is increasing at a very high rate and it is expected to reach up to 10 billion by 2050 (Hoseinifar et al. 2020a, b). With the increase in population, the demand for animal protein is also increasing. Considered the fastest-growing food-producing sector in the world, aquaculture is recognized as a possible sustainable solution for food security and increased dietary nutrition in developing countries (Adewole 2014). Fish and shellfish are important animal-based proteins and account for 16.7% of total animal-based protein intake globally. Captured fisheries are declining in past decades and many resources are almost exploited, therefore, the intensification and expansion of aquaculture are required to produce enough animal-based protein to meet increasing demand (Delgado et al. 2003).

Tilapia is the common name used for three genera (*Tilapia, Sarotherodon*, and *Oreochromis*) belonging to the family Cichlidae. Tilapias are important species in aquaculture with growing consumer preference (Ahmad et al. 2009). Aquaculture has observed an increase in tilapia culture in the last few years as these species are distributed worldwide (El-Sayed 2006) and are considered as most important fish species of the 21st century. Currently, more than 135 countries are doing tilapia culture (FAO 2014). Global annual production of tilapia was 4.5 million tons in 2013 and by 2030 it's expected to grow to 6.6. Million tons (Kuebutornye and Abarike 2020). Annual tilapia culture makes up about 12.2% of total finfish culture globally (El-Sayed 2019). Highly cultured species of tilapia belong to the genera *Oreochromis*. Mozambique tilapia (*Oreochromis mossambicus*) and Nile tilapia (*Oreochromis niloticus*) are widely cultured. Around 90% of tilapia culture is of *O. niloticus* (FAO 2016). Tilapia are among a few fish species that can be cultured in almost all types of culture systems viz. pond culture, static pond culture, integrated fish culture, recirculation or closed cycle systems in temperate countries, etc. (Fitzsimmons 2000). Recently the intensive cage culture of tilapia is gaining popularity around the globe.

With the rising global demand, the intensive culture of tilapia is gaining much attraction. However, diet quality, infectious diseases, and resistance to various therapeutics are major constraints in the intensification of tilapia culture (Ahmad et al. 2011; Elkafrawy 2020). Moreover, intensification is stressful to fish and leads to the weakening of the immune system hence leading to infections (Roosta and Hoseinifar 2014). To manage the losses resulting from disease and pathogen infections, farmers are extensively using different chemicals mainly antibiotics

(Hoseinifar et al. 2017a, b, c, d; Nawaz et al. 2018). Antibiotics have been used successfully to treat fish diseases but indiscriminate use of antibiotics has resulted in resistance and accumulation of antibiotic residues in the environment and the fish tissues leading to the emergence of multiple antibiotic-resistant (MAR) strains (Seyfried et al. 2010; Rossolini et al. 2014; Santos and Ramos 2016; Hoseinifar et al. 2017a, b, c, d; Nawaz et al. 2018). Widespread use of vaccines and antibiotics is expensive and impractical in fish farms and a single vaccine works against only one type of pathogen (Harikrishnan et al. 2011; Plant and LaPatra 2011). Therefore, fish farmers are advised to improve fish husbandry and the water quality of ponds and use sustainable dietary supplements to strengthen the immune system of fish (Daniels and Hoseinifar 2014). Organic fish farming or green aquaculture relies mainly on eco-friendly approaches to avoid disease and infection. Feed additives *e.g.* Pro- and pre-biotics and medicinal plants are now being used as an alternative to antibiotics in aquaculture. Among these feed additives, medicinal plants provide the most sustainable and eco-friendly alternative to antibiotics (Abarike et al. 2018; Awad and Awaad 2017) and are used to strengthen the fish immune system (Srichaiyo et al. 2020; Mohammadi et al. 2020a, b).

In the current chapter, a detailed review of existing literature about the use of medicinal plants in tilapia aquaculture is provided along with the possible mode of action. Furthermore, future perspectives and areas to be focused on are highlighted.

7.2 Importance of Medicinal Herbs/Plants in Aquaculture

Medicinal plants include parts of plants (flowers, seeds, leaves, and roots), plant-derived products, spices, herbs, and seaweed (Hai 2015). The use of medicinal and edible plants, vegetable herbs, and seeds dates back in human history (Davidson-Hunt 2000). Traditional medicines and drugs were made by using medicinal plants and they enhanced the utilization of dietary energy and increased performance efficiency. Moreover, medicinal plants act as natural tonics and restoratives (Götti et al. 2014). Herbs and spices were used to improve storage stability and impart flavors to a variety of foods (Ahmad et al. 2011). Medicinal plants/herbs are used in the treatment of different ailments (*e.g.* lowering blood cholesterol and prevention of cancer) by activating the immune system (Desai et al. 2008; Greenwell and Rahman 2015). Antimicrobial substances were used to treat bacterial disease in fish (Van Doan et al. 2020; Ashour et al. 2020; Ceballos-Francisco et al. 2020; Abdel-Tawwab and El-Araby 2021).

Medicinal plants are inexpensive, usually readily available, and have no or minimal effects on fish health and they don't contribute negatively to the environment (Na-Phatthalung et al. 2018). Different parts (leaves, roots, seeds, skin, flowers, and fruit) are used because of their rich nutritional value and therapeutic properties (Akinyemi et al. 2018). Medicinal plants contain a variety of biologically active compounds like carotenoids, flavonoids, terpenoids, polyphenols, tocopherols, glycosides, alkaloids, lectins, saponins, cinnamic acid, polypeptides, quinone, tannins, folic acid, biogenic amines, steroids, phytoandrogens, and phytosterols (Mulat et al. 2020). Medicinal plants also contain many vitamins (riboflavin, thiamine, niacin, and ascorbic acid), and trace minerals (calcium, sodium, zinc, iron, magnesium, and potassium) (Hussain et al. 2011; Ibrahim et al. 2019). All these compounds enhance fish growth, nutrient digestibility, gut health, gut microbiota, digestive enzymes, immunity, and anti-oxidant status (Citarasu 2010; Mohiseni et al. 2017; Astuya et al. 2017; Faheem et al. 2020).

There are almost 17,000 known species of plants and around 8,000 species of plants are being used in traditional medicines in many countries like Pakistan, India, Sri Lanka, China, Japan, and Thailand (Singh 2015; Hai 2015). Medicinal plants are used in traditional medicines as antioxidants (Guardiola et al. 2018; Parham et al. 2020), as antimicrobial, antifungal, antiparasitic, and antiviral agents (Tagboto and Townson 2001; Dhama et al. 2018; Kalsoom et al. 2020; Parham et al. 2020). Medicinal plants also have anti-inflammatory and stress-releasing properties (Reyes-Becerril et al. 2019; Shourbela et al. 2020). The use of medicinal plants in traditional medicines as an appetizing agent, hepatoprotective, gastro-protective, nephroprotective, hematoprotective, and immunostimulatory agent (Bai et al. 2009; Ayaz et al. 2017; Okaiyeto et al. 2018; Asnaashari et al. 2018; Esmail and Ali 2019; Abdel-Tawwab and Hamed 2020; Faheem et al. 2020) also very popular. For example, Cinnamon possesses potent antioxidants, and anti-inflammatory and anti-mutagenic compounds (Ahmad et al. 2011). Parsley is of high nutritive value as the extract can counteract the toxic effects of aflatoxin in fish. Parsley contains a high amount of vitamins A, C, riboflavin, and niacin and is also a rich source of minerals like sodium, calcium, magnesium, iron, and zinc (Peter 2001). Both seeds of *Nigella sativa* and seed oil are used as an immune booster and confer many positive health benefits (Mohammed and Arias 2016; Latif et al. 2020).

7.3 Bioactive Compounds of Medicinal Plants

Medicinal plants have various bioactive compounds, and phytochemicals, which are mainly responsible for their positive health effects. Various types of phytochemicals *e.g.* flavonoids, alkaloids, essential oils, phytosterols, etc. are present in many plants and plant parts. Some anti-nutrient factors are also present in plants along with phytochemicals. These anti-nutrient factors, sometimes decrease the effectiveness of medicinal plants. These anti-nutrient factors are now decreased or eliminated by making extracts or by using enzymes.

7.3.1 Flavonoids

Flavonoids are secondary plant metabolites. Structurally, flavonoids are di-phenylpropanes, a three-carbon ring molecule having two benzene rings attached on both sides of the carbon chain. Flavonoids are present in plants in various forms like flavanols, flavanone, flavones, isoflavones, flavan-3-ol, and anthocyanins. When a different functional group or compound like methyl group, hydroxyl group, sugar, or oxygen attaches to the 3-carbon chain, various classes of flavonoids are formed.

Many plants and herbs contain flavonoids e.g., tea is rich in catechin, epigallocatechin, and epicatechin while onions, olives, grapefruit, barriers, and red wine are rich in quercetin, tamarixetin, myricetin, and kaempferol. Citrus fruits like lemons, grapefruit, and oranges contain a variety of flavanones like hesperidin, taxifolin, naringenin, and naringin. Isoflavones, e.g., daidzin and genistin, are present in soybean. The skin of many fruits is also a source of flavonoids, the skin of tomato and red pepper is loaded with rutin, chrysin, luteolin, apigenin, and luteolin glucosides. Cherry, strawberry, and raspberry contain anthocyanidin and cyanidin. Flavonoids are known for their antioxidant property. However, the antioxidative potential of flavonoids is dependent on the presence and configuration of functional groups. Flavonoids having B-ring configuration have potent radical scavenging potential, they produce stable flavonoid radicals by donating hydrogen and electron to reactive oxygen species like hydroxyl, peroxyl, and peroxynitrite. Flavonoids also have other pharmacological properties like antiviral, antiallergic, antibacterial, anti-inflammatory, and vasodilatory.

7.3.2 Alkaloids

Alkaloids are the organic metabolites of plants having basic properties. Structurally, they are heterocyclic compounds. The pharmacologically active class of alkaloids is isoquinoline. Major and important isoquinoline is sanguinarine which belongs to benzo[c]phenanthridine alkaloid and has immunomodulatory, anti-inflammatory, and antibacterial properties. Sanguinarine also promotes growth by enhancing feed intake and lowering the degradation of amino acids.

7.3.3 Essential Oils

Aromatic plants produce essential oils that are hydrophobic concentrates having a strong odor. Essential oils have antimicrobial, anti-inflammatory, and analgesic properties. The essential oil extracted from oregano, *Origanum heracleoticum*, have thymol, carvacrol, and monoterpenoid phenolic compounds as major phytochemicals. An *in vitro* study with phytochemicals like cymene and carvacrol has revealed that they have bactericidal activities against *Salmonella typhi* (Rattanachaikunsopon and Phumkhachorn 2009). Therefore, it is believed that phenolic compounds present either in plant extracts or essential oil have anti-bacterial properties, that can modulate gut microflora leading to better utilization of nutrients and hence better growth performance in fish.

7.3.4 *Triterpenoids*

Triterpenoids are hydrocarbons having six isoprene units with substitution or displacement of the methyl group with oxygen. They are present in a variety of medicinal plants with antitumor antibacterial, and antiviral potential, role in glucose metabolism and diabetes. Examples of triterpenoids include lupol, glycyrrhizin, botulin, oleanolic and glycyrrhetinic, ursolic, maslinic, and betulinic acid. Triterpenoids belonging to ursane, oleanane, and lupane (betulin, betulinic acid, and lupeol) had anti-cancerous properties because of their anti-inflammatory potential. Rainbow trout fed with varying levels of maslinic acid showed improved growth.

7.3.5 *Phytoandrogens*

Many plants possess bioactive compounds that have similar effects as testosterone, such bioactive compounds are called phytoandrogens as they are produced from plants. Seeds of fenugreek have a bioactive compound called diosgenin which is a steroid-based sapogenin, similarly, soy has an isoflavone compound called daidzein and gutta-percha tree has triterpenoids that have androgenic properties.

7.4 Effect of Medicinal Plants on Fish Physiology and Growth

Fish growth may be defined as the increase in fish weight and length with time. It is a somatic trait and is characterized by an increase in fish weight due to enhanced synthesis of protein. The growth of fish depends mainly on stocking density, water quality parameters, and quality and quantity of food. While the endogenous cues also have an important role in fish growth. Both endogenous and exogenous signals are received and integrated by the central nervous system and regulate fish growth, appetite, and metabolism. In teleost, the primary regulators of fish growth are insulin-like growth factors and growth hormones.

Upon receiving the exogenous signals, the brain produces growth hormone releasing hormone and dopamine that act on the pituitary. Somatotrophs that are located in the anterior portion of the pituitary are responsible for the production of

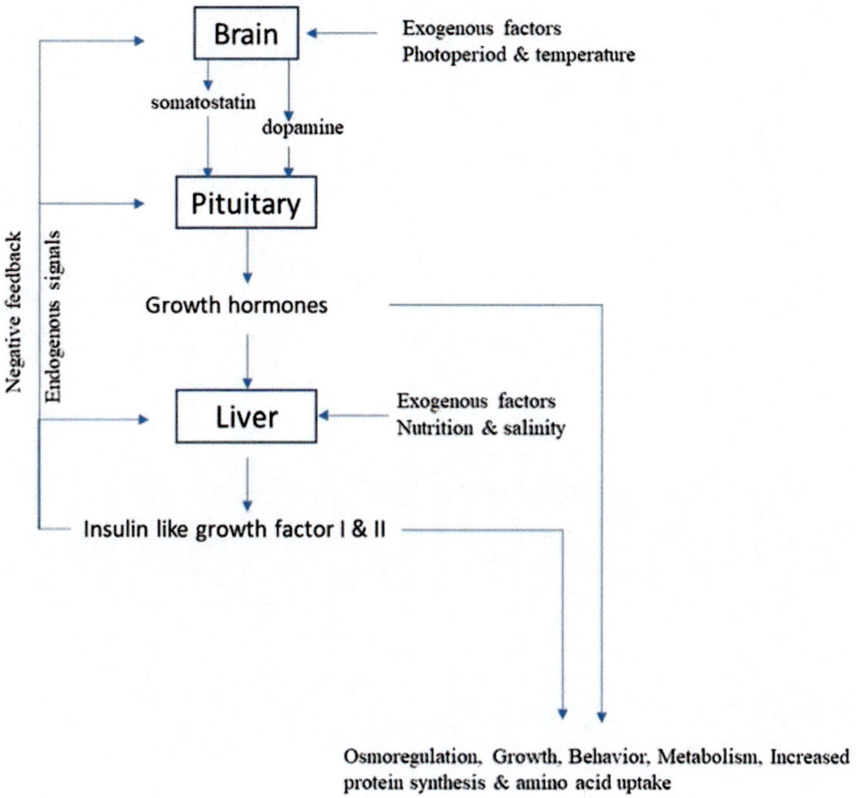

Fig. 7.1 Growth axis in a teleost adapted from

growth hormones (GHs). Once released in the blood circulation, these GHs bind to their receptors, present in various tissues and organs mainly the liver, and activate many metabolic pathways including the production of insulin-like growth factors. Both growth hormones and insulin-like growth factors are responsible for growth, metabolism, osmoregulation, and reproduction. The growth axis in a teleost is elaborated in Fig. 7.1. A balanced feed with an optimum quantity of essential nutrients that enhance fish growth is one of the major challenges in aquaculture. The addition of plants/herbs that can stimulate fish growth is an active area of research and many researchers are working on the addition of plant-based supplements to the diet that enhances the growth of tilapia. Details of such studies are listed in Table 7.1. Medicinal plants can directly act on the pituitary and enhance the synthesis and release GHs or on the liver to stimulate the release of insulin-like growth factors. Supplementing tilapia diets with 0.5% curcumin resulted in enhanced mRNA expression of GHs in the brain and IGFs in the liver, indicating that medicinal plants might directly act on the growth axis. Similarly, when tilapia feed is supplemented with 4 and 6 ppm of D-limonene, it resulted in higher mRNA levels of both GHs and IGF-I. Medicinal plants also stimulate growth by enhancing

Table 7.1 Effect of medicinal plants on growth performance of tilapia

Fish	Plant used	Dietary inclusion	Experimental duration	Observed effects	Optimal dose	Reference
Oreochromis niloticus	*Origanium vulgare* (Oregano) Ethanolic extracts	0.5%, 1%, 1.5%	10 weeks	Growth performance (↑*), Feed consumption (↑*), Total protein (↑*), Serum albumin (↑*), Globulin (↑*)		Ahmad et al. (2009)
Oreochromis niloticus	*Caraway seed meal (Carum carvi)*	0.0, 5, 10, 15, and 20 g/kg	12 weeks	Survival (95–100%), growth performance (↑*), feed utilization (↑*), and whole body composition (+)	10 g/kg	Ahmad and Abdel-Tawwab (2011)
Oreochromis niloticus	Cinnamon	0, 0.5, 1.0, and 1.5%	90 days	Specific growth rate (↑), feed conversion ratio (FCR) (↑), feed efficiency ratio (↑), protein efficiency ratio (↑), apparent protein utilization (APU) (↑), and energy utilization (EU) (↑)	1%	Ahmad et al. (2011)
Oreochromis niloticus	*Ocimum gratissimum* Essential oil	0, 0.5, 1.0, and 1.5%	55 days	Specific growth rate ↓* and feed conversion ratio ↑*	0.5%	Brum et al. (2017)
	Zingiber officinale Essential oil			Specific growth rate ↓* and feed conversion ratio ↑*	1.5%	
Gift tilapia	*Aloe vera*	0.5, 1, 2, and 4 %	8-weeks	Weight gain ↑*, specific growth rate ↑*, feed conversion ratio ↑*	2 & 4%	Gabriel et al. (2015)
Oreochromis mossambicus	*Psidium guajava*	1, 5, and 10 mg/g aqueous and ethanolic extracts	30-days	Final body weight ↑*, specific growth rate ↑* and feed conversion ratio ↓	10 mg/g	Gobi et al. (2016)

Oreochromis niloticus	*Cinnamomum camphora*	2 g/kg	90 days	Weight gain (↑*) and specific growth (↑*), feed conversion ratio (↑*) and condition factor (↑*)	2 g/kg	Kareem et al. (2016)
	Euphorbia hirta			Weight gain (ns) and specific growth (ns), feed conversion ratio (↑*), and condition factor (↑*)		
	Azadirachta indica			Weight gain (ns), specific growth (ns), feed conversion ratio (ns), and condition factor (ns)		
	Carica papaya			Weight gain (↑*), specific growth (↑*), feed conversion ratio (↑*),and condition factor (↑*)		
Oreochromis niloticus	*Astragalus membranaceus*	1500 mg/kg	Astragalus polysaccharides (APS)	Gain (↑*) , specific growth rate (↑*), feed conversion ratio (↑*) and feed intake (↑*)	1500 mg/kg	Zahran et al. (2014)
Hybrid tilapia (O. niloticus × O. aureus) juvenile,	Garlic (*Allium sativum*)	0, 0.5, 1 g/kg	4-weeks	Weight gain ↓* in 0.5% and ns in 1%)	1%	Ndong and Fall (2011)
Oreochromis niloticus	*St John's-wort (Hypericum perforatum)*	Ethanolic extract (0%, 0.2%, and 0.5)	60 days	FBW (ns), WG%,(ns) SGR (ns), and FCR(ns)	0.5%	Mohammadi et al. (2020a, b)
	Lemon balm (Melissa officinalis)			FBW↑* , WG%↑*, SGR↑*, and FCR↑*		
	Oregano (Origanum vulgare)			FBW↑* , WG%↑*, SGR↑*, and FCR↑*		

(continued)

Table 7.1 (continued)

Fish	Plant used	Dietary inclusion	Experimental duration	Observed effects	Optimal dose	Reference
Oreochromis niloticus	*Licorice (Glycyrrhiza glabra*	Root powder (0, 5, 10, and 20 g/kg)	60-days	Weight gain↑*, growth rate↑*, and feed intake↑*	10 g/kg of feed	Abdel-Tawwab and El-Araby (2021)
Oreochromis niloticus	*Assam tea (Camellia sinensis)*	0, 1, 2, 4, and 8 g/kg	4 & 8 weeks	Final body weight ↑*, weight gain ↑*, and specific growth rate ↑*, feed conversion ratio↓	2 g/kg	Van Doan et al. (2019)
Oreochromis niloticus	*Elephantopus scaber*	0, 2.5, 5, 10, and 20 g kg	4 & 8 weeks	Final body weight ↑*, weight gain ↑*, and specific growth rate ↑*	5 g/kg	Van Doan et al. (2019)
Oreochromis niloticus	Thai ginseng *(Boesenbergia rotunda)*	0, 5, 10, 20, 40 g/kg	8 weeks	Final weight ↑*, weight gain ↑*, specific growth rate ↑*, feed intake ↑*, and feed conversion ratio ↑*	10 g/kg	Van Doan et al. (2019)
Oreochromis niloticus	*Berberine*	0, 1, 3, 6, and 9 g/kg	4 & 8 weeks	Final body weight ↑*, weight gain ↑*, and specific growth rate ↑*	1 g/kg	Van Doan et al. (2020)
Oreochromis niloticus	Seaweed (TrueAlgaeMax)	0, 0.5, 1, 1.5, and 2%	70-days	Weight gain↑*, average daily weight gain ↑*, specific growth rate ↑* and achieved the best feed conversion ratio ↑*	2%	Ashour et al. (2020)
Oreochromis niloticus	Watermelon rind powder	0 20, 40, 80, 160 g/kg	8-weeks	Final weight↑*, weight gain ↑*, and specific growth rate ↑*	40 g/kg	Van Doan et al. (2020)
Oreochromis niloticus	White button mushroom	0, 0.5, 1, 2, and 4%	60-days	Final body weight ↑*, weight gain ↑*, specific growth rate↑*, and feed intake↑*, feed conversion ratio ↓	2%	Dawood et al. (2020)

Arrows indicate an increase (↑), a decrease (↓), a significant increase (↑*), and a significant decrease (↓*)

the appetite of fish. The addition of medicinal plants increases the attraction towards food, increased feed acceptance, and palatability.

Intensive aquaculture results in stress on fish which leads to detrimental effects on fish health through improper feed utilization, changes in intestinal functions, and inability to digest consumed feed (Hegazi and Hasanein 2010; Yilmaz 2019). The intestine of fish is the main entrance to the fish body, and various pathogens can alter the local intestinal immunity that results in an increased probability of pathogen infection (Abid et al. 2013; Standen et al. 2016). Many feed additives are used in fish feed, and bioactive components of medicinal plants improve intestinal functions by increasing nutrient digestibility, enzyme secretion, feed utilization, and increasing healthy gut microflora (Van Kooyk and Rabinovich 2008). Natural growth promoters are present in medicinal plants that were found to improve body weight, survival rate, and feed conversion ratio in fish (El-Dakar 2004; Shalaby 2004).

Growth promoting effects of medicinal herbs may be due to the nutrient digestibility and thus leading to enhanced absorbance of nutrients. Many spices stimulate digestion by increasing the secretions of bile and digestive enzymes (Platel et al. 2002). Important compounds present in herbs and spices act as attractants for fish leading to better feeding and improved growth. Growth of *Tilapia zillii* improved after adding olfactory stimulants to feed (Adams et al. 1988).

Organium vulgare extract was found to increase body weight, weight gain%, SGR, and survival rate due to the presence of antioxidants (tocopherols, total polar compounds, fatty acids, and oxidative stability) and antimicrobial compounds (Ahmad et al. 2009; de Souza Pereira et al. 2017). Diets containing feed additives (0.5% cinnamon and 1% marjoram) showed better growth performance when fed to tilapia (Abdel Wahab et al. 2007; Abdel El-Maksoud et al. 1999). 1% supplementation of cinnamon was reported to act as a natural growth promoter in the diets of tilapia. The presence of cinnamon in the diet enhances the sensory stimulation of fish which may result in improved appetite and thus lead to improved growth (Ahmad et al. 2011). Similarly, dietary supplementation of clove and basil at 0.5% improved the feed conversion ratio in tilapia (Brum et al. 2017). An increase in body weight gain was observed after feeding Nile tilapia with *A. vera* (Gabriel et al. 2015). Dietary inclusion of licorice root powder, aqueous, and methanolic extracts of licorice root powder improve growth performance and feed utilization in Nile Tilapia (Abdel-Tawwab and El-Araby 2021).

Many active compounds like phenols, flavonoids and isoflavonoids, saponins, and coumarins present in medicinal plants and herbs improve nutrient digestibility leading to better growth and weight gain in fish (Poolsawat et al. 2020; Bostami et al. 2021).

7.5 Medicinal Plants to Improve Pathogen Resistance in Tilapia

Mortalities due to pathogen infection are huge in tilapia culture and lead to substantial losses. There is a dire need to look for sustainable solutions and prophylactics to manage pathogen resistance and damage caused by pathogens. Many bacterial, fungal, and viral pathogens are responsible for disease outbreaks in tilapia culture. Important pathogens of tilapia are *Branchiomyces*, *Dactylogyrus*, *Saprolegnia*, *Trichodina*, *Dactylogyrus*, *Iridovirus*, *Francisella*, *Nocardia*, *Flavobacterium*, and *Edwarsiella* (Absali and Mohamad 2010; MSD Animal Health 2012). Major bacterial pathogens of tilapia culture belong to the genera *Streptococcus* especially *Streptococcus iniae and Streptococcus agalactoe.* Almost 82% of pathogens isolated from farmed tilapia belonged to *Streptococcus agalactoe* (MSD Animal Health 2012; Li et al. 2014). Other economically important bacterial pathogens of tilapia are *Aeromonas hydrophila, Francisella* spp., and *Edwardsiella* spp.

7.5.1 Streptococcus agalactiae

Streptococcus agalactiae is a gram-positive bacteria of cultured fish both in freshwater and marine water (Kümmerer 2009). Infections with *Streptococcus agalactiae* occurs both in juvenile and adults with a prevalence rate of 40-60% (Pretto-Giordano et al. 2010). *Streptococcus agalactiae* cause infection in tilapia, rainbow trout, channel catfish, and silver sea bream in tropical and temperate climates (Ye et al. 2011). Once infected with bacteria, fish show erratic swimming, injuries in the skin, loss of appetite, swelling in the abdominal region, paleness of the liver, loss of coordination in movement, and softening of the head region (Figueiredo et al. 2006; Pretto-Giordano et al. 2010; Chen et al. 2012). *S. agalactiae* can survive for long periods in pond water and mud; however, water is the primary route of transmission for *S. agalactiae*. Around 90-100% mortalities were recorded in tilapia after infection with *S. agalactiae*. Virulence of this bacteria in tilapia depends upon CFU and strain. The 72-h lethal dose (LD_{50}) of *S. agalactiae* in tilapia is 1.12×10^6 CFU/ml (Bondad-Reantaso et al. 2005; Mian et al. 2009) while the concentration ranging from 1×10^1 to 1×10^5 CFU caused 80-100% mortalities in cultured Nile Tilapia (Hernández et al. 2009).

7.5.2 Aeromonas hydrophila

Genus *Aeromonas* is gram-negative bacteria that belongs to the family Aeromonadaceae which has four species *i.e. Aeromonas hydrophila*, *Aeromonas. salmonicida*, *Aeromonas sobria*, and *Aeromonas piscicola*. These are facultative pathogens and cause dermal red plague or hemorrhagic septicemia syndrome. These pathogens were isolated from tilapia, goldfish, channel catfish, rainbow trout, and salmon (Molnár and Csaba 2005; Rattanachaikunsopon and Phumkhachorn 2009; Austin and Austin 2012). Members of the genus *Aeromonas* are naturally present in many habitats including freshwater, the degree of infection caused by the pathogens depends greatly on environmental conditions, dissolved oxygen, ammonia levels, and species of fish (Beaz-Hidalgo and Figueras 2012). Among four species of the genus *Aeromonas*, *Aeromonas hydrophila* is the most important with a prevalence of up to 80% in freshwater (Beaz-Hidalgo et al. 2012; Soto-Rodríguez et al. 2013) it is present on the skin of fish and in the digestive system (Harikrishnand and Balasundaram 2005). Loss of appetite, weakness, hemorrhages in gills and skin, and necrosis of vital organs are major signs of *Aeromonas* infection (Bastardo et al. 2012). Other symptoms include loss of scales, accumulation of fluid in the body, and loss of vision (Yardimci and Aydin 2011).

7.5.3 **Francisella** ***spp.***

Members of the genus *Francisella* are gram-negative and facultative aerobic bacteria and have two species i.e. *Francisella noatunensis orientalis* occurs in tropical zones and *Francisella noatunensis noatunensis* (Colquhoun and Duodu 2011). Both species are distributed globally with a prevalence of up to 90% in freshwater aquaculture systems (Ottem et al. 2007; Jeffery et al. 2010). Francisella is transmitted horizontally in the freshwater culture system and it has been isolated from Nile tilapia, white and striped bass, salmon, and Atlantic cod (Ostland et al. 2006; Ottem et al. 2007; Kamaishi et al. 2010). Fish infected with *Francisella* spp. shows loss of appetite, erratic swimming pattern, hemorrhage around the pectoral fin, and anemia (Soto et al. 2009). Severe infection results in gill hyperplasia, enlargement of the kidney and spleen, and necrosis of vital organs (Mauel et al. 2007; Ottem et al. 2009).

7.5.4 **Edwardsiella** ***spp.***

Genus *Edwardsiella* is a group of gram-negative pathogens and has four species *i.e. Edwardsiella tarda*, *Edwardsiella Ictaluri*, *Edwardsiella hoshinae*, and *Edwardsiella anguillimortifera.*

Edwardsiella ictaluri is commonly present but *Edwardsiella tarda* is an economically important species that also cause zoonosis (Bastardo et al. 2012). *E. tarda* has been isolated from channel catfish, Nile tilapia, hybrid tilapia, turbot, red seabream, and Japanese eel (Burr et al. 2012). Infection with *E. trada* depends mainly on temperature, the presence of organic matter in the pond, and water quality parameters. General symptoms of infection include bulging of eyes, spiral swimming, swelling of the abdominal region, and pale eyes.

The use of vaccines and antibiotics to control pathogen infection is common practice among tilapia fish farmers around the globe. Uncontrolled use of antibiotics resulted in the emergence of resistance, therefore plants and plants-derived chemicals are now being used in tilapia culture to avoid the spread of disease (Table 7.2).

Tilapia fed with different medicinal plants showed improved growth and resistance to pathogens (Table 7.2). Tilapia fed with Echinacea showed improved survival rates against *P. fluorescens* infections (Aly et al. 2008). Similarly, the addition of Cinnamon (0.5%) to the diets of tilapia resulted in improved gut microbial content and feed absorption. Moreover, guts of tipalia fed with cinnamon containing diet had a smaller number of harmful bacteria like *Campylobacter jejuni*, *Bacillus* sp., *Enterobacter* sp., *Escherichia coli*, *Listeria monocytogenes*, *Klebsiella pneumoniae*, *Staphylococcus aureus*, *etc.* (Abdel Wahab et al. 2007). Similarly, the addition of *Cinnamomum camphora* to the diet increased the survival rate in tilapia when exposed to *Streptococcus agalactiae* (Kareem et al. 2016). Several medicinal plants *e.g. Lonicera japonica* (Japanese honeysuckle), *Nyctanthes arbortristis* (night jasmine), *Panax quinquefolium* (American ginseng), *Tinospora cordifolia, Glycyrrhiza glabra* (Licorice) protect tilapia against A. hydrophila challenge (Sudhakaran et al. 2006; Ardo et al. 2008; Kirubakaran et al. 2010; Abdel-Tawwab and El-Araby 2021). Supplementing feed of tilapia with leaf extract and dried leaves of rosemary increases survival rate when challenged with *Streptococcus iniae* (Abutbul et al. 2004). Tilapia showed increased resistance against *Streptococcus agalactiae* when fed with varying levels of *Sophora flavescens* (Wu et al. 2013).

In addition to protection against bacterial pathogens, medicinal plants, and herbs also protect tilapia against parasitic infections. Indian almond (*Terminalia catappa*), garlic (*Allium sativum*), and extracts of green tea protect tilapia against Trichodina sp. infections (Chitmanat et al. 2005; El-Deen 2010).

Many bioactive compounds found in plants have medicinal properties. Active compounds like Thymol and Carvacrol have lipophilic properties and can cause the breakdown of the bacterial cell wall (Lambert et al. 2001; Ultee et al. 1998; Conner 1993; Helander et al. 1998). Other compounds like ρ-cymene present in Oregano cause swelling of the bacterial cell membranes and hence destroy the bacterial cell (Ultee et al. 2002). The addition of herbs increases mucus lysozyme activity which can break down peptidoglycans of the bacterial cell wall (Magnadottir 2010). It is believed that the addition of medicinal plants to the diet of tilapia improved immune response and hence provided better resistance to pathogens.

Table 7.2 Role of different medicinal plants to improve pathogen resistance in tilapia

Fish	Plant used	Inclusion type	Inclusion ratio	pathogen	Route of pathogen administration	Resistance	Optimal dose	Reference
Oreochromis mossambicus	*Tinospora cordifolia*	Ethanolic extract	0.8, 8, 80 mg/kg	*Aeromonas hydrophila*	Intraperitonial 10^9 cells/fish	Decreased mortality	80 mg/kg ether and ethanolic extract	Sudhakaran et al. (2006)
Oreochromis mossambicus	*Solanum trilobatum*	Water and hexane soluble fractions	0, 4, 40, 400 mg/kg	*Aeromonas hydrophila*	Intraperitonial 1×10^9 cells/fish	80% mortality in the control 23% in 400 mg water extract	4 mg/kg hexane soluble fraction & 400 mg/kg water-soluble fraction	Divyagnaneswari et al. (2007)
Oreochromis mossambicus	*Eclipta alba*	Leaf extracts	0, 0.01, 0.1, or 1%	*Aeromonas hydrophila*	Intraperitonial 1×10^8 cells/fish	Decreased mortality	1% aqueous extract-supplemented feed	Christybapita et al. (2007)
Oreochromis mossambicus	*Toona sinensis*	Hot water extract	4 and 8 μg/g injection.	*Aeromonas hydrophila*	Intraperitonial 5×10^7 cells/fish	Decreased mortality	8 μg/g	
Oreochromis niloticus	Cinnamon *C. zeylanicum*	Powder	0., 0.5, 1.0, and 1.5%	*Aeromonas hydrophila*	Intraperitoneal 5×10^5 CFU	Decreased mortality	1%	Ahmad et al. (2011)
Oreochromis niloticus	*Viscum album coloratum*	Aqueous leaf extract	0, 10, 50 g, and 200 mg /kg	*Aeromonas hydrophila*	Intraperitoneal 2×10^5 cells/fish	Decreased mortality	50 mg/kg	Park and Choi (2012)
Oreochromis niloticus	The herbal mixture of *Astragalus*, *angelica*, hawthorn,	Powder	0.0, 0.5, 1.0, 1.5, and 2.0%	*Aeromonas hydrophila*	Intraperitoneal 10^8 cells/ml	Decreased mortality	2%	Tang et al. (2014)

(continued)

Table 7.2 (continued)

Fish	Plant used	Inclusion type	Inclusion ratio	pathogen	Route of pathogen administration	Resistance	Optimal dose	Reference
	Licorice root and honeysuckle							
Gift tilapia	*Aloe vera*	Powder	0.5, 1, 2, and 4 %	*Streptococcus iniae*	Intraperitonial 7.7×10^6 cells/mL	Decreased mortality	0.5%	Gabriel et al. (2015)
Oreochromis mossambicus	*Citrus limon*	Essential oils of the peel	0, 0.5%, 0.75% and 1%	*Edwardsiella tarda*	Intraperitonial 1.2×10^8 cells/mL	Decreased mortality	0.5%	Baba et al. (2016)
Oreochromis niloticus	*Cinnamomum camphora*	Ethanolic extract	2 g/kg	*Streptococcus agalactiae*	Intramuscular 1×10^6 cells/mL	Decreased mortality	2 g/kg Cinnamomum camphora	Kareem et al. (2016)
	Euphorbia hirta							
	Azadirachta indica							
	Carica papaya							
Oreochromis mossambicus	*Psidium guajava*	Aqueous and ethanolic leaf extracts	1, 5, and 10 mg/g	*Aeromonas hydrophila*	Intraperitonial 10^7 cells/ml	Decreased mortality	10 mg/g ethanolic extracts	Gobi et al. (2016)
Oreochromis mossambicus	*Nyctanthes arbortristis*	Methanolic extracts of seeds	2, 20 or 200 mg/kg (Intraperitonal)	*Aeromonas hydrophila*	Intraperitoneal 10^8 cells/mL	Decreased mortality	20 mg/kg	Kirubakaran et al. (2010)
Oreochromis niloticus	*Zingiber officinale*	Essential oil	0, 0.5, 1, 1.5%	*Streptococcus agalactiae*	Gavage 10^8 CFU g/fish	Decreased mortality	0.5% ginger	Brum et al. (2017)
	Ocimum gratissimum							
Oreochromis mossambicus	*Moringa oleifera*	Leaf powder	0, 3, 6, 9, and 12%,	*Aeromonas hydrophila*	0, 1×10^6, 1.5×10^6, 3×10^6 and 4×10^6 cells. mL			Mbokane and Moyo (2018)

Oreochromis niloticus	Assam tea (*Camellia sinensis*)	Ethanolic extracts	0, 1, 2, 4, and 8 g/kg	*Streptococcus agalactiae*	Intraperitonial 10^7 CFU/ml	Decreased Mortality	2 g/kg	Van Doan et al. (2019)
Oreochromis niloticus	*Elephantopus scaber*	ethanolic extracts	0, 2.5, 5, 10, and 20 g/kg	*Streptococcus agalactiae*	Intraperitonial 10^7 CFU/ml	Decreased Mortality	5 g/kg	Van Doan et al. (2019)
Oreochromis niloticus	Thai ginseng (*Boesenbergia rotunda*)	Powder	0, 5, 10, 20, 40 g/kg	*Streptococcus agalactiae*	10^8 CFU/ml	Decreased Mortality	10 g/kg	Van Doan et al. (2019)
Hybrid tilapia *Oreochromis niloticus* × *O. mossambicus*	*Psidium guajava*	Leaf powder	0, 1.5, and 3%	*Vibrio harveyi*	10^4 cells/mL	Decreased Mortality	1.5%	Ceballos-Francisco et al. (2020)
Oreochromis niloticus	Berberine	Powder	0, 1, 3, 6, and 9 g/kg	*Streptococcus agalactiae*	Intraperitonial 10^7 CFU mL^{-1}	Decreased mortality	1 g/kg	Van Doan et al. (2020)
Oreochromis niloticus	Seaweed (TrueAlgaeMax)	Aqueous extracts	0, 0.5, 1, 1.5, and 2%	*Aeromonas hydrophila*	Intraperitonial 1 × 108 CFU/mL	Decreased mortality	2%	Ashour et al. (2020)
Oreochromis niloticus	Watermelon rind powder	Powder	0 20, 40, 80, 160 g/kg	*Streptococcus agalactiae*	10^7 CFU/mL	Decreased mortality	40 g/kg	Van Doan et al. (2020)
Oreochromis niloticus	Licorice (*Glycyrrhiza glabra*)	Root powder (0, 5, 10, and 20 g/kg)	60-days	*Aeromonas hydrophila*	Intraperitoneal With LD_{50} was 1×10^6 cells/ml	No mortality	10–20 g/kg diet	Abdel-Tawwab and El-Araby (2021)

7.6 Medicinal Plants and Improved Immune System: Possible Mechanism of Action

The immune system is a complex physiological, chemical, and metabolic process that protects the organism against foreign materials such as bacteria, viruses, protozoa, fungi, *etc.*. The immune system is usually divided into two categories, innate (non-specific or natural) immune system, and adaptive (specific or acquired) immune system.

7.6.1 Innate Immune System

The innate immune system is present in almost 98% of all multicellular organisms and from a phylogenetic point of view, and it is the oldest immune system. The innate immune system act as the first line of defense against infection and humoral and cellular components of this system is present even before exposure to foreign material. As it is non-specific, the reaction mounted against a foreign material is always the same.

Components of the innate immune system include

i. Chemical and physical barriers *e.g.* skin and mucus
ii. Cellular and humoral components *e.g.* blood proteins (complement & acute phase, cytokines, natural killer cells, eosinophils, and phagocytic cells (macrophages and neutrophils))

The innate immune system of fish is highly robust (Rombout et al. 2005). Toll-like receptors (TLR) and complement components are the first lines of defense. Toll-like receptors are the principal inducer of innate immune response and complement components of the innate adaptive immune response.

Scales, skin, and mucus along with gills and the gastrointestinal tract provide physical protection against pathogens (Alexander and Ingram 1992). The innate immune response is initiated when there is tissue damage, or any pathogen is sensed by leukocytes. Inflammation is the first response of the innate immune system to evade pathogen replication and initiate the process of cellular repair. Many leukocytes i.e. Neutrophils, macrophages, monocytes, etc. are recruited to the site of inflammation. Each type has a different function but they work in precise coordination to subside the effects of pathogens.

The innate immune system of teleosts works through precise mechanisms to destroy pathogens. These mechanisms include degranulation, diapedesis, phagocytosis, oxidative burst, and opsonic and hemolytic serum activities.

Upon detection of the foreign substance, a large amount of anti-microbial molecules and inflammatory mediators are released from neutrophils and mast cells. This process is known as degranulation. These chemicals induce diapedesis or chemotaxis i.e. movement of leukocytes towards the site of infection so the pathogen and foreign molecule can be eliminated. Upon reaching the site of infection, leukocytes start the process of phagocytosis i.e. engulfing the foreign particles or pathogens. Pattern recognition receptors (PRRs) present on immune cells detect the conserved pathogen-associated molecular patterns (PAMPs) present on microbes and foreign bodies and make a PRR-PAMP complex which is engulfed by lysosomes.

Production and release of reactive oxygen species (ROS) and reactive nitrogen species (RNS) from leukocytes cause respiratory burst activity (Swindle et al. 2002). These ROS/RNS takes part in degranulation and diapedesis and also help in the removal of phagolysosomes. Along with these mechanisms, the innate immune response also modifies the foreign bodies/microbes/virus to increase their interaction with phagocytes and cytotoxic cells. This process is called Opsonization. The complement component of innate immune response can neutralize viruses by enveloping them (Whyte 2007).

In fish, cells of innate immune response include neutrophils, macrophages/monocytes, mast cells, non-specific cytotoxic cells, and other nuclear leukocytes.

Neutrophils are round with large granular vesicles. They are mainly present in hematopoietic tissues and the peritoneal exudate. Neutrophils are also present in the blood, gills, gonads, submucosa of the intestine, and skin (Mulero et al. 2007). In teleost, the primary function of neutrophils is phagocytosis (Sepulcre et al. 2002). Neutrophils also release reactive oxygen species, produce cytokines, and express a broad range of toll-like receptors (Sepulcre et al. 2002).

Macrophages/monocytes are horseshoe-shaped or round cells mainly present in the head kidney, spleen, thymus, gills, connective tissue, and peritoneal exudate (Lieschke et al. 2001). Macrophages sometimes form melanomacrophage centers, a dark-colored cluster in hematopoietic tissue (Herraez and Zapata 1986). The main function of macrophages is phagocytosis along with neutrophils, they also release ROS and various cytokines in fish (Uribe et al. 2011).

Mast cells, also known as eosinophilic granular cells (EGCs) in fish are very large, round-granulated cells. They are mainly present in the head kidney, spleen, epidermis, intestinal epithelial, and gastrointestinal-associated hematopoietic tissue (Da'as Uribe et al. 2011; Prykhozhij and Berman 2014). In fish, mast cells produce anti-microbial peptides that induce activation and chemotaxis of neutrophils (Prykhozhij and Berman 2014; Crivellato et al. 2015).

Non-specific cytotoxic cells are small agranulocytes and are present in the kidney, spleen, thymus, blood, and body fluid (Rubio-Godoy 2010; Uribe et al. 2011). They have cytotoxic activity against viruses, and protozoa and also produce a variety of cytokines (Utke et al. 2008).

Humoral components of innate immune response include lysozymes, antimicrobial peptides, protease, complement components, eicosanoids, reactive oxygen and nitrogen species (ROS/RNS), and cytokines.

Cytokines are regulatory molecules of the innate immune system and include tumor necrosis factor α (TNFα), interferons (IFN), interleukin (IL), and chemokines.

Tumor necrosis factor α (TNFα) is produced by leukocytes, it activates macrophages that leads to increased respiratory burst activity and phagocytosis (Whyte

2007; Uribe et al. 2011). It also induces the expression of cyclooxygenase 2 and leukocytes (IL-1β, IL-8) (Zou et al. 2003). Tumor necrosis factor-alpha (TNF-α) is capable of mounting a powerful antimicrobial response. It activates many host-responsive genes, induces apoptosis, and inhibits intracellular replication of the pathogen (Reyes-Cerpa et al. 2012). It also mediates the NF-κB signaling pathway (Rahman and McFadden 2006). Interleukin-8 is a chemokine that is produced by macrophages in response to pathogens and other cytokines like TNF-α.

Interferons (IFN-α and IFN-β) inhibit the viral replication within infected cells and also modify the expression of various immune genes thus providing protection against viral infection in infected and non-infected cells (de Veer et al. 2001; Samuel 2001; Uribe et al. 2011). Another important type of interferon is IFN-γ which is produced by T-cells and non-specific cytotoxic cells. IFN-γ in fish serves the same function as IFN-γ in mammals (Zou and Secombes 2016).

Interleukins include important pro-inflammatory and inflammatory cytokines. Interleukins are classified into 6 sub-categories (IL-1, IL-2, IL-6, IL-10, IL-12, IL-17) (Savan and Sakai 2006). IL-1 includes pro-inflammatory cytokines IL-1β and IL-18. Inteleukin-1β is expressed in macrophages and neutrophils and elicits a humoral response (Savan and Sakai 2006). Interleukin-18 is present in many different cells like T and B cells, and macrophages and induces secretion of interferon-γ (Whyte 2007). IL-2 derives lymphocyte differentiation; production of IFN-γ is also modulated by IL-2 (Díaz-Rosales et al. 2009; Corripio-Miyar et al. 2007). Il-6 and IL-12 are important inflammatory cytokines (Zante et al. 2015) while IL-10 is an anti-inflammatory cytokine that limits the production of pro-inflammatory cytokines (Karan et al. 2016).

Chemokines are a superfamily of cytokines that include CXC, CC, C, and CX3C classes. They are produced by many different cells and their major function is to attract and engage cells at the site of infection and convert innate immune response to adaptive response.

7.6.2 *Adaptive Immune Response*

The adaptive or specific immune response is specific for each pathogen and has a memory so upon successive encounters with the same pathogen; a more pronounced response is mounted (Male and Roitt 2001).

The main components of the adaptive immune response are

i. The cellular component includes lymphocytes (B and T-cells)
ii. The humeral component includes immunoglobulins like IgM, IgD, and IgT/Z.

T-cells are mainly produced in the thymus, spleen, and head kidney (Nakanishi et al. 2015). T-cells have also been detected in blood, gut, and skin (Rombout et al. 2005). In line mammals, T-cells are divided into T-cytotoxic cells and T-helper cells. Cytotoxic T-cells have the same cell-mediated cytotoxic function as their mammalian counterpart (Fisher et al. 2006).

In fish, B-cells are mainly produced by the head kidney and thymus, and spleen (Crowhurst et al. 2002; Salinas et al. 2011) and propagate in the intestine and peritoneal cavity (Rombout et al. 2005; Parra et al. 2015). B-cells produce immunoglobulins which precipitate and agglutinate the antigens for better phagocytosis. Antibodies also activate complement components (Rubio-Godoy 2010).

Antibodies are present in blood, skin, gills, intestine, and bile (Cain et al. 2000; Hatten et al. 2001). Different types of immunoglobulins are present in fish e.g. IgM, IgD, IgT/Z. IgM is the most common and predominant antibiotic found in fish (Rubio-Godoy 2010).

The immune system of organisms is affected by various endogenous and exogenous factors that lead to either stimulation or suppression of the immune system (Trichet 2010; Nawaz et al. 2018). The immune system of fish comprises an innate and adaptive immune system. Goblet cells present in fish skin secrete mucus (Sanahuja et al. 2019). Mucus is the first line of defense and provides a biological, physical, and immunological barrier against pathogens and other exogenous chemicals (Esteban 2012; Cordero et al. 2016). Many enzymes like proteases and antiproteases, lysozyme, peroxidases, phosphatases, and esterases made major components of fish skin mucus. Many immunoglobulins, complement components, and antimicrobial peptides are part of fish skin mucus (Adel et al. 2018; Mohammadi et al. 2020a, b).

In fish, major organs that possess hematopoietic activities are the kidney (head kidney), thymus, and spleen (Reyes-Cerpa et al. 2012). The major cells that mount the innate immune response by secreting cytokines are neutrophils, mast cells, macrophages, and non-specific killer cells (Awad and Awaad 2017). Some important pro-inflammatory cytokines secreted by the cells of the innate immune response are tumor necrosis factor-alpha (TNF-α) and interleukin-8 (IL-8). Interleukin-8 is responsible for attracting lymphocytes and neutrophils and also induces a powerful respiratory burst activity (Laing et al. 2002). Interferon-γ (IFN-γ) is produced by T-lymphocytes and natural killer cells and is responsible for inducing synthesis of nitrous oxide, enhanced respiratory burst activity, and enhanced macrophage-mediated phagocytosis of the pathogen (Arts et al. 2010; Grayfer et al. 2010).

Herbs and medicinal plants can elicit the fish immune system by activating cells of the innate and adaptive immune system, phagocytosis, and complement system which result in protection against pathogens (Watanuki et al. 2006; Hoseinifar et al. 2020a, b). Different medicinal plants have been used in tilapia culture to improve the immune response (Table 7.3). Herbs interact with lymphoid organs associated with the skin (SALT), gills (GIALT), and gut (GALT) to enhance mucus immunity (Vallejos-Vidal et al. 2016; Martin and Krol 2017). The immunostimulatory properties of medicinal plants can also be attributed to various bioactive compounds. Polyphenols such as catechins can modulate the immune system by interacting with immune cells and regulating the secretions of chemokines and proinflammatory cytokines. Moreover, medicinal plants stimulate the immune system through respiratory burst activity. Hydrogen peroxide & superoxide anions produced as a result of respiratory burst activity provide a powerful antimicrobial activity; moreover, these

Table 7.3 Role of medicinal plants as immune-modulators in tilapia aquaculture

Fish	Plant used	Feeding rate	Duration	Endpoints studied	Observed effects	Optimal dose	Reference
Oreochromis mossambicus	*Tinospora cordifolia* (Ethanolic extract & petroleum ether extracts)	0.8, 8, or 80 mg/kg body weigh	2-days post-immunization	Serum immunity	Neutrophil activity (↑*), antibody production (↑*)	All doses of ethanolic extract. 0.8 mg/kg ether extract	Sudhakaran et al. (2006)
Oreochromis mossambicus	*Eclipta alba*	0, 0.01, 0.1, or 1%	1, 2 and 3 weeks	Cellular and humoral immunity	Lysozyme↑*, antiprotease↑* and complement ↑, myeloperoxidase content↑*, production of reactive oxygen and nitrogen species↑*	1%	Christybapita et al. (2007)
Oreochromis mossambicus	Solanum trilobatum	0, 4, 40, or 400 mg kg1 body weight of water or hexane soluble fraction		Nonspecific immune	Lysozyme activity (↑*), reactive oxygen species (↑*), reactive nitrogen species (↑*),	40 mg/kg water-soluble fraction 4 & 400hexanesoluble fraction	Divyagnaneswari et al. (2007)
Oreochromis mossambicus	*Toona sinensis* Hot water extract	4 and 8 μg/g injection.	2-weeks	Serum immunity	Respiratory burst activity ↑*, phagocytic activity ↑*, and lysozyme activity ↑*, total immunoglobulins (ns)	8 μg/g	
Oreochromis niloticus	*Cinnamon C. zeylanicum*	0.5%, 1.0%, or 1.5%		Blood plasma profile	Hemoglobin (HB)↑ Red blood cells (RBCs) ↑ Hematocrit (PCV) ↑ Total protein ↑ Total lipids ↑ creatinine ↓, urea ↓, aspartate	1% cinnamon	Ahmad et al. (2011)

					aminotransferase (AST) ↓, alanine aminotransferase (ALT) ↓, and glucose↓		
Hybrid tilapia (*O. niloticus* × *O. aureus*)	Garlic (*Allium sativum*)	0, 0.5, 1 g/kg	4-weeks	Cellular and humoral response	Total leucocyte count (↑*), Respiratory burst activity (↑*). Phagocytic activity (↑*), phagocytic index (↑*), Lysozyme activity (↑*)	0.5% garlic	Ndong and Fall (2011)
Oreochromis niloticus	*Viscum album coloratum*	0, 10, 50, and 200 mg /kg	80-days	Serum immunity	Respiratory burst activity ↑*, lysozyme activity ↑*, alternative complement haemolysis activity ↑* phagocytic Activity ↑*	50 mg/kg	Park and Choi (2012)
Oreochromis niloticus	Astragalus membranaceus	Astragalus polysaccharides (APS)	0.15%	Immunological parameters	Phagocytic activity (↑), the respiratory burst activity (↑), plasma lysozyme (↑), bactericidal activity (↑), superoxide dismutase (↑), glutathione peroxidase (↑), and amylase activity (↑), serum nitric oxide (ns) or Malondialdehyde (ns)		Zahran et al. (2014)
Oreochromis niloticus	An herbal mixture of *Astragalus, angelica*, hawthorn, Licorice	0.0, 0.5, 1.0, 1.5, and 2.0%	4-weeks	Serum immunity and Immune-related gene expression	Serum lysozyme, superoxide dismutase TNFα ↑*, IL-1β ↑*	2%	Tang et al. (2014)

(continued)

Table 7.3 (continued)

Fish	Plant used	Feeding rate	Duration	Endpoints studied	Observed effects	Optimal dose	Reference
	root and honeysuckle						
Gift tilapia	*Aloe vera*	0.5, 1, 2, and 4 %	8-weeks	Hematological parameters	Red blood cells ↑*, hematocrits ↑*, hemoglobin ↑*, white blood cells ↑*, neutrophils ↑*, monocytes ↑*, eosinophils↑* and serum total protein↑*	2 & 4%	Gabriel et al. (2015)
Oreochromis mossambicus	*Citrus limon* Essential oils of the peel	0, 0.5, 0.75, and 1%	60-days	Immunological & biochemical parameters	Total white blood cell ↑*, serum lysozyme activity↑* and myeloperoxidase activity ↑*, total proteins↑* , glucose ↓, cholesterol ↓ and triglycerides ↓	0.75%	Baba et al. (2016)
Oreochromis mossambicus	*Psidium guajava*	1, 5, and 10 mg/g aqueous and ethanolic extracts	30-days	Serum immunity	Serum myeloperoxidase activity ↑*, lysozyme activity ↑*, serum antiprotease activity ↑*, alternate complement haemolytic activity ↑*	10 mg/g ethanolic extracts	Gobi et al. (2016)
				Serum antioxidants	Superoxide dismutase ↑, Catalase ↑*, Glutathione peroxidase ↑*		
Oreochromis mossambicus	*Nyctanthes arbortristis*	2, 20, or 200 mg/kg injection of seed's	Not specified	Serum immune parameters	Lysozyme activity↑*, Myeloperoxidase activity ↑*, antiprotease activity ↑*	2 & 20 mg/kg	Kirubakaran et al. (2010)

		methanolic extracts					
Oreochromis niloticus	*Cinnamomum camphora*	2 g/kg	90-days	Biochemical parameters	Plasma protein (ns), creatine, albumin (ns) and AST (ns), ALT (↑*), plasma protein (ns), creatinine (ns), albumin (ns)	2 g/kg	Kareem et al. (2016)
	Euphorbia hirta				Plasma protein (ns), creatine, albumin (ns) and AST (ns), plasma protein (ns), creatinine (ns), albumin (ns)		
	Azadirachta indica				Plasma protein (ns), creatine, albumin (ns), AST (ns), ALT (↓*), plasma protein (ns), creatinine (ns), albumin (ns)		
	Carica papaya				Plasma protein (ns), creatine, albumin (ns) and AST (ns), plasma protein (ns), creatinine (ns), albumin (ns)		
Oreochromis niloticus	*Zingiber officinale Essential oil*	0, 0.5, 1, 1.5%	55-days	Hematological parameters	Ethtrocytes (ns, ↓, Hematocrit ↓, Hemoglobin ↓, Neutrophils, Monocytes	0.5 % ginger essential oil	Brum et al. (2017)
	Ocimum gratissimum Essential oil			Hematological parameters	Erythrocytes ↓, Hematocrit ↓, Total leukocytes ↑, Thrombocytes ↑,		

(continued)

Table 7.3 (continued)

Fish	Plant used	Feeding rate	Duration	Endpoints studied	Observed effects	Optimal dose	Reference
					Neutrophils ↑, Monocytes ↓		
Oreochromis mossambicus	*Moringa oleifera*	0, 3, 6, 9, and 12%,	45-days	Hematological parameters	WBC↑*, RBC ↑*, HGB ↑* and HCT ↑*		Mbokane and Moyo (2018)
Oreochromis niloticus	St John's-wort (*Hypericum perforatum*)	Ethanolic extract (0%, 0.2 %, and 0.5)	60 days	Digestive enzymes, hepatic antioxidants	Amylase (ns), Lipase (ns), Protease (ns)	0.5%	Mohammadi et al. (2020a, b)
	Lemon balm (*Melissa officinalis*)				Amylase↑*, Lipase ↑*, SOD ↑*, Catalase		
	Oregano (*Origanum vulgare*)				Protease ↑*, SOD ↑*, Catalase↑*		
Oreochromis niloticus	Assam tea (Camellia sinensis)	0, 1, 2, 4, and 8 g kg	4 & 8 weeks	Serum & mucus immunity	Serum lysozyme↑*, peroxidase↑*, alternative complement (ACH50) ↑*, phagocytosis↑*, and respiratory burst activities↑* mucus lysozyme ↑ and peroxidase activities ↑	2 g/kg	Van Doan et al. (2019)
Oreochromis niloticus	*Elephantopus scaber*	0, 2.5, 5, 10, and 20 g kg	4 & 8 weeks	Serum & Mucus immunity	Serum lysozyme↑* , peroxidase ↑*, alternative complement↑*, phagocytosis ↑*, and respiratory burst ↑*,	5 g/kg	Van Doan et al. (2019)

					mucus lysozyme ↑and skin peroxidase ↑		
Oreochromis niloticus	Thai ginseng (*Boesenbergia rotunda*)	0, 5, 10, 20, 40 g/kg	8 weeks	Serum & Mucus immunity	Mucus lysozyme and peroxidase activity ↑*, serum lysozyme ↑*, serum peroxidase ↑*, alternative complement ↑*, phagocytosis index ↑*, and respiratory burst activities↑*	10 g/kg	Van Doan et al. (2019)
Hybrid tilapia *Oreochromis niloticus* × *O. mossambicus*	*Psidium guajava*	0, 1.5, and 3%	21-days	Skin mucus immunity and gene expression	Mucus Protease ↑, antiprotease (ns), peroxidase ↑, and lysozyme activity (ns)	1.5%	Ceballos-Francisco et al. (2020)
Oreochromis niloticus	Berberine	4 & 8 weeks	0, 1, 3, 6, and 9 g/kg	Skin mucus	Skin mucus lysozyme ↑* and peroxidase activities ↑*; as well as improved serum lysozyme ↑, peroxidase ↑, alternative complement (ACH50) ↑, phagocytosis ↑, and respiratory burst activities ↑	1 g/kg	Van Doan et al. (2020)
Oreochromis niloticus	Seaweed (TrueAlgaeMax)	0, 0.5, 1, 1.5, and 2%	70-days	Serum immune response	Serum protein ↑*, lysozyme ↑*and respiratory burst activity ↑*	2%	Ashour et al. (2020)
Oreochromis niloticus	Watermelon rind powder	0 20, 40, 80, 160 g/kg	8-weeks	Serum & Mucus immunity	Mucus lysozyme & peroxidase activity ↑* serum lysozyme ↑*,	40 g/kg	Van Doan et al. (2020)

(continued)

Table 7.3 (continued)

Fish	Plant used	Feeding rate	Duration	Endpoints studied	Observed effects	Optimal dose	Reference
					peroxidase ↑*, phagocytosis ↑*		
Oreochromis niloticus	White button mushroom	0, 0.5, 1, 2, and 4%	60-days	Heamatological & gene expression	Hb ↑*, RBCs ↑*, and WBCs↑* INF-γ, TNF-α, and IL-1β	2%	Dawood et al. (2020)
				Antioxidants	Catalase ↑*, superoxide dismutase ↑*, glutathione peroxidase (ns)		
Oreochromis niloticus	Gotu kola (*Centella asiatica*)	0, 5, 10, and 20 g/kg	61-days	Serum & Mucus immunity	Skin mucus lysozyme ↑* and skin mucus peroxidase activities ↑* serum lysozyme ↑* and serum peroxidase activities ↑*, alternative complement ↑*, phagocytosis ↑*, and respiratory burst activities↑*	10 g/kg	Srichaiyo et al. (2020)
Nile tilapia, Oreochromis niloticus	Licorice (Glycyrrhiza glabra	Root powder (0, 5, 10, and 20 g/kg)	60-days	Digestive enzymes, anti-oxidants & immune parameters	Total protein↑*, albumin↑*, and globulin↑*	10–20 g/kg of feed	Abdel-Tawwab and El-Araby (2021)
					catalase↑*, superoxide dismutase↑*, and glutathione peroxidase↑*		
					Lysozyme activity↑*, respiratory burst↑*, and total immunoglobulin↑*		

Arrows indicate an increase (↑), a decrease (↓), a significant increase (↑*), and a significant decrease (↓*)

ions also influence the release of inflammatory cytokines (Klebanoff 1992; Rieger et al. 2010).

Medicinal plants/ herbs are well known for their immunomodulatory potential. Many herbs have immunostimulatory properties while many have anti-inflammatory properties. The immunoregulatory action of medicinal plants is dependent on the bioactive components. Medicinal plants exhibit anti-inflammatory action by inhibiting or activating many immune pathways. One possible mode of action may be the inhibition of enzymes involved in the arachidonic acid pathway. Arachidonic acid is formed from the lipid of the cellular membranes upon the action of phospholipase A2. This arachidonic acid produces inflammatory mediators. Many anti-inflammatory drugs block phospholipase A2 which leads to the inhibition of arachidonic acid synthesis and ultimately the formation of inflammatory mediators. Bioactive compounds present in the plants act as natural inhibitors of phospholipase A2 e.g., quercetin, a flavonoid, was the first plant-based inhibitor of phospholipase A2. Many other plants such as pimientillo (*Xylopia frutescens*), onion (*Allium cepa*), garlic (*Allium sativum*), and turmeric (*Curcuna longa*) have been reported to have phospholipase inhibitory activity. Medicinal plants also inhibit 15-lipoxygenases (LOX), an enzyme that is involved in the synthesis of leukotrienes from arachidonic acid. Leukotrienes are important mediators of pro-inflammatory and allergic reactions. Inhibition of LOX activity is linked with a decreased production of pro-inflammatory cytokines. Ethanolic and methanolic extracts of many plants can inhibit 15-LOX activity. Ethanolic extracts of *Ajuga genevensis* were found to inhibit 15-LOX up to 30% while methanolic extracts of *Petroselinum crispum* (parsley) and *Apium graveolens* (celery) inhibit LOX activity up to 50%. This inhibition activity may be attributed to the bioactive components like luteolin, luteolin-O-glucoside, caffeic acid, and apigenin present in plants.

Medicinal plants also exert their immunomodulatory action through the inhibition of prostaglandins. Prostaglandins are also known as eicosanoids. These are active lipid compounds produced from arachidonic acid and regulate blood flow, blood clotting, and inflammation. Cyclooxygenases (COX1 and COX-2) are responsible for the synthesis of prostaglandins from arachidonic acid. Non-steroidal anti-inflammatory drugs (NSAIDs) are used to block COX's active sites, lowering the synthesis of prostaglandins and thus decreasing inflammation. Bioactive compounds of many medicinal plants have NSAID-like modes of action that can block the activity of COX. Bioactive compounds, acetyl-β-boswellic acid present in *Boswellia carterii* and betulinic acid present in *Glycyrrhiza uralensis*, are active inhibitors of COX-1 while senkyunolide-O present in *Ligusticum chuanxiong* and cryptotanshinone, present in *Salvia miltiorrhiza* activity inhibit COX-2. Similarly, phenethyl-trans-ferulate present in *Notopterygium incisum* and roburic acid present in *Gentiana macrophylla* can actively inhibit both isoforms of COX.

Plants and their bioactive compounds exert their cellular regulation by activating or inhibiting different protein kinases e.g. protein kinase C and mitogen-activated protein kinase that are involved in various signal transduction pathways. These medicinal plants can directly or indirectly influence the production of pro-inflammatory cytokines. The DNA-binding capacity of different transcription factors (such as activator protein-1 or nuclear factor-kappa B) is also influenced by these compounds resulting in the regulation of downstream targeted genes.

7.6.3 *Immunostimulatory Mode of Action*

Immunostimulant is a term used for any molecule/compound that can stimulate the immune system of the organism. This immunostimulation can be specific through vaccines and antigens or can be non-specific through drugs or nutrients. Active compounds in plants and herbs can stimulate nonspecific immune responses. In the last decade, a lot of research has been done on finding the plant/herb or their active compounds that can stimulate the immune response in fish so the need for vaccines can be avoided. The use of herbal parts and their extracts are beneficial as they are cheap, can stimulate nonspecific immune responses leave no memory, are sustainable, and have almost no hazardous footprints on the environment.

Phagocytes play an important role in the non-specific immune response and exert an anti-microbial response through NADPH phagocyte oxidase or inducible nitric oxide synthase (iNOS) pathways.

Inducible nitric oxide synthase (iNOS) pathway can be activated by chemokines e.g. interferon gamma (IFN-γ) and tumor necrosis factor-α (TNF-α). Upon activation, this led to the production of reactive nitrogen species by macrophages. In an in vitro study with rat hepatocytes, it was reported that IFN-γ and TNF- α (alone or in combination with LPS) stimulate the inducible nitric oxide synthase (iNOS) pathway. Reactive nitrogen species also activate the NF-kB signaling pathway in the mononuclear cells of peripheral blood which is an important factor for proinflammatory cytokine transcription.

Another possible mechanism through which medicinal plants activate the immune system is through toll-like receptors (TLRs). These are transmembrane proteins and are the first to detect parasites, bacteria, fungi, and viruses. Around 17 TLRs have been identified in teleost. These TLRs may be located in endosomal and lysosomal compartments or maybe cell surface orientated. These TLRs are associated with adaptor proteins, each of which initiates a different signaling pathway. Two important adaptors that initiate immune-related signaling pathways are TRIF-related adaptor molecule (TRAM) and Myeloid differentiation primary-response protein 88(MyD88)> The end product of these signaling pathways may be the production of cytokines or activation of other pathways like MAPK and NF-kB pathway.

Toll-like receptors recognize the pathogen through pathogen-associated molecular patterns and activate the respective adaptor protein (either TRAM or MyD88). These activated adaptors in turn activate transcriptional factors which bind with the DNA and upregulate the synthesis of pro-inflammatory cytokines, interferons, chemokines, and anti-microbial peptides (Li et al. 2014). Medicinal plants and herbs influence the TLR pathway and stimulate the immune response. Supplementing the diet of yellow catfish with *Glycyrrhiza uralensis* stimulated the

TLR pathway and upregulate the expression of TLRs and MyD88 resulting in more production of interleukins. Similarly, supplementing the fish diet with a mixture of Chinese herbs resulted in the activation of TLR5 and MAPK pathways.

Different medicinal plants have been used in tilapia aquaculture as immune stimulants, details of which have been mentioned in Table 7.3.

7.7 Conclusion

The present chapter provides in-depth knowledge about the use of medicinal plants in tilapia aquaculture. In the light of literature reviewed here, it is suggested that the use of medicinal plants provides a sustainable, less costly, and eco-friendly alternative to antibiotics. Moreover, the use of medicinal plants has positive effects on the fish's physiological system by improving gut health, increase in digestive enzymes and antioxidants, and stimulating the immune system hence improving growth and disease resistance. Most importantly, supplementation of herbs/spices improves feed conversion ratio and feed utilization/absorption by fish. However, it must be kept in mind that the addition of a large number of medicinal plants may cause unwanted effects in fish and these effects may be species-specific. Therefore, the optimal range of medicinal plants inclusion and duration of feeding must be determined before the addition of that herb in tilapia feed on a large scale.

References

Abarike ED, Jian J, Tang J, Cai J, Yu H, Lihua C, Jun L (2018) Influence of traditional Chinese medicine and Bacillus species (TCMBS) on growth, immune response and disease resistance in Nile tilapia, *Oreochromis niloticus*. Aquac Res 49(7):66–75

Abdel El-Maksoud AMS, Aboul-Fotouh GE, Allam SM, Abou Zied RM (1999) Effect of marjoram leaves *(Majorana hortensis* L.) as a feed additive on the performance of Nile tilapia *(Oreochromis niloticus)* fingerlings. Egypt J Nutr Feeds 2:39–47

Abdel Wahab AM, Hassouna MME, Abd El-Maksoud AMS, Abd El Tawab A, Abu-Seef AMM (2007) Cinnamon as a feed supplemented in Nile tilapia *(Oreochromis niloticus)* diets reared in earthen ponds. Egypt J Nutr Feeds 10(2):881–890

Abdel-Tawwab M, El-Araby DA (2021) Immune and antioxidative effects of dietary licorice (*Glycyrrhiza glabra* L.) on performance of Nile tilapia, *Oreochromis niloticus* (L.) and its susceptibility to Aeromonas hydrophila infection. Aquaculture 530:735828

Abdel-Tawwab M, Hamed HS (2020) Antagonistic effects of dietary guava (*Psidium guajava*) leaves extract on growth, hemato-biochemical, and immunity response of cypermethrin-intoxicated Nile tilapia, *Oreochromis niloticus*, fingerlings. Aquaculture 529:735668

Abid A, Davies SJ, Waines P, Emery M, Castex M, Gioacchini G, Carnevali O, Bickerdike R, Romero J, Merrifield DL (2013) Dietary synbiotic application modulates Atlantic salmon *(Salmo salar)* intestinal microbial communities and intestinal immunity. Fish Shellfish Immunol 35:1948–1956

Absali H, Mohamad S (2010) Immune response of common carp (*Cyprinus carpio*) fed with herbal immunostimulants. J Anim Vet Adv 9(13):1839–1847

Abutbul S, Golan-Goldhirsh A, Barazani O, Zilberg D (2004) Use of *Rosmarinus officinalis* as a treatment against *Streptococcus iniae* in tilapia (*Oreochromis sp.*). Aquaculture 238:97–105

Adams MA, Johansen PB, Hong Z (1988) Chemical enhancement of feeding for the herbivorous fish Tilapia zillii. Aquaculture 72:95–107

Adel M, Safari R, Soltanian S, Zorriehzahra MJ, Esteban MÁ (2018) Antimicrobial activity and enzymes on skin mucus from male and female caspian kutum (*Rutilus frisii kutum Kamensky*, 1901) specimens. Slo Vet Res 55:235–243

Adewole AM (2014) Effects of Roselle as dietary additive on growth performance and production economy of *Clarias gariepinus*. J Emerg Trends Eng Appl Sci 5:1–8

Ahmad M, El Mesallamy A, Samir F, Zahran F (2011) Effect of Cinnamon (Cinnamomum zeylanicum) on growth performance, feed utilization, whole-body composition, and resistance to Aeromonas hydrophila in Nile Tilapia. J Appl Aquac 23(4):289–298

Ahmad MH, Abdel-Tawwab M (2011) The use of caraway seed meal as a feed additive in fish diets: growth performance, feed utilization, and whole-body composition of Nile tilapia, *Oreochromis niloticus* (L.) fingerlings. Aquaculture 314:110–114

Ahmad MH, El-Gamal RM, Hazaa MM, Hassan SM, Araby DA (2009) The use of *Origanium vulgare* extract in practical diets as a growth and immunity promoter for Nile tilapia, *Oreochromis niloticus* (L.) fingerlinges challenged with pathogenic *Pseudomonas aeruginosa* and *Pseudomonas flourscence*. J Exp Biol (Zool.) 5:457–463

Akinyemi O, Oyewole SO, Jimoh KA (2018) Medicinal plants and sustainable human health: a review. Hortic Int J 2(4):194–195

Alexander JB, Ingram GA (1992) Noncellular nonspecific defence mechanisms of fish. Annu Rev Fish Dis 2:249–279

Aly SM, Fathi M, John G (2008) Echinacea as immunostimulatory agent in Nile Tilapia (*Oreochromis niloticus*) via Earthen pond experiment. Proceedings of the 8th International Symposium on Tilapia in Aquaculture 1033–1041

Ardo L, Yin G, Xu P, Váradi L, Szigeti G, Jeney Z, Jeney G (2008) Chinese herbs (*Astragalus membranaceus* and *Lonicera japonica*) and boron enhance the non-specific immune response of Nile tilapia (*Oreochromis niloticus*) and resistance against *Aeromonas hydrophila*. Aquaculture 275:26–33

Arts JAJ, Tijhaar EJ, Chadzinska M, Savelkoul HFJ, Verburg-van Kemenade BML (2010) Functional analysis of carp interferon-γ: evolutionary conservation of classical phagocyte activation. Fish Shellfish Immunol 29:793–802

Ashour M, Mabrouk MM, Ayoub HF, El-Feky MMMM, Zaki SZ, Hoseinifar SH, Rossi W, Van Van DH, El-Haroun E, Goda AMAS (2020) Effect of dietary seaweed extract supplementation on growth, feed utilization, hematological indices, and non-specific immunity of Nile Tilapia, *Oreochromis niloticus* challenged with *Aeromonas hydrophila*. J Appl Phycol 32:3467–3479

Asnaashari S, Dastmalchi S, Javadzadeh Y (2018) Gastroprotective effects of herbal medicines (roots). Int J Food Prop 21:902–920

Astuya A, Ziehe J, Rivera A, Ortiz S, Ulloa V, Roeckel M, Aspé E, Fernández K (2017) Antioxidant and anti-inflammatory activities of *Pinus radiata* bark extract in salmonid cell lines. Aquac Res 48:3568–3578

Austin B, Austin DA (2012) Aeromonadaceae representatives (motile aeromonads). Bacterial fish pathogens-diseases of farmed and wild fish, 5th edn. Springer, Dordrecht, The Netherlands

Awad E, Awaad A (2017) Role of medicinal plants on growth performance and immune status in fish. Fish Shellfish Immunol 67:40–54

Ayaz M, Sadiq A, Junaid M, Ullah F, Subhan F, Ahmed J (2017) Neuroprotective and anti-aging potentials of essential oils from aromatic and medicinal Plants. Front Aging Neurosci 30(9):168

Baba E, Acar Ü, Öntaş C, Kesbiç OS, Yılmaz S (2016) Evaluation of *Citrus limon* peels essential oil on growth performance, immune response of Mozambique tilapia *Oreochromis mossambicus* challenged with *Edwardsiella tarda*. Aquaculture 465:13–18

Bai D, Wang F, Guo Y, Zhu G, Mao H, Sun J, Yan S (2009) Primary study of Chinese herbs as feed attractants in Tinca. J Tianjin Agric Univ 16(1):5–8

Bastardo A, Ravelo C, Castro N, Calheiros J, Romalde JL (2012) Effectiveness of bivalent vaccines against *Aeromonas hydrophila* and *Lactococcus garvieae* infections in rainbow trout *Oncorhynchus mykiss* (Walbaum). Fish Shellfish Immunol 32:756–761

Beaz-Hidalgo R, Figueras MJ (2012) Molecular detection and characterization of *furunculosis* and other aeromonas fish infections. In: Carvalho E, David GS, Silve RJ (eds) Health and environment in aquaculture. Intech Open, pp 97–132

Beaz-Hidalgo R, Romaldo JL, Figueras MJ (2012) Gene expression in fish due to infections caused by *Aeromonas spp*. Revista AquaTIC 37:113–122

Bondad-Reantaso MG, Subasinghe RP, Arthur JR, Ogawa K, Chinabut S, Adlard R, Tan Z, Shariff M (2005) Disease and health management in Asian aquaculture. Vet Parasitol 132: 249–272

Bostami ABMR, Khan MRI, Rabbi AKMZ, Siddiqui MN, Islam MT (2021) Boosting animal performance, immune index and antioxidant status in post-weaned bull calves through dietary augmentation of selective traditional medicinal plants. Vet Anim Sci 14:100197

Brum A, Pereira SA, Owatari MS, Chagas EC, Chaves FCM, Mouriño JLP, Martins ML (2017) Effect of dietary essential oils of clove basil and ginger on Nile tilapia *(Oreochromis niloticus)* following challenge with *Streptococcus agalactiae*. Aquaculture 468:235–243

Burr SE, Goldschmidt-Clermont E, Kuhnert P, Frey J (2012) Heterogeneity of Aeromonas populations in wild and farmed perch, Perca *fluviatilis* L. J Fish Dis 35:607–613

Cain KD, Jones DR, Raison RL (2000) Characterisation of mucosal and systemic immune responses in rainbow trout (*Oncorhynchus mykiss*) using surface plasmon resonance. Fish Shellfish Immunol 10:651–666

Ceballos-Francisco D, Castillo Y, De La Rosa F, Vásquez W, Reyes-Santiago R, Cuello A, Cuesta A, Esteban MÁ (2020) Bactericidal effect on skin mucosa of dietary guava *(Psidium guajava* L.) leaves in hybrid tilapia (*Oreochromis niloticus* × *O. mossambicus*). J Ethnopharmacol 259:112838

Chen M, Li LP, Wang R, Liang WW, Huang Y, Li J, Lei AY, Huang WY, Gan X (2012) PCR detection and PFGE genotype analyses of streptococcal clinical isolates from tilapia in China. Vet Microbiol 159:526–530

Chitmanat C, Tongdonmuan K, Nunsong W (2005) The use of crude extracts from traditional medicinal plants to eliminate Trichodina sp. in tilapia (*Oreochromis niloticus*) fingerlings. Songklanakarin J Sci Technol 27(Suppl. 1):359–364

Christybapita D, Divyagnaneswari M, Michael DR (2007) Oral administration of *Eclipta alba* leaf aqueous extract enhances the non-specific immune responses and disease resistance of *Oreochromis mossambicus*. Fish Shellfish Immunol 23:840–852

Citarasu Y (2010) Herbal biomedicines: a new opportunity for aquaculture industry. Aquac Int 18: 403–414

Colquhoun DJ, Duodu S (2011) Francisella infections in farmed and wild aquatic organisms. Vet Res 42:47

Conner DE (1993) Naturally occuring compounds. In: Davidson PM, Branen AL (eds) Antimicrobials in foods. Marcel Dekker, New York, pp 441–468

Cordero H, Cuesta A, Meseguer J, Esteban MÁ (2016) Changes in the levels of humoral immune activities after storage of gilthead seabream (*Sparus aurata*) skin mucus. Fish Shellfish Immunol 58:500–507

Corripio-Miyar Y, Bird S, Tsamopoulos K, Secombes CJ (2007) Cloning and expression analysis of two pro-inflammatory cytokines, IL-1 beta and IL-8, in haddock (*Melanogrammus aeglefinus*). Mol Immunol 44(6):1361–1373

Crivellato E, Travan L, Ribatti D (2015) The phylogenetic profile of mast cells. Methods Mol Biol 1220:11–27

Crowhurst MO, Layton JE, Lieschke GJ (2002) Developmental biology of zebrafish myeloid cells. Int J Dev Biol 46:483–492

Da'as Uribe C, Folch H, Enríquez R, Moran G (2011) Innate and adaptive immunity in teleost fish: a review. Vet Med (Praha) 10:486–503

Daniels C, Hoseinifar SH (2014) Prebiotic applications in shellfish. In: Merrifield D, Ringø E (eds) Aquaculture nutrition: gut health, probiotics and prebiotics, 1st edn. https://doi.org/10.1002/9781118897263.ch15

Davidson-Hunt I (2000) Ecological ethno botany: stumbling toward new practices and paradigms. MASA J 16:1–13

Dawood MAO, Eweedah NM, El-Sharawy ME, Awad SS, Van Doan H, Paray BA (2020) Dietary white button mushroom improved the growth, immunity, antioxidative status and resistance against heat stress in Nile tilapia (*Oreochromis niloticus*). Aquaculture 523:735229

Delgado LC, Wada N, Rosegrant WM, Meijer S, Ahmed M (2003) Fish to 2020: supply and demand in changing global markets. IFPRI, Washington, D.C. (USA), p 236

Desai AG, Qazi GN, Ganju RK, El-Tamer M, Singh J, Saxena AK, Bedi YS, Taneja SC, Bhat HK (2008) Medicinal plants and cancer chemoprevention. Curr Drug Metab 9(7):581–591

Dhama K, Karthik K, Khandia R, Munjal A, Tiwari R, Rana R, Khurana SK, Ullah S, Khan RU, Alagawany M, Farag MR, Dadar M, Joshi SK (2018) Medicinal and therapeutic potential of herbs and plant metabolites/extracts countering viral pathogens—current knowledge and future prospects. Curr Drug Metab 19(3):236–263

Díaz-Rosales P, Arijo S, Chabrillón M, Alarcón FJ, Tapia-Paniagua ST, Martínez-Manzanares E, Balebona MC, Moriñigo MA (2009) Effects of two closely related probiotics on respiratory burst activity of Senegalese sole (*Solea senegalensis*, Kaup) phagocytes, and protection against *Photobacterium damselae* subsp. piscicida. Aquaculture 293:16–21

Divyagnaneswari M, Christybapita D, Michael RD (2007) Enhancement of nonspecific immunity and disease resistance in *Oreochromis mossambicus* by *Solanum trilobatum* leaf fractions. Fish Shellfish Immunol 23:249–259

El-Dakar AY (2004) Growth response of hybrid tilapia, *O. niloticus x O. auraus*, fingerlings to diets supplemented with different levels of caraway seeds. Mansoura Univ J Agric Sci 29: 6083–6094

El-Deen A (2010) Green tea extract role in removing the Trichodina sp. on *Oreochromis niloticus* fry in the Egyptian fish hatcheries. Rep Opinion 2:77–81

Elkafrawy IA (2020) Impacts of feeding restriction regimes on economic and productive performance of Nile Tilapia fish. JCVR 2(2):48–54

El-Sayed AFM (2006) Tilapia culture. CABI Publishing, CABI International, Wallingford, Oxfordshire, UK

El-Sayed AFM (2019) Tilapia culture, 2nd edn. Academic Press, p 358

Esmail AA, Ali TT (2019) A review of medicinal plants with nephroprotective effects. GSC Biol Pharm Sci 08(01):114–122

Esteban MÁ (2012) An overview of the immunological defenses in fish skin. Int Sch Res Notices:1–29

Faheem M, Khaliq S, Mustafa N, Rani S, Lone KP (2020) Dietary *Moringa oleferia* leaf meal induce growth, innate immunity and cytokine expression in grass carp, *Ctenopharyngodon Idella*. Aquac Nutr 26(4):64–72

FAO (2014) The state of world fisheries and aquaculture. Food and Agriculture Organization of the United Nations, Rome

FAO (2016) The state of world fisheries and aquaculture 2016. Food and Agriculture Organization of the United Nations Contributing to food security and nutrition for all. Rome. 200 pp

Figueiredo CP, Carneiro DO, Faria FC, Costa GM (2006) *Streptococcus agalactie* associated to meningoencefalitis and systemic infection from tilapia *(Oreochromis niloticus)* in Brazil. Arq Bras Med Vet Zootec 58:678–680

Fisher DT, Chen Q, Appenheimer MM, Skitzki J, Wang WC, Odunsi K, Evans SS (2006) Hurdles to lymphocyte trafficking in the tumor microenvironment: implications for effective immunotherapy. Immunol Invest 35(3-4):251–277

Fitzsimmons K (2000) Future trends of Tilapia aquaculture in the Americas. In: Costa-Pierce BA, Rakocy JE (eds) Tilapia aquaculture in the Americas, Vol. 2. The World Aquaculture Society, Baton Rouge, Louisiana, United States, pp 252–264

Gabriel NN, Qiang J, He J, Ma XY, Kpundeh MD, Xu P (2015) Dietary *Aloe vera* supplementation on growth performance, some haemato-biochemical parameters and disease resistance against *Streptococcus iniae* in tilapia (GIFT). Fish Shellfish Immunol 44:504–514

Gobi N, Ramya C, Vaseeharan B, Malaikozhundan B, Vijayakumar S, Murugan K, Benelli G (2016) *Oreochromis mossambicus* diet supplementation with *Psidium guajava* leaf extracts enhance growth, immune, antioxidant response and resistance to *Aeromonas hydrophila*. Fish Shellfish Immunol 58:572–583

Götti RP, Melzer J, Saller R (2014) An approach to the concept of tonic: suggested definitions and historical aspects. Forsch Komplementmed 21:413–417

Grayfer L, Garcia EG, Belosevic M (2010) Comparison of macrophage antimicrobial responses induced by type II interferons of the Goldfish (*Carassius auratus* L.). J Biol Chem 285:23537–23547

Greenwell M, Rahman PK (2015) Medicinal plants: their use in anticancer treatment. Int J Pharm Sci Res 6(10):4103–4112

Guardiola FA, Bahi A, Jiménez-Monreal AM, Martínez-Tomé M, Murcia MA, Esteban MA (2018) Dietary administration effects of fenugreek seeds on skin mucosal antioxidant and immunity status of gilthead seabream (Sparus aurata L.). Fish Shellfish Immunol 75:357–364

Hai NV (2015) The use of medicinal plants as immunostimulants in aquaculture: a review. Aquaculture 446:88–96

Harikrishnan R, Balasundaram C, Heo M-S (2011) Impact of plant products on innate and adaptive immune system of cultured finfish and shellfish. Aquaculture 317:1–15

Harikrishnand R, Balasundaram C (2005) Modern trends in *Aeromonas hydrophila* disease management with fish. Rev Fish Sci Aquac 13:281–320

Hatten F, Fredriksen A, Hordvik I, Endresen C (2001) Presence of IgM in cutaneous mucus, but not in gut mucus of Atlantic salmon, *Salmo salar*. Serum IgM is rapidly degraded when added to gut mucus. Fish Shellfish Immunol 11:257–268

Hegazi MM, Hasanein SS (2010) Effects of chronic exposure to ammonia concentrations on brain monoamines and ATPases of Nile tilapia *(Oreochromis niloticus)*. Comp Biochem Physiol C Toxicol Pharmacol 151:420–425

Helander IM, Alakomi H-L, Latva-Kala K, Mattila-Sandholm T, Pol I, Smid EJ, Gorris LGM, von Wright A (1998) Characterization of the action of selected essential oil components on Gram-negative bacteria. J Agric Food Chem 46:3590–3595

Hernández E, Figueroa J, Iregui C (2009) Streptococcosis on a red tilapia, Oreochromis sp., farm: a case study. J Fish Dis 32(3):247–252

Herraez MP, Zapata AG (1986) Structure and function of the melanomacrophage centers of the goldfish *Carassius auratus*. Vet Immunol Immunopathol 12:117–126

Hoseinifar SH, Ahmadi A, Raeisi M, Hoseini SM, Khalili M, Behnampour N (2017a) Comparative study on immunomodulatory and growth enhancing effects of three prebiotics (galactooligosaccharide, fructooligosaccharide and inulin) in common carp (*Cyprinus carpio*). Aquac Res 48(7):3298–3307

Hoseinifar SH, Dadar M, Khalili M, Cerezuela R, Esteban MA (2017b) Effect of dietary supplementation of palm fruit extracts on the transcriptomes of growth, antioxidant enzyme and immune-related genes in common carp (*Cyprinus carpio*) fingerlings. Aquac Res 48(7): 3684–3692

Hoseinifar SH, Mirvaghefi A, Amoozegar MA, Merrifield DL, Ringø E (2017c) In vitro selection of a synbiotic and *in vivo* evaluation on intestinal microbiota, performance and physiological response of rainbow trout *(Oncorhynchus mykiss)* fingerlings. Aquac Nutr 23(1):111–118

Hoseinifar SH, Safari R, Dadar M (2017d) Dietary sodium propionate affects mucosal immune parameters, growth and appetite related genes expression: insights from zebrafish model. Gen Comp Endocrinol 243:78–83

Hoseinifar SH, Sun YZ, Zhou Z, Van Doan H, Davies SJ, Harikrishnan R (2020b) Boosting immune function and disease bio-control through environment-friendly and sustainable approaches in finfish aquaculture: herbal therapy scenarios. Rev Fish Sci Aquac 28(3):303–321

Hoseinifar SH, Sunb Y-Z, Zhou Z, Van Doan HV, Davies SJ, Harikrishnan R (2020a) Boosting immune function and disease bio-control through environment- friendly and sustainable approaches in finfish aquaculture: herbal therapy scenarios. Rev Fish Sci Aquac 28:10. https://doi.org/10.1080/23308249.2020.1731420

Hussain J, Khan FU, Ullah R, Muhammad Z, Rehman NU, Shinwari ZK, Khan IU, Zohaib M, Imad-ud-din HAM (2011) Nutrient evaluation and elemental analysis of four selected medicinal plants of Khyber Pakhtoon Khwa, Pakistan. Pak J Bot 43(1):427–434

Ibrahim RE, El-Houseiny W, Behairy A, Mansour MF, Abd-Elhakim YM (2019) Ameliorative effects of *Moringa oleifera* seeds and leaves on chlorpyrifos- induced growth retardation, immune suppression, oxidative stress, and DNA damage in *Oreochromis niloticus*. Aquaculture 505:225–234

Jeffery KR, Stone D, Feist SW, Verner-Jeffreys D (2010) An outbreak of disease caused by Francisella spp. in Nile tilapia *Oreochromis niloticus* at a recirculation fish farm in the UK. Dis Aquat Organ 91:161–165

Kalsoom R, Haider MS, Chohan S (2020) Phytochemical analysis and antifungal activity of some medicinal plants against alternaria specie isolated from onion. J Anim Plant Sci 30(2):454–460

Kamaishi T, Miwa S, Goto E, Matsuyama T, Oseko N (2010) Mass mortality of giant abalone *Haliotis gigantea* caused by a Francisella sp. bacterium. Dis Aquat Organ 89:145–154

Karan S, Dash P, Kaushik H, Sahoo PK, Garg LC, Dixit A (2016) Structural and functional characterization of recombinant interleukin-10 from Indian major carp *Labeo rohita*. J Immunol Res 2016:1–11

Kareem ZH, Abdelhadi YM, Christianus A, Karim M, Romano N (2016) Effects of some dietary crude plant extracts on the growth and gonadal maturity of Nile tilapia (*Oreochromis niloticus*) and their resistance to *Streptococcus agalactiae* infection. Fish Physiol Biochem 42:757–769

Kirubakaran CJW, Alexander CP, Michael RD (2010) Enhancement of non-specific immune responses and disease resistance on oral administration of *Nyctanthes arbortristis* seed extract in *Oreochromis mossambicus* (Peters). Aquac Res 41(11):1630–1639

Klebanoff SJ (1992) Myeloperoxidase: occurrence and biological function. In: Evers J, Evers KE, Grisham M (eds) Peroxidases in chemistry and biology, vol 1. CRC Press, FL, pp 2–35

Kuebutornye FKA, Abarike ED (2020) The contribution of medicinal plants to tilapia aquaculture: a review. Aquaculture International 1-19

Kümmerer K (2009) Antibiotics in the aquatic environment-a review-part I. Chemosphere 75:417–434

Laing KJ, Zou JJ, Wang T, Bols N, Hirono I, Aoki T, Secombes CJ (2002) Identification and analysis of an interleukin 8-like molecule in rainbow trout *Oncorhynchus mykiss*. Dev Comp Immunol 26:433–444

Lambert RJW, Skandamis PN, Coote PJ, Nychas G-JE (2001) A study of the minimum inhibitory concentration and mode of action of oregano essential oil, thymol and carvacrol. J Appl Microbiol 91:453–462

Latif M, Faheem M, Asmatullah, Hoseinifar SH, Van Doan H (2020) Dietary black seed effects on growth performance, proximate composition, antioxidant and histo-biochemical parameters of a culturable fish, Rohu (*Labeo rohita*). Animals (Basel) 11(1):48

Li YW, Liu L, Huang PR, Fang W, Luo ZP, Peng HL, Wang YX, Li AX (2014) Chronic *streptococcosis* in Nile tilapia, *Oreochromis niloticus* (L.), caused by *Streptococcus agalactiae*. J Fish Dis 37:757–763

Lieschke GJ, Oates AC, Crowhurst MO, Ward AC, Layton JE (2001) Morphologic and functional characterization of granulocytes and macrophages in embryonic and adult zebrafish. Blood 98(10):3087–3096

Magnadottir B (2010) Immunological control of fish diseases. Mar Biotechnol 12:361–379

Male D, Roitt I (2001) Introduction to the immune system. In: Roitt I, Brostoff J, Male D (eds) Immunology. Mosby, London, pp 1–12

Martin SAM, Krol E (2017) Nutrigenomics and immune function in fish: new insights from omics technologies. Dev Comp Immunol 75:86–98

Mauel MJ, Soto E, Moralis JA, Hawke J (2007) A piscirickettsiosis-like syndrome in cultured Nile tilapia in Latin America with Francisella spp. as the pathogenic agent. J Aquat Anim Health 19: 27–34

Mbokane EM, Moyo NAG (2018) Alterations of haemato-biochemical parameters pre and post-challenge with *Aeromonas hydrophila* and survival of *Oreochromis mossambicus* fed *Moringa oleifera*-based diets. Fish Shellfish Immunol 83:213–222

Mian GF, Godoy DT, Leal CA, Yuhara TY, Costa GM, Figueiredo HCP (2009) Aspects of the natural history and virulence of *S. agalactiae* infection in Nile tilapia. Vet Microbiol 136:180–183

Mohammadi G, Rafiee G, El Basuini MF, Van Doan H, Ahmed HA, Dawood MAO, Abdel-Latif HMR (2020a) Oregano (*Origanum vulgare*), St John's-wort (*Hypericum perforatum*), and lemon balm (*Melissa officinalis*) extracts improved the growth rate, antioxidative, and immunological responses in Nile tilapia (*Oreochromis niloticus*) infected with *Aeromonas hydrophila*. Aquac Rep 18:100445

Mohammadi G, Rashidian G, Hoseinifar SH, Naserabad SS, Van Doan H (2020b) Ginger (*Zingiber officinale*) extract affects growth performance, body composition, haematology, serum and mucosal immune parameters in common carp (*Cyprinus carpio*). Fish Shellfish Immunol 99: 267–273

Mohammed HH, Arias CR (2016) Protective efficacy of *Nigella sativa* seeds and oil against columnaris disease in fishes. J Fish Dis 39(6):693–703

Mohiseni M, Sepidnameh M, Bagheri D, Banaee M, Nematdust HB (2017) Comparative effects of Shirazi thyme and vitamin E on some growth and plasma biochemical changes in common carp (*Cyprinus carpio*) during cadmium exposure. Aquacult Res 48(9):4811–4821

Molnár K, Csaba G (2005) Sanitary management in Hungarian aquaculture. Vet Res Commun 29 (Suppl 2):143–146

MSD Animal Health (2012) Technical Bulletin: *Streptococcus* in the Tilapia environment. http://aqua.merckanimalhealth.com/binaries/AquaVac_StreSa_tech01_0Nov12_050STREP_tcm56-35564.pdf

Mulat M, Khan F, Muluneh G, Pandita A (2020) Phytochemical profile and antimicrobial effects of different medicinal plant: current knowledge and future perspectives. Curr Tradit Med 6:24

Mulero I, García-Ayala A, Meseguer J, Mulero V (2007) Maternal transfer of immunity and ontogeny of autologous immunocompetence of fish: a mini review. Aquaculture 1-4:244–250

Nakanishi T, Shibasaki Y, Matsuura Y (2015) T cells in fish. Biology 4:640–663

Na-Phatthalung P, Teles M, Voravuthikunchai SP, Tort L, Fierro-Castro C (2018) Immunomodulatory effects of *Rhodomyrtus tomentosa* leaf extract and its derivative compound, rhodomyrtone, on head kidney macrophages of rainbow trout (*Oncorhynchus mykiss*). Fish Physiol Biochem 44:543–555

Nawaz A, Bakhsh Javaid A, Irshad S, Hoseinifar SH, Xiong H (2018) The functionality of prebiotics as immunostimulant: evidences from trials on terrestrial and aquatic animals. Fish Shellfish Immunol 76:272–278

Ndong D, Fall J (2011) The effect of garlic (*Allium sativum*) on growth and immune responses of hybrid tilapia (*Oreochromis niloticus x Oreochromis aureus*). J Clin Immunol Immunopathol Res 3:1–9

Okaiyeto K, Nwodo UU, Mabinya LV, Okoh AI (2018) A review on some medicinal plants with hepatoprotective effects. Pharmacogn Rev 12:186–199

Ostland VE, Stannard JA, Creek JJ, Hedrick RP, Ferguson HW et al (2006) Aquatic Francisella-like bacterium associated with mortality of intensively cultured hybrid striped bass *Morone chrysops x M. saxatilis*. Dis Aquat Organ 72:135–145

Ottem KF, Nylund A, Karlsbackk E, Friis-Moller A, Krossoy B et al (2007) New species in the genus Francisella (Gammaproteobacteria; Francisella); *Francisella piscicida* sp. nov. isolated from cod (*Gadus morhua*). Arch Microbiol 188:547–550

Ottem KF, Nylund A, Karlsbakk E, Friis-Møller A, Kamaishi T (2009) Elevation of Francisella philomiragia subsp. noatunensis Mikalsen et al. (2007) to Francisella noatunensis comb. nov.

[syn. Francisella piscicida Ottem et al. (2008) syn. nov.] and characterization of Francisella noatunensis subsp. orientalis subsp. nov., two important fish pathogens. J Appl Microbiol 106: 1231–1243

Parham S, Kharazi AZ, Bakhsheshi-Rad HR, Nur H, Ismail AF, Sharif S, RamaKrishna S, Berto F (2020) Antioxidant, antimicrobial and antiviral properties of herbal materials. Antioxidants 9: 1309

Park KH, Choi SH (2012) The effect of mistletoe, *Viscum album coloratum*, extract on innate immune response of Nile tilapia (*Oreochromis niloticus*). Fish Shellfish Immunol 32:1016–1021

Parra D, Reyes-Lopez FE, Tort L (2015) Mucosal immunity and B cells in teleosts: effect of vaccination and stress. Front Immunol 6:1–12

Peter KV (2001) Handbook of herbs and spices, vol 3. Woodhead publishing Limited, Abington. ISBN: 185573 5628

Plant KP, LaPatra SE (2011) Advances in fish vaccine delivery. Dev Comp Immunol 35:1256–1262

Platel K, Rao A, Saraswathi G, Srinivasan K (2002) Digestive stimulant action of three Indian spice mixes in experimental rats. Nahrung 46:394–398

Poolsawat L, Yu Y, Li X, Zhen X, Yao W, Wang P, Luo C, Leng X (2020) Efficacy of phytogenic extracts on growth performance and health of tilapia (*Oreochromis niloticus* × *O. aureus*). Aquac Fish. https://doi.org/10.1016/j.aaf.2020.08.009

Pretto-Giordano LG, Eckehard-Müller E, De Freitas JC, Gomes da Silva V (2010) Evaluation of the pathogenesis of Streptococcus agalactiae in Nile tilapia *(Oreochromis niloticus)*. Braz Arch Biol Technol 53:87–92

Prykhozhij SV, Berman JN (2014) The progress and promise of zebrafish as a model to study mast cells. Dev Comp Immunol 46:74–83

Rahman MM, McFadden G (2006) Modulation of tumor necrosis factor by microbial pathogens. PLoS Pathog 2:e4

Rattanachaikunsopon P, Phumkhachorn P (2009) Potential of Chinese chive oil as a natural antimicrobial for controlling *Flavobacterium columnare* infection in Nile tilapia *Oreochromis niloticus*. Fish Sci 75:1431–1437

Reyes-Becerril M, Angulo C, Sanchez V, Vázquez-Martínez J, López MG (2019) Antioxidant, intestinal immune status and anti-inflammatory potential of *Chenopodium ambrosioides* L. in fish: in vitro and in vivo studies. Fish Shellfish Immunol 86:420–428

Reyes-Cerpa S, Maisey K, Reyes-Lpez F, Toro-Ascuy D, Mara A, Imarai M (2012) Fish cytokines and immune response, In: New advances and contributions to fish biology. InTech.

Rieger AM, Hall BE, Barreda DR (2010) Macrophage activation differentially modulates particle binding, phagocytosis and downstream antimicrobial mechanisms. Dev Comp Immunol 34: 1144–1159

Rombout JH, Huttenhuis HBT, Picchietti S, Scapigliati S (2005) Phylogeny and ontogeny of fish leucocites. Fish Shellfish Immunol 19:441–455

Roosta Z, Hoseinifar SH (2014) The effects of crowding stress on some epidermal mucus immune parameters, growth performance and survival rate of tiger barb *(Pentius tetrazona)*. Aquac Res 47(5):1682–1686

Rossolini GM, Arena F, Pecile P, Pollini S (2014) Update on the antibiotic resistance crisis. Curr Opin Pharmacol 18:56e60

Rubio-Godoy M (2010) Teleost fish immunology. Rev Mex Cienc Pecu 1:47–57

Salinas I, Zhang Y-A, Sunyer JO (2011) Mucosal immunoglobulins and B cells of teleost fish. Dev Comp Immunol 35:1346–1365

Samuel CE (2001) Antiviral actions of interferons. Clin Microbiol Rev 14:778–809

Sanahuja I, Fernández-Alacid L, Ordóñez-Grande B, Sánchez-Nuño S, Ramos A, Araujo RM, Ibarz A (2019) Comparison of several non-specific skin mucus immune defences in three piscine species of aquaculture interest. Fish Shellfish Immunol 89:428–436

Santos L, Ramos F (2016) Analytical strategies for the detection and quantification of antibiotic residues in aquaculture fishes: a review. Trends Food Sci Technol 52:16e30

Savan R, Sakai M (2006) Genomics of fish cytokines. Comp Biochem Physiol Part D Genomics Proteomics 1(1):89–101

Sepulcre MP, Pelegrín P, Mulero V, Meseguer J (2002) Characterisation of gilthead seabream acidophilic granulocytes by a monoclonal antibody unequivocally points to their involvement in fish phagocytic response. Cell Tissue Res 308(1):97–102

Seyfried EE, Newton RJ, Rubert KF 4th, Pedersen JA, McMahon KD (2010) Occurrence of tetracycline resistance genes in aquaculture facilities with varying use of oxytetracycline. Microb Ecol 59(4):799–807

Shalaby SM (2004) Response of Nile tilapia, *Oreochromis niloticus*, fingerlings to diets supplemented with different levels of fenugreek seeds (Hulba). Mansoura Univ J Agric Sci 29(5):2231–2242

Shourbela RM, El-Hawarry WN, Abd El-Latif AM, Abo-Kora SY (2020) Potentiality of *Moringa oleifera* aqueous extract as a growth modulator and antistress in acute hypoxic Nile tilapia *Oreochromis niloticus*. Iran J Fish Sci 19(1):67–84

Singh R (2015) Medicinal plants: a review. J Plant Sci 3(1):50–55

Soto E, Hawke JP, Fernandez D, Morales JA (2009) Francisella sp., an emerging pathogen of tilapia, *Oreochromis niloticus* (L.), in Costa Rica. J Fish Dis 32:713–722

Soto-Rodríguez SA, Cabanillas-Ramos J, Alcaraz U, Gómez-Gil B, Romalde JL (2013) Identification and virulence of *Aeromonas dhakensis*, *Pseudomonas mosselii* and *Microbacterium paraoxydans* isolated from Nile tilapia, *Oreochromis niloticus*, cultivated in Mexico. J Appl Microbiol 115:654–662

de Souza Pereira JJ, Pereira AP, Jandú JJ, da Paz JA, Crovella S, Dos Santos Correia MT, de Azevêdo Silva J (2017) *Commiphora leptophloeos* phytochemical and antimicrobial characterization. Front Microbiol 24(8):52

Srichaiyo N, Tongsiri S, Hoseinifar SH, Dawood MAO, Jaturasitha S, Esteban MÁ, Ringø E, Van Van DH (2020) The effects gotu kola (*Centella asiatica*) powder on growth performance, skin mucus, and serum immunity of Nile tilapia (*Oreochromis niloticus*) fingerlings. Aquac Rep 16: 100239

Standen BT, Peggs DL, Rawling MD, Foey A, Davies SJ, Santos GA, Merrifield DL (2016) Dietary administration of a commercial mixed-species probiotic improves growth performance and modulates the intestinal immunity of tilapia, *Oreochromis niloticus*. Fish Shellfish Immunol 49:427–435

Sudhakaran SD, Srirekha P, Devasree LD, Premsingh S, Michael DR (2006) Immunostimulatory effect of *Tinospora cordifolia* Miers leaf extract in *Oreochromis mossambicus*. Indian J Exp Biol 44(9):726–732

Swindle EJ, Hunt JA, Coleman JW (2002) A comparison of reactive oxygen species generation by rat peritoneal macrophages and mast cells using the highly sensitive real-time chemiluminescent probe pholasin: inhibition of antigen-induced mast cell degranulation by macrophage-derived hydrogen peroxide. J Immunol 169:5866–5873

Tagboto S, Townson S (2001) Antiparasitic properties of medicinal plants and other naturally occurring products. Adv Parasitol 50:199–295

Tang J, Cai J, Liu R, Wang J, Lu Y, Wu Z, Jian J (2014) Immunostimulatory effects of artificial feed supplemented with a Chinese herbal mixture on *Oreochromis niloticus* against *Aeromonas hydrophila*. Fish Shellfish Immunol 39:401–406

Trichet VV (2010) Nutrition and immunity: an update. Aquac Res 41:356e72

Ultee A, Bennik MHJ, Moezelaar R (2002) The phenolic hydroxyl group of carvacrol is essential for action against the food-borne pathogen *Bacillus cereus*. Society 68(4):1561–1568

Ultee A, Gorris LG, Smit EJ (1998) Bactericidal activity of carvacrol towards the food-borne pathogen *Bacillus cereus*. J Appl Microbiol 85:211–218

Uribe C, Folch H, Enriquez R, Moran G (2011) Innate and adaptive immunity in teleost fish: a review. Vet Med 56:486–503

Utke K, Kock H, Schuetze H, Bergmann SM, Lorenzen N, Einer-Jensen K, Kollner B, Dalmo RA, Vesely T, Ototake M, Fischer U (2008) Cell mediated immune responses in rainbow trout after

DNA immunization against the viral hemorrhagic septicemia virus. Dev Comp Immunol 3:239–252

Vallejos-Vidal E, Reyes-López F, Teles M, MacKenzie S (2016) The response of fish to immunostimulant diets. Fish Shellfish Immunol 56:34–69

Van Doan H, Hoseinifar SH, Jaturasitha S, Dawood MAO, Harikrishnan R (2020) The effects of berberine powder supplementation on growth performance, skin mucus immune response, serum immunity, and disease resistance of Nile tilapia (*Oreochromis niloticus*) fingerlings. Aquaculture 520:734927

Van Doan H, Hoseinifar SH, Sringarm K, Jaturasitha S, Khamlor T, Dawood MAO, Esteban MÁ, Soltani M, Musthafa MS (2019) Effects of elephant's foot (*Elephantopus scaber*) extract on growth performance, immune response, and disease resistance of nile tilapia (*Oreochromis niloticus*) fingerlings. Fish Shellfish Immunol 93:328–335

Van Kooyk Y, Rabinovich GA (2008) Protein-glycan interactions in the control of innate and adaptive immune responses. Nat Immunol 9(6):593–601

de Veer MJ, Holko M, Frevel M, Walker E, Der S, Paranjape JM, Silverman RH, Williams BR (2001) Functional classification of interferon stimulated genes identified using microarrays. J Leukoc Biol 6:912–920

Watanuki H, Ota K, Tassakka ACMAR, Kato T, Sakai M (2006) Immunostimulant effects of dietary *Spirulina platensis* on carp, *Cyprinus carpio*. Aquaculture 258:157–163

Whyte SK (2007) The innate immune response of finfish-a review of current knowledge. Fish Shellfish Immunol 6:1127–1151

Wu Y-R, Gong Q-F, Hong F, Liang W-W, Chen M, He R-J (2013) Effect of *Sophora flavescens* on non-specific immune response of tilapia (GIFT *Oreochromis niloticus*) and disease resistance against *Streptococcus agalactiae*. Fish Shellfish Immunol 34:220–227

Yardimci B, Aydin Y (2011) Pathological findings of experimental *Aeromonas hydrophila* infection in Nile tilapia *(Oreochromis niloticus)*. Vet Fak Derg 58:47–54

Ye X, Li J, Lu M, Deng G, Jiang X, Tian Y, Quan Y, Jian Q (2011) Identification and molecular typing of *Streptococcus agalactiae* isolated from pond-cultured tilapia in China. Fish Sci 77: 623–632

Yilmaz E (2019) Effects of dietary anthocyanin on innate immune parameters, gene expression responses, and ammonia resistance of Nile tilapia *(Oreochromis niloticus)*. Fish Shellfish Immunol 93:694–701

Zahran E, Risha E, Abdel Hamid F, Mahgoub HA, Ibrahim T (2014) Effects of dietary Astragalus polysaccharides (APS) on growth performance, immunological parameters, digestive enzymes, and intestinal morphology of Nile tilapia (*Oreochromis niloticus*). Fish Shellfish Immunol 38: 149–157

Zante MD, Borchel A, Brunner RM, Goldammer T, Rebl A (2015) Cloning and characterization of the proximal promoter region of rainbow trout (*Oncorhynchus mykiss*) interleukin-6 gene. Fish Shellfish Immunol 43:249–256

Zou J, Clark MS, Secombes CJ (2003) Characterisation, expression and promoter analysis of an interleukin 10 homologue in the puffer fish, *Fugu rubripes*. Immunogenetics 55:325–335

Zou J, Secombes CJ (2016) The function of fish cytokines. Biology (Basel) 5(2):23

Chapter 8
Alternative Protein Sources for Sustainable Tilapia Farming

Sherine Ragab, Mohmed Hassaan, Kevin Fitzsimmons, and Ehab El-Haroun

Abstract During the past 20 years, there has been a growing concern about how best to replace fish meals with more sustainable, cost-effective, and environment-friendly ingredients. A global crisis of high feed prices, expansion of the aquaculture sector, and the increasing proportion of farmed fish have run into a decrease in the supply of fishmeal products. Therefore, scientists have redoubled efforts to discover alternatives and more renewable ingredients for aquafeeds worldwide. The goal has been to protect ecosystems and increase economic viability based on enhanced efficiency of production facility outputs and nutritional efficiency. A key challenge though is sourcing sustainable, renewable protein ingredients. This chapter reviews the advances, novel ingredients, and new techniques that serve the mission for the future of aquafeed research on alternatives for fishmeal to achieve global aquaculture sustainability, with a focus on partial or full replacement of fishmeal, especially in the feed of tilapia.

Keywords Aquaculture · Sustainability · Plant-based protein · Animal-based Protein

---

S. Ragab
Ministry of Agriculture, Giza, Egypt

M. Hassaan
Animal Production Department, Fish Nutrition Research Laboratory, Faculty of Agriculture, Benha University, Banha, Egypt

K. Fitzsimmons
Department of Soil Water and Environmental Science, College of Agriculture and Life Sciences, University of Arizona, Tucson, AZ, USA

E. El-Haroun (✉)
Fish Nutrition Research Laboratory, Faculty of Agriculture, Cairo University, Giza, Egypt

© The Author(s), under exclusive license to Springer Nature Switzerland AG 2023
S. H. Hoseinifar, H. Van Doan (eds.), *Novel Approaches Toward Sustainable Tilapia Aquaculture*, Applied Environmental Science and Engineering for a Sustainable Future, https://doi.org/10.1007/978-3-031-38321-2_8

8.1 Introduction

The world population is predicted to reach 9.8 billion by 2050, which consequently will increase our need for protein by 30%–80% (Hua et al. 2019). However, the aquaculture industry faces big challenges to provide balanced amino acid proteins and other essential nutrients (SOFIA 2020; Barrows et al. 2007). Aquaculture will be considered a major source of animal protein in the near future to satisfy the nutritional needs of a rising human population (Gamboa-Delgado and Marquez-Reyes 2018). Already, animal nutrition plays a main role in aquaculture (Kaushik et al. 2011). Feed costs make up a significant fraction of total fish production costs, representing 40–70% of operating costs (USB 2008). Protein is the most essential nutrient needed for growth to be efficiently utilized and is relatively expensive. Thus, fishmeal (FM) has been considered an essential of aquafeeds, supplying over 60% of the protein total. Therefore, it has been the most costly ingredient in aquafeeds and has been the favorite dietary protein source for many fish feeds due to its essential amino acid profile, vitamin-mineral content, high digestibility, and palatability (Miles and Chapman 2006). The production of fishmeal fluctuates depending on several items including global environmental conditions including El Niño—La Niña oscillation, government fishing subsidies, and availability of fishing crews. The general situation for the last 30 years has been decreasing fishmeal supply as stocks are overfished and fishers resort to capturing juveniles of non-forage fishes., Therefore, the development of alternative sources of protein has become a priority of research in aquaculture. Affordable and sustainable alternative ingredients would compensate for the shortages of fishmeal and fish oil and help to achieve aquaculture sustainability and increase industry profitability (Ayadi et al. 2012; Nathaly et al. 2018; Alsager et al. 2018). The development of alternative ingredients would also address the competition for limited feed resources with other animal livestock, especially in the poultry industry (Hardy 2010). The aquafeed industry currently relies on derived ingredients to formulate cost-effective diets with a target by 2025 to supply 40 million tons of aquafeeds (Aas et al. 2019; Willora et al. 2020; Hua et al. 2019). Hence, FM is increasingly replaced with ingredients such as plant and terrestrial animal proteins, agriculture by-products, and microbial proteins to partially or replace FM without compromising the growth performance of the fish or feed utilization efficiency. In addition, the industry has learned to remove many Anti-Nutrition Factors in certain animal and plant-based ingredients by soaking, cooking, fermentation, chemical and enzyme treatment, and deficiency of particular amino acids by using additives (Archit et al. 2019). In this chapter, we review the beneficial use of alternative protein sources in aquafeeds for a more sustainable and secure future for the aquaculture industry, with the potential to partially or fully substitute for fish feed, especially in the feed of tilapia (Table 8.1).

Table 8.1 Summary of alternative proteins (animal, plant, and microbial) in aquafeeds

Source (specification)	Species fish	Levels (%) inclusion	Supported inclusion level (%)	Species (weight, g)	Type replacement	References
Animal protein sources						
Meat and bone meal	*Oreochromis niloticus*	0, 25, 50, 75, and 100%	50	1.80	Fish meal(FM)	Gomaa and El Moghazy (2014)
Poultry by-products	*O. niloticus*	10, 20, 30, or 40%	Up to 20	—	Soybean meal (SBM)	Dawood et al. (2020)
Poultry by-product	*O. niloticus*	50, 75, and 100	100	1.5	FM	Yones and Metwalli (2015)
Feather meal	Hybrid red tilapia (*Oreochromis niloticus x Oreochromis mossambicus*)	10% and 15	15	37	FM	Yong et al. (2018)
Feather meal	Hybrid red tilapia (*O.niloticus x O.mossambicus*)	10 and 15	15	18	SBM	Farahiyah et al. (2018)
Hydrolyzed chicken feather meal	Red tilapia	3, 6, 9, and 12	Up to 12	24.09	FM	Nursinatrio and Nugroho (2019)
Hydrolyzed feather meal	*O. niloticus*	33, 66, and 100	Up to 33–66	2.30	FM	Suloma et al. (2014)
Chicken feathers	*O. niloticus*	0, 12.5, 25, 37.5, and 50%	12.5	3.21	FM	Diana and Istiyanto (2019)
Black soldier fly larvae meal	Hybrid tilapia, Nile x Mozambique (*O. niloticus x O. Mozambique*)	0, 5, 10, 20, and 30%	30	2.6	SBM	Yildirim-Aksoy et al. (2020)
Black soldier fly maggot meal (BSFM)	*O. niloticus*	0, 25, 50,75, and 100%	Up to 50%	3.07 g	FM	Muin et al. (2017)
Plant protein sources						
Soybean meal	*O. niloticus*	0,32, 42, and 50,9%	Up to 42%	2.09	FM	Obirikorang et al. (2020)

(continued)

Table 8.1 (continued)

Source (specification)	Species fish	Levels (%) inclusion	Supported inclusion level (%)	Species (weight, g)	Type replacement	References
Soybean meal	*O. niloticus*	0%, 25%, 50%, and 75%	50%	15.58	FM	Ahmad et al. (2020)
Soybean meal	*O. niloticus*	0, 22, 44, 66, and 88%	Up to 22%	18.51	FM	Amer et al. (2019)
Soybean meal	*O. niloticus*	0, 5, 10, 15, and 20%	20%	3.22	FM	Sangeera and Reddy (2019)
Soybean meal	*O. niloticus x O. aureus*	0, 10, 20, 30, and 40%)	Up to 30%	32.47	Rubber seed meal	Deng et al. (2015)
Soybean meal	Hybrid tilapia (*O. niloticus x O. aureus*)	0, 33, 50, 67, and 100%	Up to 50	1.54	Shrimp shell meal	Fall et al. (2012)
Soybean meal	*O. niloticus × O. aureus*	25, 50, 75, and 100%	50%	4	FM	Lin and Luo (2011)
Soybean meal	Hybrid tilapia (*O.niloticus x O. aureus)*	0, 15, 30, 45, 60, and 100%	30%	56.3	Canola meal	Zhou and Yi-Rong (2010)
Fermented soybean meal (FSBM)	(*O. niloticus,* L.1758)	0, 20, 30, 40, and 50%	Up to 40%	0.74	FM	Yigit (2016)
Cottonseed meal	*O. niloticus*	12, 16, and 23%	12%	11.62	FM	Hassaan et al. (2019)
Cottonseed meal	*O. niloticus*	0, 25, 50, and 75%	50%	4.24	FM	Agbo et al. (2011)
Cottonseed meal	*O. niloticus*	0, 25, 50, 75, and 100%	Up to 25%	1.43	FM	Soltan et al. (2011)
Cottonseed meal	*O. niloticus*	0, 25, 50, 75, and 100%	25%	4.79	SBM	El-Saidy and Saad (2011)
Cottonseed meal	Hybrid tilapia (*O. niloticus x O. aureus*)	15, 30, 45, 60, and 100%	Up to 30%	6.24	SBM	Yue and Zhou (2008)

Sunflower seed meal	*O. niloticus*	0, 25, 50, or 75%	25	10.30	FM	Soltan et al. (2015)
Sunflower seed meal	Tilapia rendalli	10,20,30,40, and 50%	Up to 20	0.93 g	FM	Olvera-Novoa et al. (2002)
Fermented sunflower meal	*O. niloticus*	0, 25, 50, and75%	25	10.28 g	FM	Hassaan et al. (2018)
Canola meal	*O. niloticus*	0, 12.5, 25, 37.5, and 50%	25	0.1	FM	Mohammadi et al. (2019)
Canola meal	*O. niloticus*	0, 15, 30, 45, 60, and 75%	75%	20	FM	Luo et al. (2012)
Canola meal	*O. niloticus*	0,10,20,30,40, and 50%	Up to 10%	1.21	FM	Yigit and Olmez (2009)
Canola seed meal (CSM)	*O. niloticus*	0,25, 50, 75, and 100%	50%	8.42	FM	Soltan (2005)
Fermented canola meal	*O. niloticus*	0, 25, 50,75, and 100%	Up to 50	10	SBM	Pichet and Yakupitiyage (2014)
Lupin seed meal (LSM)	*O. niloticus*	0, 15, 30, 45, and 60%	30%	3	FM	Srour (2007)
Lupin seed meal (LSM)	Hybrid tilapia (*O. niloticus xO. Aureus*)	0, 33%, 67%, and 100%	Up to 33%	0.3	SBM	Chien and Chiu (2003)
Linseed protein	*O. niloticus*	0, 25, 50, 75, and 100%	Up to 75	12.4	FM	El-Saidy and Gaber (2001)
Microbial biomass (single-cell protein)						
Single-cell protein	*O. niloticus*	25, 50, and 75%	Up to 50%	15.39	FM	Al-Hafedh and Alam (2013)
Single-cell protein	*O. niloticus*	0, 10, 20, 30, 40, and 50%	Up to 50%	8	FM	Bob-Manuel and Alfred-Ockiya (2011)

(continued)

Table 8.1 (continued)

Source (specification)	Species fish	Levels (%) inclusion	Supported inclusion level (%)	Species (weight, g)	Type replacement	References
Microalga spirulina	*O. niloticus*	20, 40, 60, 80, and 100%,	40	26.7	FM	Olvera-Novoa et al. (2008)
Spirulina (Arthrospira platensis)	*O. niloticus*	0, 30, 45, 60, and 75%	30	0.89	FM	Stephanie et al. (2016)
Spirulina (Spirulina platensis)	*O. niloticus*	0, 5, 10, 15, and 20%	Up to 20	—	SBM	Abo El-Ward et al. (2016)
Spirulina (Arthrospira platensis)	Hybrid red tilapia (*O.niloticus x O. mossambicus*)	0, 50, 75, and 100%	75	0.23	FM	El-Sheekh et al. (2014)
Green seaweed Ulva	*O. niloticus*	0, 10, 20, and 30%	10	3.5	FM	Suryaningrum and Samsudin (2019)
Seaweed Ulva	*O. niloticus*	0, 10, 20, and 30%	10	3.42	FM	Lusi and Reza (2020)

8.1.1 Animal-Based Protein Sources

Animal by-products are well known as valuable sources of protein in aquafeeds and are recognized for their high content of protein, comparable amino acid profiles, ready availability within the market, and competitive price (Hassaan et al. 2018; Moutinho et al. 2017). The types of animal protein mainly are meat and bone meal (MBM), poultry by-product meal (PBM), feather meal, and insect meal (Bureau et al. 2000; Metts et al. 2011; Wu et al. 2018).

8.1.1.1 Meat and Bone Meal

Meat and bone meal (MBM) is typically 48–52% protein, 33–35% ash, 8–12% fat, and 4–7% water. Meat and bone meal (MBM) has several advantages including (i) high protein content, minerals, good amino acid profile, lack of anti-nutritional factors, and a reasonable price usually less than fish meal (Ferouz, et al. 2012; Suloma et al. 2013). The nutritive value of meat and bone meal depends on the quality of raw materials and the processing techniques (Kureshy et al. 2000). However, the occurrence of bone, high levels of ash and mineral content can limit the amount used in fish diets (Bureau et al. 1999). A recent research stated that meat and bone meal could positively substitute for fish meal for up to 75% of diets for Mozambique tilapia or seabream without compromising performance (Gomaa and El Moghazy 2014; Davies et al. 2020). In addition, Yang et al. (2004) and Hu et al. (2008) found that MBM supplemented with lysine and methionine successfully replaced up to 50% of the fishmeal enhancing growth performance for gibel carp (*Carassius auratus gibelius*).

8.1.1.2 Poultry by-Products

Poultry by-product meal (PBM) is often considered the most important animal protein source for replacing fish meal (Meeker and Hamilton 2006). The AAFCO (USA) defines poultry by-product meal as the ground, rendered, clean parts of the carcass of slaughtered poultry such as necks, heads, feet, undeveloped eggs, gizzards, and intestines (provided their content is removed), exclusive of feathers (Watson 2006). Whole poultry carcass meal can also be obtained from culled laying hens (spent hen meal), notably in areas where there is no market for culled hens (Hertrampf and Piedad-Pascual 2000). Poultry by-product meal (PBM) has been shown to substitute for 80–90% fishmeal in fish feed. With high protein content (55–75%), and an essential amino acids profile similar to FM's AAs profile, except for lower lysine and methionine levels, PBM can lower costs in fish diets (Gupta et al. 2020; Gonzalez-Rodrıguez et al. 2016; Heuzé et al. 2015). However, the composition of PBM depends on processing conditions and drying techniques, and the original raw materials (Johnson and Parsons 1997). Previous trials have been conducted to investigate the effects of replacing FM by PBM for several fish species including Nile tilapia, redbelly tilapia (*Coptodon zillii)*, and hybrid striped bass (Metwalli 2008; El-Husseiny et al. 2006; Yones and Metwalli 2015; Dawood et al. 2020; Yildrim et al. 2009; Rawles et al. 2009). Tacon et al. recommended that 10–30% of PBM could replace FM depending on the type of fish and processing technique. Recent research conducted by Yones and Metwalli (2015) found that up to 100% of FM could be replaced by PBM with positive impacts on the growth of *O. niloticus*. El-Haroun and Burea (2007); El-Haroun et al. (2009) found using blood meal, or a combination of animal protein including PBM up to 30% in rainbow trout diets, could replace fish meal with growth matching the control diet with fish meal as the main source of protein. Furthermore, Dawood et al. (2020) found that PBM could replace up to 20% of FM content, without any negative impacts on the performance and feed utilization efficiency of Nile tilapia. NRC (2011) concluded that two main factors could limit the high inclusion of PBM in aquatic feeds; (i) high lipid content, and (ii) the deficiency of DHA and EPA in lipids of PBM.

8.1.1.3 Feather Meal

Feather meal (FEM) is a by-product of poultry processing, made from poultry feathers by partially grinding them under pressure and heat, and then additional grinding and drying. Although total protein content levels are high, the bioavailability of this nitrogen may be low due to the difficulty of digestion and assimilation from the presence of beta-keratin and fibrous fiber components (Adelina et al. 2019). This has been a constraint of using FEM in aquafeeds. Several protocols and solutions have been developed to mitigate these challenges and enhance- diet quality, flavor, and improvement in terms of digestibility (Nursinatrio Nugroho 2019) including (i) hydrolysis and; (ii) fermentation techniques improving protein digestibility and palatability (Isika et al. 2006; Madigan and Martinko 2005). Laporte et al. (2007) and Davies et al. (2009) demonstrated that using these techniques improved the coefficients of crude protein (CP) from 67 to 87% indicating a significant improvement in the digestibility of feather meal. These enhancements could be attributed to good manufacturing techniques and the optimization of the processing conditions (Poppi et al. 2011). Thus, the inclusion of feather meals in diets can decrease the cost of feed and thus help in the exploitation of these by-products (Bishop et al. 1995). Kumari and Sundarabarathy (2011); Suloma et al. (2014) and Abwao et al. (2017) found that FEM could replace up to 65% of FM in diets for Nile tilapia without negative impacts on performance or feed utilization. Similarly, Farahiyah et al. (2018) and Yong et al. (2018) found that FM could be replaced at 15–30% with feather meal without compromising the growth performance, feed utilization, and protein efficiency in juvenile red hybrid tilapia and Nile tilapia.

Hydrolyzed feather meal (HFM) is another type of feather meal with keratins being hydrolyzed, that is commonly used due to its high crude protein content (74%–91%) (Campos et al. 2017; Divakala et al. 2009; Munguti et al. 2014; Suloma et al.

2014) and is highly digestible ~80% (Davies et al. 2020). Recent research confirmed this idea as Yu et al. (2020) and Suloma et al. (2014) found that up to 25–45% or 66%, respectively, of FM, could be replaced by HFM without affecting growth rate or feed utilization on *O. niloticus*. Nursinatrio Nugroho (2019) and Yones and Metwalli (2015) reported that 12% of hydrolyzed feather meal incorporated into red tilapia and juvenile Nile tilapia diets enhanced growth compared to the control with fish meal as the sole source of protein.

8.1.1.4 Insect Meal

Historically, the European Union, the US, and other countries had strict regulations regarding insects and insect parts allowed in feed ingredients or finished feed. With the advent of farmed insect meals specifically as an ingredient, regulations are being rewritten. In 2021, there are about ten insect species that have received approval for use in animal feeds in major markets and are considered non-pathogenic (Wang and Shelomi 2017). The most heavily researched are the black soldier fly, the common housefly, and the yellow mealworm. Larval meals from these as well as crickets, grasshoppers, and silkworms all have high levels of crude protein (similar to soymeal and fishmeal) and lipids. Insect meals show promise as an ingredient in aquafeed due to their increasing availability and superior protein content of up to 60–70%, depending on the stage and species of insect (Prachom et al. 2021) and balanced amino acid profiles, vitamins, and minerals similar to that of FM (Barroso et al. 2014; Henry et al. 2015; Gasco et al. 2019). Their amino acid composition is comparable to FM and protein digestibility is high, and these can substitute for 25–100% of soybean meal or fishmeal protein in aquafeeds. High oil content in some insect meals can be extracted to produce another value-added product, biodiesel; and the defatted insect meals have protein content even higher than soybean meal or fishmeal. Several insect meals are recognized for providing high productivity and low FCR (Berggren et al. 2019; IPIFF 2018; Bruni et al. 2018). The main challenge of using certain insect meals is their unfavorable fatty acids (FA')s profiles, which can be altered by the feed offered to the insect larvae or by removal of FAs using different processing and drying techniques (Lock et al. 2016; Ferrer et al. 2019).

Defatted insect meal can contain up to 83% crude protein (Makkar et al. 2014a, b). Previous studies confirmed that insect meals can partially or completely replace FM (Silvia et al. 2019; Tran et al. 2015; Henry et al. 2015). Belghit et al. (2019) and Borgogno et al. (2017) found that 100% of fish meal could be replaced with different types of insect meals that could help to boost growth performance, disease resistance, and immune response (Ido et al. 2019). Confirming this hypothesis (Gasco et al. 2016; Iaconisi et al. 2018; Magalhães et al. 2017; Renna et al. 2017; Bruni et al. 2018; Devic et al. 2018; Wang et al. 2017) found that up to 70% substitution of fish meal with the black soldier fly meal obtained growth performance and feed efficiency matching basal diets with the fish meal as the main source of protein. Kurniawan et al. (2018) reported using BSF that up to 14% replacement of fish meal improved growth and FCR ($P \leq 0.05$) in Nile tilapia with replacement of fish meal in the control diet. Successful 75% FM substitution with *Musca domestica* (MD) was reported in Nile tilapia (Ogunji et al. 2007) and in hybrid tilapia up to 30% (Yildirim-Aksoy et al. 2020). The sum of this research highlights the importance of FM replacement with BSF and variation with larval meal level depending on fish species and processing techniques (Henry et al. 2015).

8.1.2 Plant-Based Protein Sources

Modified plant proteins treated with exogenous enzymes or solid-state fermentation (SSF) are finding their way into aquafeeds, as a complete or partial replacement for FM due to their low cost and increasing availability. Modified plant protein sources including soybean oil meal (SBM), sunflower meal, and cottonseed meal are considered promising ingredients to partially or totally replace fish meal in aquafeeds (Puja et al. 2019). The main challenges of using these untreated ingredients are (i) low initial protein content, (ii) high fiber content (iii) the presence of antinutritional factors (ANFs), (iv) unbalanced profile of EAA, and (v) deficit of essential AAs lysine and methionine, and low palatability which negatively impact their use in fish diets (Henry et al. 2015; Naylor et al. 2009; Oliva-Teles et al. 2015). Different strategies are used in ingredient processing to mitigate and avoid the negative impacts of ANF on plant ingredients (Puja et al. 2019). Soaking, heat treatment, fermentation, and enzymatic treatment are available as detoxification methods (Aguihe et al. 2017). Modified plant proteins are providing partial or total replacement of FM in many fish species (Hassaan et al. 2019; Bowyer et al. 2020), based on new advances in feed technology processing and gene technology. Many studies to evaluate alternatives to FM have been conducted with plant proteins, which have achieved success in fish diets (Glencross et al. 2011; Ngo et al. 2016; Agbo et al. 2015; Al-Thobaiti et al. (2017). These studies reported the substitution of FM by a mixture of plant protein sources (soya bean meal, cottonseed meal, groundnut cake, corn gluten meal, and wheat gluten meal) or using new technology including solid state fermentation (SSF) was able to replace FM (50%) in the diets for Nile tilapia (*O. niloticus*) and 75–100% in common carp and seabream respectively without affecting growth performance or feed utilization. The results showed that fish meals can be replaced up to 100% by a mixture of plant proteins without a negative impact on growth, feed utilization, and fish health.

8.1.2.1 Soybean Meal

Soybean meal (SBM) is certainly the most common plant protein source to replace FM protein in fish diets with its high protein content (45–55%) (Ozkan et al. 2015). A challenge of using SBM in fish diets is ANFs (Cabral et al. 2011; Gerile and Pirhonen 2017). To maximize the incorporation of SBM in fish diets it is necessary to eliminate or inactivate its ANFs content (Sales 2009). Different processing techniques have been adapted to remove ANFs to adjust and enhance the nutritive value of SBM including soaking, solvent-extracting, or fermenting processes (Hotz and Gibson 2007). Selective breeding and advanced genetic tools have been employed to increase the level of essential AAs that were otherwise limited in older soybean varieties to mimic FM (Llanes and Toledo 2011; Hassaan et al. 2015).

Feeding trials with several species of fish reported that improved versions of SBM could enhance performance, digestibility, and immune function compared to traditional SMB and FM (Bansemer et al. 2015; Ferrara et al. 2015; Liu et al. 2017; Grisdalle-Helland et al. 2002; Bonaldo et al. 2008; Zhou et al. 2011). Choi et al. (2019) reported that fermented SBM was used to replace 40% of FM without a negative impact on the growth performance and feed utilization of rainbow trout. Ye et al. found that up to 75% of FM could be replaced by a complementary mixture of plant protein (soybean meal, corn gluten meal, and cottonseed meal) without significant differences in growth and feed utilization of hybrid Nile tilapia. Also, Obirikorang et al. (2020) found that up to 42% of fish meals could be replaced with SBM in Nile tilapia and up to 30% in hybrid tilapia. However, as the price for fermented SBM is increasing there is interest in less-traditional plant protein sources to improve the sustainability of aquafeeds, aquaculture, and farmers' profitability (Fawole et al. 2016).

8.1.2.2 Cottonseed Meal (CSM)

The more common end-efficient technique used for the production of cottonseed meal is the extraction of oil from cottonseed meal. Cottonseed meal (CSM) has been used as an alternative ingredient to both fishmeal and soybean meal due to its lower cost, being readily available in some countries, particularly in the USA, China, India, and Egypt. Cottonseed meal has advantages including (i) nutritive value, (ii) good palatability. (iv) low cost and (vi) good protein content which is variable (23–53%) contingent on the processing (Yue and Zhou 2008; Ng and Romano 2013; El-Saidy and Saad 2011). The main challenge of using cottonseed meal in aquafeed could be summarized: (i) low levels of EAAs such as methionine and lysine., (ii) CSM contains high concentrations of the ANFs phytic acid and gossypol, and (iii) CSM contains a toxic pigment that decreases lysine availability (Ayadi et al. 2012). Previous trials reported that CSM was used to replace FM and SBM in the diets of *O. niloticus* without effects on growth performance (El-Saidy et al. 2012; Agbo et al. 2011). Sun et al. (2015) reported that CSM replaced up to 16% of FM in a feed trial for black seabream without negative effects on performance and feed conversion ratio. Additional research proved the benefits of using CSM to replace FM for hybrid tilapia and Nile tilapia (Yue and Zhou 2008; Soltan et al. 2015; El-Saidy and Saad 2011).

8.1.2.3 Sunflower Seed

The nutritive value of sunflower meal is similar to that of soybean meal and contains high amounts of sulfur amino acids. However, it contains relatively high amounts of crude fiber, low lysine content, and contains high ANFs protease and arginase inhibitors which have limited the incorporation of CSM in aquafeed (Gonzalez-Salas et al. 2014; Ochieng et al. 2017). Previous research concluded that the quality and quantity of protein in the sunflower meal depended on the oil extraction technique.

Ochieng et al. (2017) reported that up to 25% of FM could be replaced with CSM in Nile tilapia diets without ($P \geq 0.05$) impacting growth. Soltan et al. (2015) found that up to 25% of FM could be replaced by fermented sunflower meal for Nile tilapia, without negative impacts on performance, feed consumption, or protein efficiency ratio. Abdel et al. (2008) confirmed the possibility of substituting up to 30% of soy protein with sunflower and sesame seed for Nile tilapia without impacts on performance and carcass composition. However, the high inclusion of sunflower could affect protein conversion in the fish. In addition, (Abdel et al. 2010; Mugo-Bundi et al. 2013) found that the protein level in the body composition tended to decrease consistently with the high ash and crude fiber content in the diets. When testing the cumulative presence of sunflower meal up to 100% replacement, increasing the inclusion of CSM led to reduced digestibility and consequently a decrease in the growth rate of fish.

8.1.2.4 Rapeseed or Canola Seed

The chemical composition of rapeseed oil meal or canola protein concentrate (CPC) contains high levels of crude protein equal to FM with high content of lysine and methionine compared to SBM (Gonzalez-Salas et al. 2014). Friedman (1996) stated that the amino acid profile of canola protein is superior compared to many other plant proteins. However, some disadvantages limit the inclusion of rapeseed meals in both fresh and marine water species such as the high amount of fiber and ANFs including glucosinolates, tannins, and phytate (Ayadi et al. 2012). Burr et al. (2013) reported that for *Salmo salar,* there was not a significant difference in growth comparing the salmon fed the 0% CPC and the 10% CPC diets ($p > 0.05$). Canola protein concentrates significantly depressed growth when included in the diet at 20%, but not at 10%, indicating that canola could be used as a minor feed ingredient when available. The nutritive value and quality of canola meal rely on the intensity of heat and drying techniques used within oil extraction. However, excessive heating during the processing leading can lead to the reduction of protein digestibility and amino acid availability, especially lysine (Ayadi et al. 2012). An early trial conducted by Lim et al. (1997) found that rapeseed meal protein was a cost-effective replacement for FM. Soares et al. (2001) stated that rapeseed meal could substitute for 49% of SBM in tilapia diets. Mohammadi et al. (2019) found that canola meal can replace up to 35% of SBM or FM for hybrid tilapia without any impact on the growth of fish. Plaipetch and Yakupitiyage (2014) found that up to 50% of yeast-fermented canola meal in *O. niloticus* could be included in a diet without a negative impact on growth, survival, or feed consumption. Though, a high inclusion level of rapeseed meal up to 75–100% had an adverse effect on nutrient digestion and protein utilization efficiency of Nile tilapia.

8.1.2.5 Lupin Meal and Pea Meal

Lupin meals (LM) are recognized to be a unique replacer of FM or SBM due to their high content of protein (up to 45%), balanced nutritional characteristics, palatability, high digestibility, and stable price in the market (Bartkiene et al. 2015; Rajeev and Bavitha 2015; Lucas et al. 2015). The major concern about the inclusion of lupins in aquafeeds is the presence of some ANFs, which can reduce nutrient availability and digestibility (Petterson 2000; Francis et al. 2001). Burel et al. (2000) found up to 45% extruded lupin meal diets could substitute for fishmeal with positive support of the growth performance of rainbow trout, and turbot (*Psetta maxima*). Abdel-Moneim Yones (2010) found that LM succeeded to replace FM in hybrid red tilapia without affecting growth performance. The appropriateness of lupin meal in aquafeeds could have a positive effect on the fish diet, as it would alleviate the dependence on imported plant-based proteins (Anwar et al. 2020). Furthermore, Vo-Binh et al. (2015) reported that fermented lupin meal succeeded to replace 70% of FM without significant changes in performance and FCR compared to the basal diets in Nile tilapia containing FM as a main source of protein. The findings of these results stated that FM and soybean meal could be replaced by lupin meal without affecting performance, survival, and nutrient utilization efficiency (Pham et al. 2020).

Øverland et al. (2009) reported that pea protein concentrate was shown to be a promising new protein ingredient for salmonids and could replace 20% of high-quality fish meal protein in the feed without any adverse effect on growth performance, carcass composition, or histology of the gut. Schulz et al. (2007) reported that for juvenile Nile tilapia inclusion of 30% protein derived from pea protein isolate in the diet resulted in a growth performance (in terms of WG and SGR) that did not differ significantly from the diet with 100% of protein derived from FM.

8.1.2.6 Linseed Meal

Linseed meal has been recognized for its high protein content (30%), after the extraction of its oil (Oomah and Mazza 1993). However, it has some components limiting its use in fish nutrition (i) high fiber content (37%) and its water retention capacity are considered responsible for the depletion of growth that reduces food intake (Ndou et al. 2017; Goulart et al. 2015), and (ii) trypsin inhibitor and total polyphenolic compounds which may impair the assimilation and availability of the nutrients (Soltan 2005; Francis et al. 2001). The high fiber content could contribute to a high capacity to engross water and increase volume thereby limiting intestinal topography capacity and degrading the nutrients (Eastwood et al. 2009). The application of protein isoelectric precipitation is a solution to mitigate the ANFs, increase the incorporation of linseed in aquafeeds and increase protein concentration and eliminate the antinutrients components in fish diets, (Lovatto et al. 2014, 2015, 2016). Soltan (2005) compared raw LSM with roasted LSM and autoclaved LSM as substitutes for SBM in tilapia diets. The trials demonstrated that there was no significant difference in growth or FCR with up to 25% replacement of SBM with roasted or autoclaved LSM but a 7% decrease in feed cost. At the level of 75% replacement, there were significant decreases in growth and FCR, but feed costs were decreased by 18%. Pianesso et al. (2020) found that up to 40% linseed protein concentrate (LPC) could replace FM in silver catfish without influence on growth and nutrient efficiency. Wei et al. (2018) found that the replacement of fish oil with linseed oil in the fish feed did not affect performance, SGR, feed efficiency, or survival rate in Nile tilapia. Hanafy (2006) tested several diet formulations including LSM and concluded that soybean could be completely replaced by linseed in Nile tilapia diets without adverse effects on the growth rate when diets supplemented with *Yucca schidigera* as an additional nutrient and functional ingredient source.

8.1.3 Microbial Biomass and Macro-Algae (Seaweeds)

Recently feed nutritionists are examining different sources of microbial proteins (bacteria, algae, and yeast) as novel ingredients for fish (Pacheco-Vega et al. 2018). Single-cell proteins (SCP) have the highest protein content (65–85%) and contain balanced essential amino acid profiles. Moreover, SCP products are rapidly increasing in volume, renewable, and constitute a variety of compounds that can be blended to substitute for protein sources such as FM and SBM (Intraco 2020; Sharif et al. 2021; Hua et al. 2019; Jones et al. 2020). The quality and quantity of single-cell proteins vary widely due to feed sources, type of bacteria, algae or yeast, conditions of fermentation, and processing technique (Øverland et al. 2010). Previous research conducted by (Zamani et al. 2020; Gamboa-Delgado et al. 2016; Zhao et al. 2015; Rosales et al. 2017; Tlusty et al. 2017; Kalhoro et al. 2018; Sun et al. 2015; Chen et al. 2020) found that SCP could replace FM of SBM up to 60% in *O. niloticus* diets without effects on growth rate feed consumption. Likewise, Aas et al. (2006) found that yeast biomass could be included in 30% of the diet instead of fishmeal with no effect on fish growth rate and nutrient absorption. The benefits of microalga biomass as a feed additive or substitute for FM or SBM in tilapia fish feed have been documented with positive impacts on growth and feed efficiency (Gamboa-Delgado and Marquez-Reyes 2018; Sarker et al. 2018). The inclusion of *Spirulina* up to 10% has been reported as a feasible alternative to enhance the nutritional value in the feed of Nile tilapia (Rosas et al. 2019; Van Vo et al. 2020). Also, up to 30% replacement of fish meal by *Arthrospira maxima* without negative impact on the performance of red tilapia has been reported (Rincón et al. 2012). Badawy et al. (2008) reported that the substitution of up to 50% FM with *Chlorella* sp. and *Scenedesmus* sp. enhanced the growth of Nile tilapia. Replacement of fish meal by *Spirulina platensis* at up to 100% did not cause negative effects on growth in various fish species Nandeesha et al. (2001).

The use of seaweed as an aquaculture feed ingredient has demonstrated positive responses to enhance growth performance, feed utilization, enhance serum parameters, immune responses, hematological parameters, and disease resistance of many fish species; gray mullet (*Mugil cephalus)*, rainbow trout, Asian sea bass (*Lates calcarifer*), Nile tilapia, rabbitfish, red tilapia, and Atlantic salmon (Zeraatpisheh et al. 2018; Akbary and Aminikhoei 2018; Akbary et al. 2018; Yangthong et al. 2016; Meurer et al. 2009; Hussein 2017; Ashour et al. 2020; Doan et al. 2017; Yengkhom et al. 2018; Hosseinifer et al. 2020; Yengkhom et al. 2018; Abdel-Aziz and Ragab 2017; El-Tawil 2010; Collins et al. 2019), without any negative effect on fish. Pratiwy et al. (2018) found that the inclusion of 8% *Sargassum* meal enhanced the growth of Nile tilapia. Younis et al. (2018) reported that up to 20% of red algae inclusion in the diet improved the performance, feed utilization, and body composition of tilapia.

References

Aas TS, Hatlen B, Grisdale-Helland B, Terjesen BF, Bakke-McKellep AM, Helland SJ (2006) Effects of diets containing a bacterial protein meal on growth and feed utilisation in rainbow trout (*Oncorhynchus mykiss*). Aquaculture 261:357–368

Aas TS, Ytrestøyl T, Åsgård T (2019) Utilization of feed resources in the production of Atlantic salmon (*Salmo salar*) in Norway: an update for 2016. Aquac Rep 15:100216

Abdel HNFM, Lashin ME, Al-Azab AA, Nazmi HM (2008) Effect of replacing soybean meal protein by other plant protein sources on growth performance and economic efficiency of mono sex Nile Tilapia (*Oreochromis niloticus*) cultured in tanks. Proceedings of the 8th International Symposium on Tilapia in Aquaculture. Cairo. Egypt pp. 739–755

Abdel TM, Mohammad H, Ahmad MH, Khattab YAE, Shalaby AME (2010) Effect of dietary protein level, initial body weight, and their interaction on the growth, feed utilization, and physiological alterations of Nile tilapia, *Oreochromis niloticus* (L.). Aquaculture 298:267–274

Abdel-Aziz MFA, Ragab MA (2017) Effect of use fresh macroalgae (seaweed) *Ulva fasciata* and *Enteromorpha flexusa* with or without artificial feed on growth performance and feed utilization of rabbitfish (*Siganus rivulatus*) fry. J Aquac Res Develop 8:482

Abdel-Moneim Yones AM (2010) Effect of lupin kernel meal as plant protein source in diets of red hybrid tilapia (*Oreochromis niloticus x O. mossambicus*), on growth performance and nutrients utilization. African J Biol Sci 6(1):1–16

Abo El-Ward A, Eid AE, Mohamed KA, Abd-elfattah B, Hasan MA (2016) Growth performance of Nile tilapia (*Oreochromis niloticus*) fingerlings fed diet supplemented with different levels of *Spirulina platensis*. Egyptian J Anim Prod 53(3):181–190

Abwao J, Safina M, Ondiba R, Ogello E, Obiero K (2017) Effect of replacing marine protein with hydrolyzed feather meal on growth, apparent digestibility and body composition of juvenile tilapias; *Oreochromis mossambicus* (peters, 1852). Int J Fish Aquatic Stud 5(6):242–250

Adelina A, Feliatra F, Siregar Suharman I, Pamukas NA (2019) Fermented chicken feathers using *Bacillus subtilis* to improve the quality of nutrition as a fish feed material. Earth Environ Sci 348(2019):012008

Agbo NW, Madalla N, Jauncey K (2015) Mixtures of oilseed meals as dietary protein sources in diets of juvenile Nile tilapia (*Oreochromis niloticus L.*). J Sci Technol 35:11–24

Agbo NW, Madalla N, Jauncey K (2011) Effects of dietary cottonseed meal protein levels on growth and feed utilization of Nile tilapia, *Oreochromis niloticus* L. J Appl Sci Environ Manag 15(2):235–239

Aguihe PC, Kehinde AS, Ospina-Rojas CI, Murakami AE (2017) Evaluation of processing methods of rubber (*Hevea brasiliensis*) seed meal for use as a feed ingredient for broiler chickens. J Poultry Res 14(1):20–27

Ahmad N, Jamal P, Siddiqui A, Khan KM, Ali A, Tahir M, Akbar NU, Amin M, Attaullah M, Boneri ID (2020) Effects of partial substitution of fishmeal by soybean meal in Nile tilapia (*Oreochromis niloticus*) diet. J Animal Plant Sci 30(2):364–370. ISSN: 1018-7081

Akbary P, Aminikhoei Z (2018) Effect of water-soluble polysaccharide extract from the green alga Ulva rigida on growth performance, antioxidant enzyme activity, and immune stimulation of grey mullet *Mugil cephalus*. J Appl Phycol 30:1345–1353

Akbary P, MollaZei E, Aminikhoei Z (2018) Effect of dietary supplementation of *Ulva rigida* C. Agardh extract on several of physiological parameters of grey mullet, *Mugil cephalus* (Linnaeus). Iranian J Aquat Animal Health 4:59–68

Al-Hafedh YS, Alam A (2013) Replacement of fishmeal by single cell protein derived from yeast grown on date (*Phoenix dactylifera*) industry waste in the diet of Nile tilapia (*Oreochromis niloticus*) fingerlings. J Appl Aquac 25(4):346–358

Alsager OA, Alnajrani MN, Alhazzaa O (2018) Decomposition of antibiotics by gamma irradiation: kinetics, antimicrobial activity, and real application in food matrices. Chem Eng J 338:548–556. https://doi.org/10.1016/j.cej.2018.01.065

Al-Thobaiti A, Al-Ghanim K, Ahmed Z, Suliman EM, Mahboob S (2017) Impact of replacing fish meal by a mixture of different plant protein sources on the growth performance in Nile tilapia (*Oreochromis niloticus L.*) diets. Braz J Biol. https://doi.org/10.1590/1519-6984.172230

Amer SA, Ahmed SAA, Ibrahim RE, Al-Gabri NA, Ali O, Mahmoud S (2019) Impact of partial substitution of fish meal by methylated soy protein isolates on the nutritional, immunological, and health aspects of Nile tilapia, *Oreochromis niloticus* fingerlings. Aquaculture 518:734871. https://doi.org/10.1016/j.aquaculture.2019.734871

Anwar A, Wan AHL, Omar S, El-Harounb E, Davies SJ (2020) The potential of a solid-state fermentation supplement to augment white lupin (*Lupinus albus*) meal incorporation in diets for farmed common carp (*Cyprinus carpio*). Aquac Rep 17:100348

Archit S, Munish K, Gyandeep G, Neeraj P, Varun M (2019) A review on replacing fish meal in aqua feeds using plant and animal protein sources. Int J Chem Stud 7(3):4732–4739

Ashour M, Mabrouk MM, Ayoub HF, El-Feky MMMM, Zaki SZ, Hoseinifar SH, Rossi W Jr, Van Doan H, El-Haroun E, Goda AMA-S (2020) Effect of dietary seaweed extract supplementation on growth, feed utilization, hematological indices, and non-specific immunity of Nile tilapia, *Oreochromis niloticus* challenged with Aeromonas Hydrophila. J Appl Phycol 32:3467–3479. https://doi.org/10.1007/s10811-020-02178-1

Ayadi FY, Rosentrater KA, Muthukumaprappan K (2012) Alternative protein sources for aquaculture feeds. J Aquac Feed Sci Nutr 4:1–26

Badawy TM, Ibrahim EM, Zeinhom MM (2008) Partial replacement of fishmeal with dried microalgae (*chlorella spp.* and *Scenedesmus spp.*) in Nile tilapia (*Oreochromis niloticus*) diets. In: Elghobashy H, Fitzsimmons K, Diab AS (eds) From the pharaohs to the future: proceedings of the 8th international symposium on tilapia in aquaculture. Egypt Ministry of Agriculture, Cairo, pp 801–810

Bansemer MS, Forder REA, Howarth GS, Suitor GM, Bowyer J, Stone DAJ (2015) The effect of dietary soybean meal and soy protein concentrate on the intestinal mucus layer and development of subacute enteritis in yellowtail kingfish (*Seriola lalandi*) at suboptimal water temperature. Aquac Nutr 21(3):300–310

Barroso FG, de Haro C, Sánchez-Muros MJ, Venegas E, Martínez-Sánchez A, Pérez-Bañón C (2014) The potential of various insect species for use as food for fish. Aquaculture 422:193–201

Barrows FT, Gaylord TG, Stone DAJ, Smith CE (2007) Effect of protein source and nutrient density on growth efficiency, histology and plasma amino acid concentration of rainbow trout (*Oncorhynchus mykiss* Walbaum). Aquac Res 38(16):1747–1758

Bartkiene E, Krungleviciute V, Juodeikiene G, Vidmantiene D, Maknickiene Z (2015) Solid state fermentation with lactic acid bacteria to improve the nutritional quality of lupin and soya bean. J Sci Food Agric 95(6):1336–13342

Belghit I, Liland NS, Gjesdal P, Biancarosa I, Menchetti E, Li Y, Waagbø R, Krogdahl ÅEJ (2019) Black soldier fly larvae meal can replace fish meal in diets of sea-water phase Atlantic salmon (*Salmo salar*). Aquaculture 503:609–619

Berggren A, Jansson A, Low M (2019) Approaching ecological sustainability in the emerging insects-as-food industry. Trends Ecol Evol 34:132–138

Bishop CD, Angus RA, Watts SA (1995) The use of feather meal as a replacement for fish meal in the diet of *Oreochromis niloticus* fry. Bioresour Technol 54:291–295

Bob-Manuel FG, Alfred-Ockiya JF (2011) Evaluation of yeast single cell protein (SCP) diets on growth performance, feed conversion and carcass composition of tilapia *Oreochromis niloticus* (L.) fingerlings. Afr J Biotechnol 10(46):9473–9478

Bonaldo A, Roem AJ, Fagioli P, Pecchini A, Cipollini I, Gatta PP (2008) Influence of dietary levels of soybean meal on the performance and gut histology of gilthead sea bream (*Sparus aurata L.*) and European sea bass (*Dicentrarchus labrax L.*). Aquac Res 39(9):970–978

Borgogno MC, Dinnella V, Iaconisi R, Fusi C, Scarpaleggia A, Schiavone E, Monteleone L, Gasco GP (2017) Inclusion of *Hermetia illucens* larvae meal on rainbow trout (*Oncorhynchus mykiss*) feed: effect on sensory profile according to static and dynamic evaluations. J Sci Food Agric 97:3402–3341

Bowyer P, El-Haroun ER, Davies SJ (2020) Benefits of a commercial solid-state fermentation (SSF) product on growth performance, feed efficiency and gut morphology of juvenile Nile tilapia (*Oreochromis niloticus*) fed different UK lupin meal cultivars. Aquaculture 523:735192

Bruni L, Pastorelli R, Viti C, Gasco L, Parisi G (2018) Characterisation of the intestinal microbial communities of rainbow trout (*Oncorhynchus mykiss*) fed with *Hermetia illucens* (black soldier fly) partially defatted larva meal as partial dietary protein source. Aquaculture 487:56–63

Bureau DP, Harris AM, Cho CY (1999) Apparent digestibility of rendered animal protein ingredients for rainbow trout. Aquaculture 180:345–358

Bureau DP, Harris AM, Bevan DJ, Simmons LA, Azevedo PA, Cho CY (2000) Feather meals and meat and bone meals from different origins as protein sources in rainbow trout (*Oncorhynchus mykiss*) diets. Aquaculture 181:281–291

Burel C, Boujard T, Tulli F, Kaushik SJ (2000) Digestibility of extruded peas, extruded lupin, and rapeseed meal in rainbow trout (*Oncorhynchus mykiss*) and turbot (*Psetta maxima*). Aquaculture 188:285–298

Burr GS, Wolters WR, Barrows FT, Donkin AW (2013) Evaluation of a canola protein concentrate as a replacement for fishmeal and poultry by-product meal in a commercial production diet for Atlantic salmon (*Salmo salar*). Int Aquatic Res 5(1):1–8

Cabral EM, Bacelar M, Batista S, Castro-Cunha M, Ozorio ROA, Valente LMP (2011) Replacement of fishmeal by increasing levels of plant protein blends in diets for Senegalese sole (*Solea senegalensis*) juveniles. Aquaculture 322(323):74–81

Campos I, Matos E, Marques A, Valente LMP (2017) Hydrolyzed feather meal as a partial fishmeal replacement in diets for European seabass (*Dicentrarchus labrax*) juveniles. Aquaculture 476: 152–165

Chen Y, Sagada G, Xu B, Chao W, Zou F, Ng W, Sun Y, Wang L, Zhong Z, Shao Q (2020) Partial replacement of fishmeal with *clostridium autoethanogenum* single-cell protein in the diet for juvenile black sea bream (*Acanthopagrus schlegelii*). Aquac Res 51:1000–1011

Chien YH, Chiu YH (2003) Replacement of soybean (*Glycine max (L.) Merrill*) meal by lupin (*Lupinus angustifolius*) seed meal in diet for juvenile tilapia (*Oreochromis niloticus x O. aureus*) reared indoors. Aquac Res 34:1261–1268

Choi D, He M, Fang H, Wang X, Li X, Leng X (2019) Replacement of fish meal with two fermented soybean meals in diets for rainbow trout (*Oncorhynchus mykiss*). Aquac Nutr 26:37–46. https://doi.org/10.1111/anu.12965

Collins K, Ravinder S, Tarek MM (2019) Brown seaweed (*AquaArom*) supplementation increases food intake and improves growth, antioxidant status and resistance to temperature stress in Atlantic salmon, *Salmo salar*. PLoS One 15:2019

Davies SJ, Gouveia A, Laporte J, Woodgate SL, Nates S (2009) Nutrient digestibility profile of premium (*category III grade*) animal protein by-products for temperate marine fish species (European sea bass, gilthead sea bream and turbot). Aquac Res 40(15):1759–1769

Davies SJ, Guroyd G, Hassaan M, El-Ajnaf SM, El-Haroun E (2020) Nutritional evaluation of a novel co-fermented apple-pomace, molasses and formic acid generated sardine based fish silages as ingredients in diets for juvenile European sea bass (*Dicentrachus labrax*). Aquaculture 521:735087

Dawood MAO, Abo-Al-Ela HG, Hasan MT (2020) Modulation of transcriptomic profile in aquatic animals: probiotics, prebiotics and synbiotics scenarios. Fish Shellfish Immunol 97:268–282. https://doi.org/10.1016/j.fsi.2019.12.054

Deng J, Kangsen M, Liqiao C, Haifeng M, Zhang L (2015) Effects of replacing soybean meal with rubber seed meal on growth, antioxidant capacity, non-specific immune response, and resistance to *Aeromonas hydrophila* in tilapia (*Oreochromis niloticus x O. aureus*). Fish Shellfish Immunol 44:436–444

Devic E, Leschen W, Murray F, Little DC (2018) Growth performance, feed utilization and body composition of advanced nursing Nile tilapia (*Oreochromis niloticus*) fed diets containing black soldier Fly (*Hermetia illucens*) larvae meal. Aquac Nutr 24:416–423

Diana R, Istiyanto S (2019) The effects of chicken feather silage substitution for fish meal in the diet on growth of saline tilapia fingerlings (*Oreochromis niloticus*). IOP Conf Series: Earth and Environmental Science 246(2019):012015

Divakala K, Chiba L, Kamalakar R, Rodning S, Welles E, Cummins K et al (2009) Amino acid supplementation of hydrolyzed feather meal diets for finisher pigs. J Anim Sci 87(4):1270–1281

Doan HV, Hoseinifar SH, Tapingkae W, Khamtavee P (2017) The effects of dietary kefir and low molecular weight sodium alginate on serum immune parameters, resistance against *Streptococcus agalactiae* and growth performance in Nile tilapia (*Oreochromis niloticus*). Fish Shellfish Immunol 62:139–146

Eastwood L, Kish PR, Beaulieu AD, Leterme P (2009) Nutritional value of flaxseed meal for swine and its effects on the fatty acid profile of the carcass. J Anim Sci 87:3607–3619

El-Haroun ER, Azevedo PA, Bureau DP (2009) High dietary incorporation levels of rendered animal protein ingredients on performance of rainbow trout *Oncorhynchus mykiss* (Walbaum, 1972). Aquaculture 290:269–274

El-Haroun ER, Burea DP (2007) Comparison of the bioavailability of lysine in blood meals of various origins to that of L-lysine HCL for rainbow trout (*Oncorhynchus mykiss*). Aquaculture 262:402–409

El-Husseiny OM, Abdul-Aziz GM, El-Haroun ER, Goda AMAS (2006) Fish meal replacer studies with Nile tilapia and mullet under polyculture conditions in Egypt. International Aquafeed 9(1): 20–29

El-Saidy DMSD, Ibrahim SA, Gaber HS, Midhat A, El-Kasheif (2012) Effect of dietary replacement of soybean meal protein with cotton seed meal protein on biochemical features of Nile tilapia, MUSCLE Egyptian. J Anim Prod 47:201–210

El-Saidy DMSD, Gaber MMA (2001) Linseed meal: its successful use as a partial and complete replacement for fish meal in practical diets for Nile Tilapia, *Oreochromis niloticus* (L.). Second Inter. Conf. On Animal Prod. & Health in Semi-Arid Areas Sept. 4–6

El-Saidy DMS, Saad AS (2011) Effects of partial and complete replacement of soybean meal with cottonseed meal on growth, feed utilization and haematological indexes for mono-sex male Nile tilapia, *Oreochromis niloticus (L.)* fingerlings. Aquac Res 42:351–359

El-Sheekh M, El-Shourbagy I, Shalaby S, Hosny S (2014) Effect of feeding *Arthrospira platensis* (spirulina) on growth and carcass composition of hybrid red tilapia (*Oreochromis niloticus x Oreochromis mossambicus).* Turk J Fish Aquat Sci 14(2):471–478

El-Tawil NE (2010) Effects of green seaweeds (Ulva sp.) as feed supplements in red tilapia (Oreochromis sp.) diet on growth performance, feed utilization and body composition. J Arab Aquacu Soc 5:179–194

Fall J, Yi-Theng T, Diegane N, Shin-Shyn S (2012) The effects of replacement of soybean meal by shrimp shell meal on the growth of hybrid tilapia (*Oreochromis niloticus x Oreochromis aureus*) reared under brackish water. Int J Fish Aquac 5:85–91

Farahiyah IJ, Mardhati M, Yong ST (2018) Effects of using puffed feather meal on the growth performance and carcass characteristics of red hybrid tilapia. Malaysian J Anim Sci 21(2): 99–107

Fawole FJ, Narottam PS, Jain KK, Subodh G, Shamn N, Vikas P, Prabu DL (2016) Nutritional evaluation of protein isolate from rubber seed in the diet of *Labeo rohita*: effects on growth performance, nutrient utilization, whole body composition and metabolic enzymes activity. Anim Feed Sci Technol 219:189–199

Ferrara E, Gustinelli A, Fioravanti ML, Restucci B, Quaglio F, Marono S, Piccolo G (2015) Histological and micro−/macro-morphological evaluation of intestine in sharp-snout seabream *(Diplodus puntazzo)* fed soybean meal-based diets added with Mos and inulin as prebiotics. Aquac Int 23:1525–1537

Ferrer P, Llagostera KZ, Reig L, Amores de Gea D (2019) The use of insect meal as a sustainable feeding alternative in aquaculture: current situation, Spanish consumers' perceptions and willingness to pay. J Clean Prod 229:10–21

Francis G, Makkar HPS, Becker K (2001) Antinutritional factors present in plant-derived alternate fish feed ingredients and their effects in fish. Aquaculture 199:197–227

Friedman M (1996) Nutritional value of proteins from different food sources. A review. J Agric Food Chem 44(1):6–29

Gamboa-Delgado J, Marquez-Reyes JM (2018) Potential of microbial- derived nutrients for aquaculture development. Rev Aquac 10:224–246

Gamboa-Delgado J, Fernández-Díaz B, Nieto-López M, Cruz-Suárez LE (2016) Nutritional contribution of torula yeast and fish meal to the growth of shrimp *Litopenaeus vannamei* as indicated by natural nitrogen stable isotopes. Aquaculture 453:116–121

Gasco L, Henry M, Piccolo G, Marono S, Gai F, Renna M, Lussiana C, Antonopoulou E, Mola P, Chatzifotis S (2016) *Tenebrio molitor* meal in diets for European sea bass (*Dicentrarchus labrax L.*) juveniles: growth performance, whole body composition and in vivo apparent digestibility. Anim Feed Sci Technol 220:34–45

Gasco L, Biasato I, Dabbou S, Schiavone A, Gai F (2019) Animals fed insect-based diets: state-of-the-art on digestibility, performance and product quality. Animals 9:170. https://doi.org/10.3390/ani9040170

Gerile S, Pirhonen J (2017) Replacement of fishmeal with corn gluten meal in feeds for juvenile rainbow trout (*Oncorhynchus mykiss*) does not affect oxygen consumption during forced swimming. Aquaculture 479:616–618

Glencross B, Rutherford N, Jones B (2011) Evaluating options for fishmeal replacement in diets for juvenile barramundi (*Lates calcarifer*). Aquac Nutr 17:722–732

Gomaa AHM, El Moghazy GM (2014) Effect of replacing fish meal with meat and bone meal in Nile tilapia (*O. Niloticus*) diets. 1-nutritional evaluation and safety assessment study. J Anim Poultry Prod, Mansoura Univ 6:365–376

Gonzalez-Rodrıguez A, Celada JD, Carral JM, Saez-Royuela M, Garcıa V, Fuertes JB (2016) Evaluation of poultry by-product meal as partial replacement of fish meal in practical diets for juvenile tench (*Tinca tinca* L.). Aquac Res 47(5):1612–1621

Gonzalez-Salas R, Romero-Cruz O, Valdivie-Navarro M, Ponce-Palafox JT (2014) Los productos y subproductos vegetales, animales y agroindustriales: Una alternativa para la alimentacion de la tilapia. Revista Biociencias 2:240–251

Goulart FR, Speroni CS, Lovatto NM, Loureiro BB, Bender AB, Pretto A, Radünz Neto J, Silva LP (2015) Alternative protein sources in nutrition and metabolism of jundiá: in nature and demucilaged linseed meal. Rev Bras Saude Prod Anim 1:244–252

Grisdalle-Helland B, Helland SJ, Baeverfjord G, Berge GM (2002) Full-fat soybean meal in diets for Atlantic halibut: growth, metabolism and intestinal histology. Aquac Nutr 8:265–270

Gupta SK, Fotedar R, Foysal MJ, Priyam M, Siddik MA, Chaklader MR (2020) Impact of varied combinatorial mixture of non-fishmeal ingredients on growth, metabolism, immunity and gut microbiota of *Lates calcarifer* (Bloch, 1790) fry. Sci Rep 10(1):1–13

Hanafy M (2006) Effect of replacement of soybean meal by linseed meal on growth performance, and body composition of the Nile tilapia, *Oreochromis niloticus* (l) cultured in concrete ponds. Egyptian J Aquatic Biol Fish 10(3):185–200

Hardy RW (2010) Utilization of plant proteins in fish diets: effects of global demand and supplies of fishmeal. Aquac Res 41(5):770–776

Hassaan MS, El-Sayed AIM, Soltan MA, Iraqi MM, Goda AM, Davies SJ, El-Haroun ER, Ramadan HA (2019) Partial dietary fish meal replacement with cotton seed meal and supplementation with exogenous protease alters growth, feed performance, haematological indices and associated gene expression markers (GH, IGF-I) for Nile tilapia, *Oreochromis niloticus*. Aquaculture 503:282–292

Hassaan MS, Soltan MA, Mohammady EY, Elashr MA, El-Haroun ER, Davies SJ (2018) Growth and physiological responses of Nile tilapia, *Oreochromis niloticus* fed dietary fermented sunflower meal inoculated with *Saccharomyces cerevisiae* and *Bacillus subtilis*. Aquaculture 495:592–601

Hassaan MS, Soltan MA, Abdel-Moez AM (2015) Nutritive value of soybean meal after solid state fermentation with Saccharomyces cerevisiae for Nile tilapia, *Oreochromis niloticus*. Anim Feed Sci Technol 201:89–98

Henry M, Gasco L, Piccolo G, Fountoulaki E (2015) Review on the use of insects in the diet of farmed fish: past and future. Anim Feed Sci Technol 203:1–22

Hertrampf JW, Piedad-Pascual F (2000) Handbook on ingredients for aquaculture feeds. Kluwer Academic Publishers, p 624

Heuzé V, Tran G, Chapoutot P, Noblet J, Renaudeau D, Lessire M, Lebas F (2015) Poultry by-product meal. Feedipedia, a programme by INRA, CIRAD, AFZ and FAO. http://www.feedipedia.org/node/214 Last updated on October 2, 2015, 11:41

Hosseinifer SM, Hoseinifar SH, Mazandarani M, Paknejad H, Van Doan H, El-Haroun ER (2020) The potential benefits of orange peels derived pectin on serum and skin mucus immune parameters, antioxidant defence and growth performance in common carp (*Cyprinus carpio*). Fish Shellfish Immunol 103:17–22

Hotz C, Gibson RS (2007) Traditional food-processing and preparation practices to enhance the bioavailability of micronutrients in plant-based diets. J Nutrition 137:1097–1100

Hu M, Wang Y, Wang Q, Zhao M, Xiong B, Qian X (2008) Replacement of fish meal by rendered animal protein ingredients with lysine and methionine supplementation to practical diets for gibel carp, Carassius auratus gibelio. Aquaculture 275(1–4):260–265

Hua K, Cobcroft JM, Cole A, Condon K, Jerry DR, Mangott A, Praeger C, Vucko MJ, Zeng C, Zenger K (2019) The future of aquatic protein: implications for protein sources in aquaculture diets. One Earth 1:316–329

Hussein EEM (2017) Effect of seaweed supplemented diets on Nile tilapia, *Oreochromis niloticus* performance. Int J Fish Aquat Stud 5:205–210

Iaconisi V, Bonelli A, Pupino R, Gai F, Parisi G (2018) Mealworm as dietary protein source for rainbow trout: body and fillet quality traits. Aquaculture 484:197–204

Ido A, Hashizume A, Ohta T, Takahashi T, Miura C, Miura T (2019) Replacement of fish meal by defatted yellow mealworm (*Tenebrio molitor*) larvae in diet improves growth performance and disease resistance in red seabream (*Pargus major*). Animals 9:100

Intraco Ltd., Antwerp, Belgium (2020) Available online: http://www.arevin.com/products/company/intraco-ltd/pl-68.html. Accessed on 10 September 2020

IPIFF (2018) International platform of insects for food and feed (IPIFF). The European insect sector today: challenges, opportunities and regulatory landscape. IPIFF, Brussels, Belgium

Isika MA, Eneji CA, Agiang EA (2006) Evaluation of sun and oven-dried broiler offal meal as replacement for fishmeal in broiler and layer rations. Int J Poult Sci 5(7):646–650

Johnson ML, Parsons CM (1997) Effect of raw material source, ash content and assay length on protein efficiency ratio values for animal protein meals. Poult Sci 76(12):1377–1387

Jones SW, Karpol A, Friedman S, Maru BT, Tracy BP (2020) Recent advances in single cell protein use as a feed ingredient in aquaculture. Curr Opin Biotechnol 61:189–197

Pacheco-Vega JM, Gamboa-Delgado J, Alvarado-Ibarra AG, Nieto-López MG, Tapia-Salazar M, Cruz-Suárez LE (2018) Research article nutritional contribution of fish meal and microalgal biomass produced from two endemic microalgae to the growth of shrimp *Penaeus vannamei*. Lat Am J Aquat Res 46(1):53–62

Kalhoro H, Zhou J, Hua Y, Ng WK, Ye L, Zhang J, Shao Q (2018) Soy protein concentrate as a substitute for fish meal in diets for juvenile *Acanthopagrus schlegelii:* effects on growth, phosphorus discharge and digestive enzyme activity. Aquac Res 49(5):1896–1906

Kaushik SC, Reddy VS, Tyagi SK (2011) Energy and exergy analyses of thermal power plants: a review. Renew Sust Energ Rev 15(4):1857–1872

Kumari MMT, Sundarabarathy TV (2011) Use of fermented feather meal in the artificial diet of gift tilapia (*Oreochromis niloticus*) fry. Annual Research Session 2011. Rajarata University of Sri Lanka. 1–4

Kureshy N, Davis DA, Arnold CR (2000) Partial replacement of fish meal with meat and bone meal, flash-dried poultry by-product meal, and enzyme-digested poultry by-product meal in practical diets for juvenile red drum. N Am J Aquac 62:266–272

Kurniawan DR, Arief M, Agustono Lamid M (2018) Effect of maggot (*Hermetia illucens*) flour in commercial feed on protein retention, energy retention, protein content, and fat content in tilapia (*Oreochromis niloticus*). In: IOP Conference Series: Earth and Environmental Science. https://doi.org/10.1088/1755-1315/137/1/012030

Laporte J , Woodgate SL, Davies SJ, Serwata R, Gouveia A, Nates SF, Laporte J, Woodgate SL, Davies SJ, Serwata R, Gouveia A, Nates SF (2007) Biotechnological process and biological value of feather meal. Evaluation of a novel protein source for the partial replacement of fish meal in aquafeeds: International Aquafeed 07:16–23

Lim C, Beames RM, Eales JG, Prendergast AF, McLeese JM, Shearer KD (1997) Nutritive values of low and high fibre canola meals for shrimp (*Penaeus vannamei*). Aquac Nutr 3(4):269–279

Lin S, Luo L (2011) Effects of different levels of soybean meal inclusion in replacement for fish meal on growth, digestive enzymes and transaminase activities in practical diets for juvenile tilapia, *Oreochromis niloticus* × *O. aureus*. Anim Feed Sci Technol 168:80–87

Liu H, Jin J, Zhu X, Han D, Yang Y, Xie S (2017) Effect of substitution of dietary fish meal by soybean meal on different sizes of gibel carp (*Carassius auratus gibelio*): digestive enzyme gene expressions and activities, and intestinal and hepatic histology. Aquac Nutr 23(1):129–147

Llanes J, Toledo J (2011) Desempe~no productivo de la tilapia del Nilo (*Oreochromis niloticus*) con la inclusion de altos niveles de harina de soya en la dieta. Revista Cubana de Ciencia Agr_ıcola 45:183–186

Lock ER, Arsiwallla T, Waagbø R (2016) Insect larvae meal as an alternative source of nutrients in the diet of Atlantic salmon (*Salmo salar*) post-smolt. Aquac Nutr 22:1202–1213

Lovatto NM, Goulart FR, Loureiro BB, Adorian TJ, Freitas ST, Pianesso D, Dalcin MO, Athayde ML, Silva LP (2016) Effects of phosphorylated protein concentrate of pumpkin seed meal on growth and digestive enzymes activity of silver catfish (*Rhamdia quelen*). Aquac Nutr 23:201–209

Lovatto NM, Goulart FR, Freitas ST, Mombach PI, Loureiro BB, Bender AB, Boligon AA, Radünz Neto J, Silva LP (2015) Nutritional evaluation of phosphorylated pumpkin seed (*Cucurbita moschata*) protein concentrate in silver catfish *Rhamdia quelen* (Quoy and Gaimard, 1824). Fish Physiol Biochem 1:1–11

Lovatto NM, Silva LP, Loureiro BB, Goulart FR, Pretto A, Speroni CS, Radünz Neto J, Loro VL (2014) Efeitos de dietas contendo concentrados proteicos vegetais no desempenho e atividade de enzimas digestivas de silver catfish (*Rhamdia quelen*). Sem Cienc Agr 35:1071–1082

Lucas MML, Stoddard FL, Annicchiarico P, Frias J, Martinez-Villaluenga C, Sussmann D, Duranti M, Seger A, Zander PM, Pueyo JJ (2015) The future of lupin as a protein crop in Europe. Front Plant Sci 6:705

Luo Z, Cai-Xia L, Hua W (2012) Effect of dietary fish meal replacement by canola meal on growth performance and hepatic intermediary metabolism of genetically improved farmed tilapia strain of Nile tilapia, *Oreochromis niloticus*, reared in fresh water. J World Aquacult Soc 43:5

Lusi HS, Reza S (2020) Nutrient digestibility of green seaweed Ulva meal and the influence on growth performance of Nile tilapia (Oreochromis niloticus). Emirates J Food Agric 32(7): 488–494

Madigan MT, Martinko JM (2005) Brock biology of microorganisms, 11th edn, Prentice Hall New Jersey, p 415

Magalhães R, Sánchez-López A, Leal RS, Martínez-Llorens S, Oliva-Teles A, Peres H (2017) Black soldier fly (*Hermetia illucens*) pre-pupae meal as a fish meal replacement in diets for European seabass (*Dicentrarchus labrax*). Aquaculture 476:79–85

Makkar HP, Tran G, Heuzé V, Ankers P (2014a) State-of-the-art on use of insects as animal feed. Anim Feed Sci Technol 197:1–33

Makkar HPS, Tran G, Heuze V, Ankers P (2014b) State-of-the-art on use of insects as animal feed. Anim Feed Sci Technol 197:1–33

Meeker DL, Hamilton CR (2006) An overview of the rendering industry. In: Meeker (ed) Essential rendering. National Renderers Association

Metts LS, Rawles SD, Brady YJ, Thompson KR, Gannam AL, Twibell RG, Webster CD (2011) Amino acid availability from selected animal and plant derived feedstuffs for market size sunshine bass (*Morone chrysops*× *Morone saxatilis*). Aquac Nutr 17:123–131

Metwalli AAA (2008) Using of poultry by-product and hatchery by-product as partial replacement for fish meal in Nile tilapia (*Oreochromis niloticus*) diets. University of Sharjah Journal of Pure & Applied Sciences 5(2):83–104

Meurer F, Costa MM, Barros DAD, Oliveira STL, Paixão PS (2009) Brown propolis extract in feed as a growth promoter of Nile tilapia (*Oreochromis niloticus* Linnaeus, 1758) fingerlings. Aquac Res 40:603–608

Miles RD, Chapman FA (2006) The benefits of fish meal in aquaculture diets. Series of the Department of Fisheries and Aquatic Sciences University of Florida. FA 122:7–8

Mohammadi M, Imani A, Farhangi M, Gharaei A, Hafezieh M (2019) Replacement of fishmeal with processed canola meal in diets for juvenile Nile tilapia (*Oreochromis niloticus*): growth performance, mucosal innate immunity, hepatic oxidative status, liver and intestine histology. Aquaculture 518:734824. https://doi.org/10.1016/j.aquaculture.2019.734824

Moutinho S, Martínez-Llorens S, Tomas-Vidal A, Jover Cerda M, Oliva-Teles A, Peres H (2017) Meat and bone meal as a partial replacement for fish meal in diets for gilthead seabream (*spares aurata*): growth, feed efficiency angry amino acid utilization, and economic efficiency. Aquaculture 468(1):271–277

Mugo-Bundi J, Oyoo-Okoth E, Ngugi CC, Manguya-Lusega D, Rasowo J, Chepkirui-Boit V et al (2013) Utilization of *Caridina nilotica* (roux) meal as a protein ingredient in feeds for Nile tilapia (*Oreochromis niloticus*). Aquac Res 2013:1–12

Muin H, Taufek NM, Kamarudin MS, Razak SA (2017) Growth performance, feed utilization and body composition of Nile tilapia, *Oreochromis niloticus* (Linnaeus, 1758) fed with different levels of black soldier fly, *Hermetia illucens* (Linnaeus, 1758) maggot meal diet. Iran J Fish Sci 16(2):2017

Munguti JM, Ogello EO, Liti D, Waidbacher H, Straif M, Zollitsch W (2014) Effects of pure and crude papain on the utilization and digestibility of diets containing hydrolysed feather meal by Nile tilapia (*Oreochromis niloticus* L.). Int J Adv Res 296:809–822

Nandeesha MC, Gangadhara B, Manissery JK, Venkataraman LV (2001) Growth performance of two Indian major carps, catla (*Catla catla*) and rohu (*Labeo rohita*) fed diets containing different levels of *Spirulina platensis*. Bioresour Technol 80(2):117–120

Nathaly MC, Enrique M, Francisco JC, Jos-e Luis C, Jos-e Antonio L, Saul R, Edgar Iv-an J, Martha ERV, Vıctor M (2018) Advances in the use of alternative protein sources for tilapia feeding. Rev Aquac 0:1–12

Naylor RL, Hardy RW, Dominique P, Chiu A, Elliott M, Farrell AP, Forster I, Gatlin DM, Goldburg RJ, Hua K, Nichols PD, Gatlin DM, Bureau DP, Naylor RL, Elliott M, Chiu A, Hardy RW, Forster I, Farrell AP, Hua K, Goldburg RJ (2009) Feeding aquaculture in an era of finite resources. Proc Natl Acad Sci Unit States Am 106:15103–15110

Ndou SP, Kiarie E, Thandapilly SJ, Ames N, Nyachoti CM (2017) Flaxseed meal and oat hulls supplementation modulates growth performance, blood lipids, intestinal fermentation, bile acids, and neutral sterols in growing pigs fed corn-soybean meal-based diets. J Anim Sci 95: 3068–3078

Ng WK, Romano N (2013) A review of the nutrition and feeding management of farmed tilapia throughout the culture cycle. Rev Aquac 5:220–254

Ngo DT, Wade NM, Pirozzi I, Glencross BD (2016) Effects of canola meal on growth, feed utilisation, plasma biochemistry, histology of digestive organs and hepatic gene expression of barramundi (Asian seabass; *Lates calcarifer*). Aquaculture 464:95–105

NRC; National Research Council (2011) Nutrient requirements of fish and shrimp. The National Academies Press, Washington, D.C 376 pp. https://doi.org/10.17226/13039

Nursinatrio Nugroho RA (2019) Hydrolyzed chicken feather meal as protein source for red tilapia (*Oreochromis sp.*) aquafeeds. Pakistan J Zool 51(4):1489–1496

Obirikorang KA, Gyamfi S, Goode ME (2020) Effect of soybean meal diets on the growth performance, ammonia excretion rates, gut histology and feed cost of Nile tilapia (*Oreochromis niloticus*) fry. Aquac Res 51:3520–3532

Ochieng OE, Kembenya EM, Githukia CM, Aera CN, Munguti JM, Nyamweya CS (2017) Substitution of fish meal with sunflower seed meal in diets for Nile tilapia (*Oreochromis niloticus L.*) reared in earthen ponds. J Appl Aquac 29:81–99

Ogunji JO, Nimptsch J, Wiegand C, Schulz C (2007) Evaluation of the influence of housefly maggot meal (magmeal) diets on catalase, glutathione S-transferase and glycogen concentration in the liver of *Oreochromis niloticus* fingerling. Comp Biochem Physiol A 147:942–947

Oliva-Teles A, Enes P, Peres H (2015) Replacing fishmeal and fish oil in industrial aquafeeds for carnivorous fish. In: Davis DA (ed) Feed and feeding practices in aquaculture. Woodhead Publishing, Oxford, pp 203–233

Olvera-Novoa MA, Domínguez-Cen LJ, Olivera-Castillo L, Carlos AM (2008) Effect of the use of the microalga *Spirulina maxima* as fish meal replacement in diets for tilapia, *Oreochromis mossambicus* (Peters), fry. Aquac Res 29:709–715. https://doi.org/10.1046/j.1365-2109.1998.29100709.x

Olvera-Novoa M, Olivera-Castillo L, Martinez CA, Palacios (2002) Sunflower seed meal as a protein source in diet for *Tilapia rendalli* (Boulanger, 1896) fingerlings. Aquac Res 33:223–229

Oomah BD, Mazza G (1993) Flaxseed proteins–a review. Food Chem 48:109–114. https://doi.org/10.1016/0308-8146(93)90043-F

Øverland M, Tauson AH, Shearer K, Skrede A (2010) Evaluation of methane-utilising bacteria products as feed ingredients for monogastric animals. Arch Anim Nutr 64:171–189

Øverland M, Sørensen M, Storebakken T, Penn M, Krogdahl Å, Skrede A (2009) Pea protein concentrate substituting fish meal or soybean meal in diets for Atlantic salmon (*Salmo salar*)—effect on growth performance, nutrient digestibility, carcass composition, gut health, and physical feed quality. Aquaculture 288(3–4):305–311

Ozkan YF, Engin K, Özlüer HA (2015) The effects of balanced diets with soy bean extract or meat and bone meal on muscle and liver tissue protein and glycogen levels of the Nile tilapia (*Oreochromis niloticus* L.) infected with *vibrio anguillarum*. J Appl Biol Sci 9:37–42

Petterson D (2000) The use of lupins in feeding systems-review. Asian-Aust J Anim Sci 13:861–882

Pham DH, Siddik MAB, Fotedar R, Foysal MJ, Chaklader MR, Nguyen CM, Sukham MK (2020) Substituting fishmeal with lupin *Lupinus angustifolius* kernel meal in the diets of cobia *Rachycentron canadum*: effects on growth performance, nutrient utilization, haemato-physiological response, and intestinal health. Anim Feed Sci Technol 267:114556

Pianesso D, Adorian TJ, Mombach PI, Dalcin MO, Loebens L, Telles YB, Roballo SS, Lovatto NM, Silva LP (2020) Nutritional assessment of linseed meal (*Linum usitatissimum L.*) protein concentrate in feed of silver catfish. Anim Feed Sci Technol 265:114517

Pichet P, Yakupitiyage A (2014) Effect of replacing soybean meal with yeast-fermented canola meal on growth and nutrient retention of Nile tilapia, *Oreochromis niloticus* (Linnaeus 1758). Aquac Res 45:1744–1753

Plaipetch P, Yakupitiyage A (2014) Effect of replacing soybean meal with yeast-fermented canola meal on growth and nutrient retention of Nile tilapia, *Oreochromis niloticus* (Linnaeus1758). Aquac Res 45:1744–1753

Poppi DA, Quinton VM, Hua K, Bureau DP (2011) Development of a test diet for assessing the bioavailability of arginine in feather meal fed to rainbow trout (*Oncorhynchus mykiss*). Aquaculture 314:100–109

Prachom N, Boonyoung S, Hassaan MS, El-Haroun E, Davies SJ (2021) Preliminary evaluation of Superworm (*Zophobas morio*) larval meal as a partial protein source in experimental diets for juvenile Asian sea bass, Lates calcarifer. Aquac Nutr 27:1304–1314

Pratiwy FM, Kohbara J, Susanto AB (2018) Effectiveness of sargassum meal as feed additive on growth performance of Nile tilapia, *Oreochromis niloticus*. Aquac Sci 66:25–31

Puja C, Abhijit M, Sarang N, Somu SL (2019) A review on alternative plant protein sources available for future sustainable aqua feed production. Int J Chem Stud 7(3):1399–1404

Rajeev R, Bavitha M (2015) Lupins-an alternative protein source for aquaculture diets. Int J Appl Res 1(3):4–8

Rawles SD, Gaylord TG, Matthew EM, Freeman DW (2009) Evaluation of poultry by-product meal in commercial diets for hybrid striped bass, *Morone chrysops x Morone saxatilis*, in pond production. J World Aquacult Soc 40:2

Renna M, Biasato I, Brugiapaglia A, Gai F, De Marco M, Schiavone A, Malfatto V, Zoccarato I, Gasco L, Biasibetti E, Lussiana C, Prearo M, Capucchio MT, Dabbou S (2017) Evaluation of the suitability of a partially defatted black soldier fly (*Hermetia illucens* L.) larvae meal as ingredient for rainbow trout (*Oncorhynchus mykiss* Walbaum) diets. J Anim Sci Biotechnol 8. https://doi.org/10.1186/s40104-017-0191-3

Rincón DD, Velásquez HA, Dávila MJ, Semprun AM, Morales ED, Hernández JL (2012) Substitution levels of fishmeal by *Arthrospira* (= *spirulina*) maxima meal in experimental diets for red tilapia fingerlings (*Oreochromis sp.*). Rev Colomb Cienc Pec 25:430–437

Rosales M, Castilloa S, Pohlenz C, Gatlin DM III (2017) Evaluation of dried yeast and threonine fermentation biomass as partial fish meal replacements in the diet of red drum Sciaenops ocellatus. Anim Feed Sci Technol 232:190–197

Rosas VT, Poersch LH, Romano LA, Tesser MB (2019) Feasibility of the use of *spirulina* in aquaculture diets. Rev Aquac 11:1367–1378

Sales J (2009) The effect of fish meal replacement by soyabean products on fish growth: a meta-analysis. Br J Nutr 102:1709–1722

Sangeera MDR, Reddy PB (2019) Partial replacement of fish meal with Soya Bean (Sbm) on the growth performance and body composition of Tilapia. https://www.researchgate.net/publication/343207433

Sarker PK, Kapuscinski AR, Bae AY, Donaldson E, Sitek AJ, Fitzgerald DS, Edelson OF (2018) Towards sustainable aquafeeds: evaluating substitution of fishmeal with lipid-extracted microalgal co-product (*Nannochloropsis oculata*) in diets of juvenile Nile tilapia (*Oreochromis niloticus*). PLoS One 13:e0201315

Schulz C, Wickert M, Kijora C, Ogunji J, Rennert B (2007) Evaluation of pea protein isolate as alternative protein source in diets for juvenile tilapia (*Oreochromis niloticus*). Aquac Res 38(5): 537–545

Sharif M, Muhammad HZ, Amjad IA, Muhammad S, Mayada RF, Mahmoud A (2021) Single cell protein: sources, mechanism of production, nutritional value and its uses in aquaculture nutrition. Aquaculture 531:735885

Silvia NM, Paola G, Damian J, Jan M, Krzysztof D, Mateusz R, Bartosz K, Agata J (2019) Insect meals in fish nutrition. Rev Aquac 11:1080–1103

Soares CM, Hayashi C, Faria ACEAD, Furuya WM (2001) Replacement of soybean meal protein by canola meal protein in diets for Nile tilapia (*Oreochromis niloticus*) in the growing phase. Rev Bras Zootec 30(4):1172–1177

SOFIA (2020) The state of the world fisheries and aquaculture 2020. FAO Fisheries and Aquaculture Department, Rome, pp 1–224

Soltan M, El-Bab AF, Saudy A-N (2011) Effect of replacing dietary fish meal by cottonseed meal on growth performance and feed utilization of the Nile tilapia, (*Oreochromis niloticus*). Egypt J Aquat Biol & Fish 15(2):17–33

Soltan MA, Hassaan MS, Abdella MS, El-Sayed FA, El-Ashry MA (2015) Yeast fermented sunflower meal as a replacer for fish meal in diets of the Nile tilapia, *Oreochromis niloticus*. Egyptian J Aquatic Biol Fish 19:65–72

Soltan MA (2005) Partial and total replacement of soybean meal by raw and heat treated linseed meal in tilapia, diets. Egyptian J Nutr Feeds 8(1):1091–1109

Srour TM (2007) Effect of dietary lupin (*Lupinus angustifolius*) wholeseed meal protein on the growth of tilapia (*Oreochromis niloticus*) reared in tanks. Egyptian Soc Anim Prod 44(1):83–96

Stephanie FV, Merab A, Rhea GA, Rex FMT, Melchor MT, Gian CFM, Janice AR (2016) Dietary spirulina (*Arthrospira platensis*) replacement enhances performance of juvenile Nile tilapia (*Oreochromis niloticus*). J Appl Phycol 28:1023–1030

Suloma A, El-Husseiny OM, Hassane MI, Mabroke RS, El-Haroun ER (2014) Complementary responses between hydrolyzed feather meal, fish meal and soybean meal without amino acid supplementation in Nile tilapia *Oreochromis niloticus* diets. Aquac Int 22:1377–1390

Suloma A, Mabroke RS, El-Haroun ER (2013) Meat and bone meal as a potential source of phosphorus in plant-protein-based diets for Nile tilapia (*Oreochromis niloticus*). Aquac Int 21:375–385

Sun H, Tang JW, Yao XH, Wu YF, Wang X, Liu Y, Lou B (2015) Partial substitution of fish meal with fermented cottonseed meal in juvenile black sea bream (*Acanthopagrus schlegelii*) diets. Aquaculture 446:30–36

Suryaningrum LH, Samsudin R (2019) Nutrient digestibility of green seaweed Ulva meal and the influence on growth performance of Nile tilapia (*Oreochromis niloticus*). Emirates J Food Agric 2020.32(7):488–494

Tlusty M, Rhyne A, Szczebak JT, Bourque B, Bowen JL, Burr G, Marx CJ, Feinberg L (2017) A transdisciplinary approach to the initial validation of a single cell protein as an alternative protein source for use in aquafeeds. Peer J 3170:1–19

Tran G, Heuzé V, Makkar HPS (2015) Insects in fish diets. Anim Front 5(2):37–44

USB (2008) Nutrient specifications and quality standards for ingredients commonly used in diets for aquaculture species. US Soya bean Export Council, USA

Van Vo B, Muhammad ABS, Ravi F, Reaz C, Abu H, Javed F, Huy QN (2020) Progressive replacement of fishmeal by raw and enzyme-treated alga, *Spirulina platensis* influences growth, intestinal micromorphology and stress response in juvenile barramundi, *Lates calcarifer*. Aquaculture 529:735–741

Vo-Binh V, Bui DP, Nguyen HQ, Fotedar R (2015) Optimized fermented lupin (*Lupinus angustifolius*) inclusion in juvenile barramundi (*Lates calcarifer*) diets. Aquaculture 444(0): 62–69

Wang P, Zhu J, Feng J, He J, Lou Y, Zhou Q (2017) Effects of dietary soy protein concentrate meal on growth, immunity, enzyme activity and protein metabolism in relation to gene expression in large yellow croaker *Larimichthys crocea*. Aquaculture 477:15–22

Wang YS, Shelomi M (2017) Review of black soldier fly (*Hermetia illucens*) as animal feed and human food. Foods 6:91

Watson H (2006) Poultry meal vs poultry by-product meal. Dogs in Canada Magazine

Wei H, YuH CX, Chao W, Zou F, Chen P, Que M (2018) Effects of *clostridium ethanolum* protein instead of soybean meal on growth performance, plasma biochemical indexes and pathology of hepatopancreas and intestinal tissue of *Ctenopharyngodon idellus*. J Anim Nutri 30(10): 401–412

Willora FP, Nadanasabesan N, Knutsen HR, Liu C, Sørensen M, Hagen Ø (2020) Growth performance, fast muscle development and chemical composition of juvenile lumpfish (*Cyclopterus lumpus*) fed diets incorporating soy and pea protein concentrates. Aquac Rep 17:100352

Wu YB, Ren X, Chai XJ, Li P, Wang Y (2018) Replacing fish meal with a blend of poultry by-product meal and feather meal in diets for giant croaker (*Nibea japonica*). Aquac Nutr 24: 1085–1091

Yang Y, Xie S, Cui Y, Lei W, Zhu X, Yang Y, Yu Y (2004) Effect of replacement of dietary fish meal by meat and bone meal and poultry by-product meal on growth and feed utilization of gibel carp, *Carassius auratus gibelio*. Aquac Nutri 10:289–294

Yangthong M, Hutadilok-Towatana N, Thawonsuwan J, Phromkunthong W (2016) An aqueous extract from sargassum sp. enhances the immune response and resistance against *streptococcus iniae* in the Asian sea bass (Lates calcarifer Bloch). J Appl Phycol 28:3587–3598

Yengkhom O, Shalini KS, Subramani PA, Michael RD (2018) Nonspecific immunity and disease resistance are enhanced by the polysaccharide fraction of a marine chlorophycean macroalga in *Oreochromis niloticus* (Linnaeus, 1758). J Appl Ichthyol 34:556–567

Yigit NO, Demir (2016) Use of fermented soybean meal with whey as a protein source for feeding juvenile tilapia (*Oreochromis niloticus*). The Israeli J Aquac-Bamidgeh, IJA 68:1340

Yigit NO, Olmez M (2009) Canola meal as an alternative protein source in diets for fry of tilapia (*Oreochromis niloticus*). The Israeli J Aquac–Bamidgeh 61(1):35–41

Yildirim-Aksoy M, Rashida E, Cole S, Benjamin HB (2020) Use of dietary frass from black soldier fly larvae, *Hermetia illucens*, in hybrid tilapia (*Nile x Mozambique, Oreochromis niloticus x O. mossambicus*) diets improves growth and resistance to bacterial diseases. Aquac Rep 17: 100373

Yildrim O, Turker A, Ergun S, Yigit M, Gulsahin A (2009) Growth performance and feed utilization of *Tilapia zillii* (Gervais, 1848) fed partial or total replacement of fish meal with poultry by-product meal. Afr Biotech 8:3092–3096

Yones AMM, Metwalli AA (2015) Effects of fish meal substitution with poultry by-product meal on growth performance, nutrients utilization and blood contents of juvenile Nile tilapia (*Oreochromis niloticus*). J Aquac Res Develop 6:389. https://doi.org/10.4172/2155-9546.1000389

Yong ST, Mardhati M, Farahiyah IJ, Noraini S, Wong HK (2018) Replacement of fishmeal in feather meal based diet and its effects on tilapia growth performance and water quality parameters. J Trop Agric and Food Sci 46(1):47–55

Younis EM, Al-Quffail AS, Al-Asgah NA, Abdel-Warith AA, Al-Hafedh YS (2018) Effect of dietary fish meal replacement by red algae, *Gracilaria arcuata*, on growth performance and body composition of Nile tilapia *Oreochromis niloticus*. Saudi J Biol Sci 25:198–203

Yu R, Cao H, Huang Y, Peng M, Kajbaf K, Kumar V, Tao Z, Yang G, Wen C (2020) The effects of partial replacement of fishmeal protein by hydrolysed feather meal protein in the diet with high inclusion of plant protein on growth performance, fillet quality and physiological parameters of Pengze crucian carp (*Carassius auratus var. Pengze*). Aquac Res 51:636–647

Yue YR, Zhou QC (2008) Effect of replacing soybean meal with cottonseed meal on growth, feed utilization, and hematological indexes for juvenile hybrid tilapia, *Oreochromis niloticus×O. aureus*. Aquaculture 284:185–189

Zamani A, Vahtera V, Sääksjärvi IE, Scherz MD (2020) The omission of critical data in the pursuit of "revolutionary" methods to accelerate the description of species. Syst Entomol 46:1–4

Zeraatpisheh F, Firouzbakhsh F, Khalili KJ (2018) Effects of the macroalga Sargassum angustifolium hot water extract on hematological parameters and immune responses in rainbow trout (*Oncohrynchus mykiss*) infected with *yersinia rukeri*. J Appl Phycol 30:2029–2037

Zhao L, Wang W, Huang X, Guo T, Wen W, Feng L, Wei L (2015) The effect of replacement of fish meal by yeast extract on the digestibility, growth and muscle composition of the shrimp *Litopenaeus vannamei*. Aquac Res 48:311–320

Zhou F, Song W, Shao Q, Peng X, Xiao J, Hua Y, Owari BN, Zhang T, Ng WK (2011) Partial replacement of fish meal by fermented soybean meal in diets for black sea bream, *Acanthopagrus schlegelii*, juveniles. J World Aquac Soc 42:184–197

Zhou Q, Yi-Rong Y (2010) Effect of replacing soybean meal with canola meal on growth, feed utilization and haematological indices of juvenile hybrid tilapia, *Oreochromis niloticus* x *Oreochromis aureus*. Aquac Res 41:982–990

Chapter 9
Sustainable Tilapia Farming, the Role of Culture Systems

Hien Van Doan

Abstract Considering environmental sustainability and vulnerability to the effects of climate change on fish production, sustainable and adaptable aquaculture systems are necessary". Biofloc technology (BFT) and recirculating aquaculture systems (RAS) are eco-friendly, water efficient, highly productive intensive farming systems, which are not associated with adverse environmental impacts, such as habitat destruction, water pollution and eutrophication, biotic depletion, ecological effects on biodiversity due to captive fish and exotic species escape, disease outbreaks, and parasite transmission. Moreover, BFT and RAS operate in an indoor controlled environment, and thus, are only minimally affected by climatic factors, including rainfall variation, flood, drought, global warming, cyclone, salinity fluctuation, ocean acidification, and sea level rise. This chapter provides into insight the application of these techniques for sustainable tilapia production, which focuses on their effects on growth performance, immune response, and disease resistance.

Keywords Tilapia · Sustainable · Biofloc technology · RAS

9.1 Introduction

Aquaculture, is one of the fastest food-producing sectors, with an average annual growth rate of 5.3% during the period 2001–2018, and production has increased by over 600% since 1990 (FAO 2020b). Global aquaculture production achieved 82.1 million tons in 2018, of which inland aquaculture produced 51.3 million tons (62%), while coastal and marine aquaculture1 reached 30.8 million tons (38%) (FAO 2020b). Aquaculture is performed in various environments and regions, employing various technologies, and cultured systems, and raising many species (Ahmad et al. 2022). Asia accounts for around 90% of global aquaculture production (FAO 2020a, b), and aquaculture, with its expansion outpacing global population growth,

H. Van Doan (✉)
Department of Animal and Aquatic Sciences, Faculty of Agriculture, Chiang Mai University, Chiang Mai, Thailand

© The Author(s), under exclusive license to Springer Nature Switzerland AG 2023

S. H. Hoseinifar, H. Van Doan (eds.), *Novel Approaches Toward Sustainable Tilapia Aquaculture*, Applied Environmental Science and Engineering for a Sustainable Future, https://doi.org/10.1007/978-3-031-38321-2_9

is playing an important role in boosting food production for contribution to food security and human nutrition (Pradeepkiran 2019). Although the expansion of aquaculture, the challenge of feeding a growing global population, which is expected to reach 9.7 billion by 2050 (Bahar et al. 2020), is a vivid reality, which is central in global political and scientific debates (Berners-Lee et al. 2018). Because of population growth and stagnation of capture fisheries, global aquaculture production is challenged to achieve remarkable targets, estimated to possibly reach 109 million tons in 2030 (FAO 2020a, b), and 140 million tons in 2050 (Waite et al. 2014).

To achieve the required further increase in global seafood production, aquaculture is envisaged as the only available solution, but it could bring additional adverse environmental effects if its expansion is not based on sustainable farming systems (Ahmed and Turchini 2021). Accordingly, the rapid growth of aquaculture has been linked to raising concerns about its environmental sustainability (Ahmad et al. 2022; Tom et al. 2021). Broadly, aquaculture has already been increasingly associated with a great variety of negative environmental impacts, including habitat destruction, water pollution and eutrophication, biotic depletion, disease and parasite transmission, and greenhouse gas (GHG) emissions (Ahmed et al. 2019; (Carballeira Braña et al. 2021; Adegbeye et al. 2019; Kosten et al. 2020). For example, some intensive aquaculture practices have been reported to cause antibiotic pollution, eutrophication, land occupation, and other environmental hazards (Dauda et al. 2019). The invasion of exotic fish species in some aquaculture systems has been reported to have potentially negative effects on biodiversity and the ecosystem; (Banha et al. 2022; Kiruba-Sankar et al. 2018). Freshwater aquaculture, particularly tilapia, has been reported to cause adverse effects on water resources with a broad range of impacts on the biodiversity, ecosystems, and societies (Bashir et al. 2020; Kaleem and Bio Singou Sabi 2021; Moyo and Rapatsa 2021). Tilapia aquaculture has also been censured for a variety of environmental issues, including water pollution and pathogen transmission to wild fish (Shaaban et al. 2021). The environment, on the other hand, has a series of impacts and imposes certain limitations on aquaculture, with climate change posing a significant threat to increasing global fish production (Baag and Mandal 2022; Maulu et al. 2021). A wide range of climatic factors, such as rainfall, flood, drought, cyclone, global warming, sea surface temperature change, salinity fluctuation, ocean acidification, and sea level rise have a significant impact on aquaculture practices (Ahmed et al. 2019). In essence, future climate change will certainly have adverse impacts on sustainable increasing aquaculture production (Ahmed et al. 2019; Boyd et al. 2020). It is therefore necessary to develop and implement adaptation strategies to cope with these challenges.

Considering the environmental concerns and impacts, as well as the vulnerability to the effects of climate change and other environmental variables of fish production in aquaculture, one of the potential and increasingly proposed adaptation strategies is the implementation of sustainable aquaculture systems, such as BFT and RAS. These systems allow for rising fish in a land-based, indoor, and controlled environment to minimize the direct interactions between the production processes and the environment (Ahmed and Turchini 2021; Khanjani et al. 2022). They offer benefits in improving aquaculture production that could contribute to the achievement of sustainable development goals (Bossier and Ekasari 2017; Nguyen et al. 2021). This technology could result in higher productivity with less impact on the environment. Furthermore, biofloc systems may be developed and performed in integration with other food production, thus promoting productive integrated systems, aiming at producing more food and feed from the same area of land with less input (Nisar et al. 2022). The chapter aims to address different aspects of BFT and RAS as alternatives for sustainable tilapia production.

9.2 Biofloc Technology

9.2.1 History of Biofloc Technology (BFT)

The biofloc system (BFT) evolved as an alternative to the conventional aquaculture system that is used in shrimp and tilapia productions (Ulloa Walker et al. 2020). It originated in the 1970s at the French Research Institute for Exploitation of the Sea (IFREMER) with Gerard Cuzon as the pioneer (Emerenciano et al. 2012; Devi and Kurup 2015). The BFT was then widely applied in commercial shrimp farming (Samocha et al. 2019). In the 1990s, different studies at pilot and commercial scales were conducted in the USA with penaeid shrimp led by J. Stephen Hopkins and with finfish at the Technion-Israel Institute of Technology led by Yoram Avnimelech (Emerenciano et al. 2013). In the mid-2000s, several studies on penaeid shrimp were conducted at the Federal University of Rio Grande-FURG (Brazil) led by Wilson Wasielesky and the Texas A&M University (Corpus Christi Campus, USA) led by Tzachi Samocha. After that, there was a significant increase in the number of scientific publications about biofloc technology worldwide. The number has increased from less than 10 in 2009 to more than 100 publications in 2018, with studies conducted mainly in Brazil, China, the United States of America, Mexico, and India (Ulloa Walker et al. 2020), helping to strengthen the technology and boost the industry. Another important factor for such progress was the wide range of courses and lectures offered in both scientific and commercial events for the scientific community, academia, and aquaculturists. However, despite the progress and benefits of BFT as reported by the scientific community and academia, there is still room for its commercial expansion. For example, in Indonesia, it is estimated that only 20–25% of shrimp production has occurred using biofloc technology (Thong 2014). Among the reasons behind such a scenario are the higher implementation and production costs compared to traditional land-based systems, and the complexities in the management and implementation of the technology, which requires greater technical knowledge and permanent monitoring of water quality (Avnimelech 2015).

9.2.2 Principles of Biofloc Technology

The BFT operates on the principle of nutrient recycling by maintaining a higher carbon: nitrogen (C: N) ratio above 15 to stimulate the mass growth of heterotrophic bacteria (Guo et al. 2020). Higher C: N ratio is maintained when more carbon sources, such as molasses, cassava, hay, sugarcane, starch, wheat bran, cellulose, etc., are sprayed on the surface of pond water with continuous aeration (Ogello et al. 2021). Under favorable BFT conditions, up to 0.5 g of heterotrophic bacterial biomass g^{-1} substrate of carbon can be produced (Eding et al. 2006). With the information that 1 g of carbon produces 0.5 g of bacteria, farmers can estimate quantities of floc in the culture systems (Nisar et al. 2022). The biofloc process stimulates the natural growth of macro-aggregates of organisms that enhance self-nitrification in the culture water (Jamal et al. 2020).

In outdoor BFT systems, the photosynthetic pathway that produces algae normally precedes the bio-flocking process (Ogello et al. 2021). The algae provide a substrate to which the bioflocs attach and are usually referred to as green bioflocs (Ahmad et al. 2017; Khanjani and Sharifinia 2020). Under indoor conditions, bioflocs are mainly bacteria and are referred to as brown bioflocs (El-Sayed 2021; Emerenciano et al. 2021). With the addition of an adequate carbon source bacterial floc stimulates a second production line that involves the degradation of organic wastes by bacteria to produce billions of bacterial cells under optimum aeration conditions (Khanjani et al. 2022). During this process, autotrophic and heterotrophic bacteria proliferate and attract billions of other cells including diatoms, fungi, algae, protozoans, and various types of plankton (Avnimelech 2007; Bossier and Ekasari 2017). The traditional aquaculture ponds lack injection of carbon source, and aeration mechanisms and thus harbor fewer and less diverse bacterial communities, as opposed to BFT (Felix and Menaga 2021). Small quantities of bacteria cannot form substantial flocs in the culture system. The sediment of traditional ponds accumulates higher quantities (49%) of nitrogenous waste while the BFT pond sediments have less (5%) nitrogenous waste (Ogello et al. 2021).

9.2.3 Biofloc in Tilapia

Tilapia has played an important role in global aquaculture in recent decades (FAO 2020a). It ranks the second most cultured fish species worldwide due to its fast growth, resistance to various environmental conditions, and capability of being produced in dense and ultra-dense forms (Menaga et al. 2019; Avnimelech 2007; Avnimelech 2015; Khanjani et al. 2021). Moreover, tilapia is an omnivore species that filter food particles and easily feeds on a rich natural food source and biofloc-dependent microorganisms (Durigon et al. 2020; Prabu et al. 2019). Considering the above considerations, tilapia is a suitable candidate for the biofloc system (Khanjani et al. 2022). The biofloc technique has been accepted in many countries over the past decade (Vyas 2020). Currently, the biofloc technique has been successfully developed in large-scale farms in Asia (China, South Korea), Latin America, and Central America, as well as on a small scale in the United States, Italy, and other countries. In addition, many research centers and academic institutions are expanding BFT, mainly in key areas such as growth, nutrition, reproduction, microbial ecology, biotechnology, and economics (Khanjani and Sharifinia 2020).

9.2.3.1 Biofloc as Growth Promoters

In situ utilization of microbial flocs generated in biofloc systems by some aquaculture organisms as well as the utilization of processed biofloc as a feed ingredient has been well documented (Kuhn et al. 2009; Kuhn et al. 2010; Anand et al. 2014). It has been demonstrated that the concentrations of free amino acids such as alanine, glutamate, arginine, and glycine, which are known attractants in the shrimp diet, are present in biofloc (Vyas 2020; Ahmad et al. 2017). Levels in biofloc were found to be comparable to that of the shrimp commercial diet suggesting that biofloc are likely to be recognized as food particles by some aquaculture organisms. Furthermore, biofloc technology application in larviculture may provide an easily accessible food source for the larvae outside the regular feeding moments, thus minimizing possible negative social interaction during feeding (Ekasari et al. 2015b).

Studies by various researchers have reported that the biofloc system improves the growth performance of the Nile tilapia (Azim and Little 2008; Luo et al. 2013; Mirzakhani et al. 2019; Khanjani et al. 2021; Table 9.1). The improved growth performance was attributed to optimum water quality and continuous production of biofloc. Biofloc contains poly beta-hydroxybutyrate (De Schryver and Verstraete 2009) and bioactive compounds, such as carotenoids, chlorophylls, and phytosteroids that promote the growth of cultured aquatic organisms. The adaptability of tilapia to biofloc consumption and digestion of microbial protein has been demonstrated in several studies (Azim and Little 2008). Avnimelech (2007) reported that the production of biofloc in fish ponds can meet 50% of tilapia's protein requirements. Khanjani et al. (2021) found that tilapia feed better on the biofloc impacted by molasses daily, resulting in the highest biomass increase and the lowest feed conversion ratio. Based on their results, the highest feed conversion ratio and the lowest protein efficiency were observed in the control treatment without biofloc. Researchers have found that a combination of biofloc and artificial foods improves feed conversion ratio and feed efficiency (Khanjani et al. 2021; Mirzakhani et al. 2019).

Biofloc includes bacterial proteins, polyhydroxy butyrate, and bacteria containing peptidoglycan or lipopolysaccharides, which can promote growth performance (Khanjani and Sharifinia 2020). In addition, biofloc have probiotic properties that help fish in digestion and absorption of artificial diets. In the study using orange peel-derived pectin (OPDP) added to the biofloc system, Doan et al. (2018) indicated that the inclusion of 10 g kg^{-1} OPDP significantly improved the growth performance and feed utilization of Nile tilapia. Similarly, a significant increase in growth

Table 9.1 Weight gain (WG), Specific growth rate (SGR), Food conversion efficiency (FCE), Food conversion ratio (FCR), Protein efficiency ratio (PER), Survival rate (SR), and digestive enzyme of Nile tilapia in biofloc and RAS systems

Studied systems	Studies parameters	Fish species	Doses and supplementations duration	Results	References
BFT	Orange peels derived pectin (OPDP)	Nile tilapia (*Oreochromis niloticus*) (9.09 ± 0.05 g)	0 (control in clear water), 0 (control in biofloc system), 5, 10, and 20 g kg^{-1} OPDP for 8 weeks	SGR, WG, and FW ↑ FCR ↓	Doan et al. (2018)
BFT	Stocking density	Tilapias (0.51 ± 0.05 g)	166 orgs m^{-3} (LD, low density), 333 orgs m^{-3} (MD, middle density) and 600 orgs m^{-3} (HD, high density) 120 days	Growth performance ↓ FCR ↑	Liu et al. (2018)
BFT	Carbon sources	Nile tilapia (*O. niloticus*) 2.7 ± 0.4 g	100% of MO, 75% of MO+ 25% of WF, 50% of MO+ 50% of WF, 25% of MO+ 75% of WF,and 100% of WF	Growth performance ↑	Mirzakhani et al. (2019)
BFT	Pizzeria by-product	Nile tilapia (*O. niloticus*) (2.90 ± 0.02 g)	0, 20, 40, 60, 80 and 100% 38 days	Growth performance ↑ Economic benefits ↑	de Sousa et al. (2019)
BFT	Dietary digestible protein and digestible energy	Nile tilapia (*O. niloticus*) (1.25 ± 0.15 g)	Digestible protein (22, 26 and 30% DP) and digestible energy (3000, 3150 and 3300 kcal kg^{-1}) 42 days	Pepsine activity ↑ Trypsine activity ↑	Durigon et al. (2019)
BFT	Carbon sources	Fingerlings	Corn flour (CF), wheat flour (WF), sugar (SU) and a control (C)	FCR and water ↓ Survival rate ↑	García-Ríos et al. (2019)
BFT	OPDP and *L. plantarum*	Nile tilapia (*O. niloticus*) (1.25 ± 0.15 g) 5.92 ± 0.08 g	Diet 1 (0 g kg^{-1} OPDP and 0 CFU g^{-1} *L. plantarum*), diet 2 (10 g kg^{-1} OPDP), diet 3 (108 CFU g^{-1} *L. plantarum*), and diet 4 (10 g kg^{-1} OPDP +108 CFU g^{-1} *L. plantarum*)	Growth performance ↑ FCR ↓	Van Doan et al. (2019)

BFT	*in-situ* and *ex-situ* biofloc	Nile tilapia (*O. niloticus*) (5.1 ± 0.05 g/fish)	T1-biofloc developed within the culture systems (*insitu*), T2-biofloc supplementation in fish feed (*exsitu*) and C- Control without biofloc	Growth performance ↑	Menaga et al. (2019)
BFT	Biochar	Nile tilapia (*O. niloticus*) (36.2 g/fish)	B; only biochar, GB; biochar + glucose, while for the control (G) only glucose	Growth performance →	Abakari et al. (2020)
BFT	Jaggery-based BFT	Nile tilapia (*O. niloticus*) (32.2 ± 10.1 g/fish)	22.5%, 27.7%, and 32.3% digestible protein (DP) and 6% lipid	Growth performance ↑	Elayaraja et al. (2020)
BFT	Chestnut polyphenols (CSP)	Nile tilapia (*O. niloticus*) (12.77 ± 0.17 g)	0, 1, 2, 4, and 8 g kg^{-1} of CSP	Growth performance ↑	H. Van Doan et al. (2020a)
BFT	Watermelon rind powder (WMRP)	Nile tilapia (*Oreochromis niloticus*) (17.14 ± 0.12 g)	0 (Diet 1- control), 20 g kg^{-1} WMRP (Diet 2), 40 g kg^{-1} WMRP (Diet 3), 80 g kg^{-1} WMRP (Diet 4), and 160 g kg^{-1} WMRP (Diet 5)	Growth performance ↑ FCR ↓	Hien Van Doan et al. (2020b)
BFT	Phosphatidylcholine	Nile tilapia (*O. niloticus*) 8.03 ± 0.03 g	0, 400, 800 and 1200 mg/kg of feed 40 days	Growth performance → Energy metabolism ↑	Sousa et al. (2020)
BFT	*Tenebrio molitor* meal (TM)	Nile tilapia (*O. niloticus*) (2.08 ± 0.19 g)	0% or control, 5, 10, 15 and 20% of TM	Productivity and survival rate ↑	Tubin et al. (2020)
BFT	Density and dietary carbon sources	Nile tilapia (*O. niloticus*) (50.47 ± 0.05 g)	Stocking densities [20, 40 and 60 fish per m^3] and fed the basal diet without carbon sources or with broken rice flour (BRF) or broken wheat grain flour (BWGF) 84 days	Growth performance ↑ Feed utilization ↑	Zaki et al. (2020)

(continued)

Table 9.1 (continued)

Studied systems	Studies parameters	Fish species	Doses and supplementations duration	Results	References
BFT	Pineapple peel powder (PAPP)	Nile tilapia (*O. niloticus*) (20.91 ± 0.11 g)	0, 10, 20, 30 and 40 g kg^{-1} PAPP 8 weeks	Growth performance ↑ FCR ↓	Van Doan et al. (2021a)
BFT	Watermelon rind powder (WMRP) + *L. plantarum*	Nile tilapia (*O. niloticus*) (16.57 ± 0.14 g)	Diet 1 (0 g kg^{-1} WMRP and 0 CFU g^{-1} L. *plantarum*) (control), Diet 2 (40 g kg^{-1} WMRP), Diet 3 (10^8 CFU g^{-1} LP), and Diet 4 (40 g kg^{-1} WMRP + 10^8 CFU g^{-1} LP) 8 weeks	Growth performance ↑ FCR ↓	Van Doan et al. (2021b)
BFT	Pineapple peel powder (PAPP) + *L. plantarum*	Nile tilapia (*O. niloticus*) (20.91 ± 0.11 g)	Diet 1 (0 g kg^{-1} PAPP and 0 CFU g^{-1} L. *plantarum*) (control), Diet 2 (10 g kg^{-1} PAPP), Diet 3 (10^8 CFU g^{-1} LP), and Diet 4 (10 g kg^{-1} PAPP + 10^8 CFU g^{-1} LP) 8 weeks	Growth performance ↑ FCR ↓	Van Doan et al. (2021d)
BFT	Amla (*Phyllanthus emblica*) fruit extract	Nile tilapia (*O. niloticus*) (10.48 ± 0.56 g)	0, 5, 10, 20, and 40 mg kg^{-1} AFE 8 weeks	Growth performance ↑ FCR ↓	Van Doan et al. (2022b)
BFT	Light levels	Nile tilapia (*O. niloticus*) (1.73 ± 0.16 g)	24 h of light (24hL), 12 h of light/12 h of darkness (12hL/12hD) and 24 h of darkness (24hD)	Carcass quality ↑ Digestive and hepatic enzymes ↑	Khanjani and Sharifinia (2021)
BFT	Symbiotics	Nile tilapia (*O. niloticus*) (30–35 g)	BFT with and without symbiotics 40 days	Growth performance ↑	Laice et al. (2021)
BFT	Stocking densities	Nile tilapia (*O. niloticus*) (133.91 g)	18.75, 37.50, 56.25, and 75.00 fish·m^{-3} 260 days	Growth performance ↑	Manduca et al. (2021)

	Beneficial bacteria	Nile tilapia (*O. niloticus*) (8.63 ± 3.35g)	T1, SR control; T2, SR + SSP; T3, SR + MSP; T4, BF + MSP; T5, BF + SSP; T6, BF control 112 days	Growth performance ↑	Mohammadi et al. (2021)
BFT	Coffee silverskin (CSS)	Nile tilapia (*O. niloticus*) (15.54 ± 0.21 g)	CSS1 (Control), CSS2 (10 g kg−1), CSS3 (20 g kg−1), CSS4 (40 g kg−1), and CSS5 (80 g kg−1) 8 weeks	Growth performance ↑ FCR ↓	Van Doan et al. (2021c)
BFT	Carbon sources and stocking densities	Nile tilapia (*O. niloticus*) (5.15 ± 1.12 g)	Low stocking density (LSD), 140 fish /m^3 and high stocking density, (HSD), 280/m^3 98 days	Growth performance ↑	El-Hawarry et al. (2021)
BFT	Salinities	Nile tilapia (*O. niloticus*) (13.78 ± 0.62 g)	T1 (5 ppt), T2 (10 ppt), T3 (15 ppt), T4 (20 ppt) and control (0 ppt) 90 days	Growth performance ↑ Survival rate ↑ Carcass quality ↑	Kumari et al. (2021)
BFT	Dietary phytase	Nile tilapia (*O. niloticus*) (29.8 g)	6-phytase (3000 FTU/kg; Quantum Blue™; P-F + Phy), and positive (C) and negative (P-F) control 154–156 days	Growth performance →	Green et al. (2021)
BFT	Spent coffee grounds (SCG)	Nile tilapia (*O. niloticus*) (15.25 ± 0.07 g)	SCG1 (control), SCG2 (10 g kg^{-1}), SCG3 (20 g kg^{-1}), SCG4 (40 g kg^{-1}), and SCG5 (80 g kg^{-1}) 8 weeks	Growth performance ↑ FCR ↓	Van Doan et al. (2022a)
BFT	Host-associated probiotic *Bacillus altitudinis* B61-34b	Nile tilapia (*O. niloticus*) (25.50 ± 0.52 g)	0 (BAA1—Control), 10^6 (BAA2), 10^7 (BAA3), 10^8 (BAA4) and 10^9 (BAA5) CFU ml^{-1} 8 weeks	Growth performance ↑ FCR ↓	Van Doan et al. (2021e)
BFT	Host-associated probiotic *Lactobacillus paracasei* l61-27b	Nile tilapia (*O. niloticus*) (25.40 ± 0.52 g)	LP1 = 0 (Control), LP2 = 10^6 CFU mL^{-1}, LP3 = 10^7 CFU mL^{-1}, LP4 = 10^8 CFU mL^{-1}, and LP5 = 10^9 CFU mL^{-1}) 8 weeks	Growth performance ↑ FCR ↓	Van Doan et al. (2021c)
BFT	Grade feeding rates	Nile tilapia (*O. niloticus*) (3.1 ± 0.1 g)	0%, 2.5%, 5.0%, 7.5%, and 10% of body weight per day 70 days	Growth performance ↑	Oliveira et al. (2021)

(continued)

Table 9.1 (continued)

Studied systems	Studies parameters	Fish species	Doses and supplementations duration	Results	References
BFT	17-α-methyltestosterone	Nile tilapia (*O. niloticus*) (Fry)	60, 90, 120, 150 and 180 mg kg^{-1}	Masculinization rates ↑	Costa e Silva et al. (2022)
BFT	Feeding levels and stocking densities	Nile tilapia (*O. niloticus*) (3.2 ± 0.05 g)	Feeding levels (0, 15, 30, 45 and 100) and 2 stocking densities (500 fish/m^3 and 1000fish/m^3)	Growth performance ↑	Sarsangi Aliabad et al. (2022)
BFT	Longan seed powder (LS)	Nile tilapia (*O. niloticus*) (13.82 ± 0.06 g)	Control (LS0), 10 (LS10), 20 (LS20), 40 (LS40), and 80 (LS80) g kg^{-1} LS 8 weeks	Growth performance ↑ FCR ↓	Wannavijit et al. (2022)
BFT	Rambutan seed (RS)	Nile tilapia (*O. niloticus*) (14.77 ± 0.80 g)	0, 5, 10, 20, and 40 g kg^{-1} of RS 8 weeks	Growth performance ↑ FCR ↓	Xuan et al. (2022)
BFT	Rambutan peel (RP)	Nile tilapia (*O. niloticus*) (17.14 ± 0.12 g)	0 g kg^{-1} (control – RP0); 10 g kg^{-1} (RP10); 20 g kg^{-1} (RP20); 40 g kg^{-1} (RP40), and 80 g kg^{-1} (RP80) 8 weeks	Growth performance ↑ FCR ↓	Le Xuan et al. (2022)
BFT	Chitosan	Nile tilapia (*O. niloticus*) (1.70 ± 0.36 g)	30, 60 and 90 ppm and 10, 20 and 30 ppm of chitosan	Growth performance ↑ Digestive enzymes ↑ FCR ↓	Chutia et al. (2022)
RAS	Light intensity and photoperiod	Nile tilapia (*O. niloticus*) (5 ± 0.9 g)	(1000, 2000, and 3000 lx) and photoperiods (12L:12D, 18L:6D, 24L:0D) 160 days	Growth performance ↑ FCR ↓ Stress →	Wang et al. (2020)

RAS	Magnetic field	Nile tilapia (*O. niloticus*) (7.16 ± 0.05 g)	0.00, 0.10, 0.15 and 0.20 T 70 days	Growth performance ↑ FCR ↓	Hassan et al. (2018)
RAS	Rearing systems and dietary probiotic	Nile tilapia (*O. niloticus*) (embryos)	FTS + control diet, RAS + control diet and RAS + *B. subtilis* coated diet (RASB) 33 days	Survival rate ↑ Beneficial bacteria in gut ↑	Deng et al. (2022)
BFT-RAS	Culture systems	Nile tilapia (*O. niloticus*) (0.17 ± 0.00 g)	Clear-water (CW), biofloc (BF), or hybrid (HY) 9 weeks	Growth performance ↑ FCR ↓	Fleckenstein et al. (2018)
RAS + BFT	Nitrogen and phosphorus budgets	Nile tilapia (*O. niloticus*) (3.54 ± 2.82 g)	(BFT) aquaculture system and a recirculation aquaculture system (RAS) during over-wintering of tilapia 64 days	Recovery rate of N in BFTSs ↑ P recovery rate →	Cao et al. (2020)
BFT-RAS	Protein levels	Nile tilapia (*O. niloticus*) (39.1 ± 2.5 g)	23, 27, 31 or 35% crude protein 9 weeks	Growth performance ↑ FCR ↓	Nguyen et al. (2021)

performance and feed utilization were observed in Nile tilapia fed dietary inclusion of OPDP and *Lactobacillus plantarum*, chestnut polyphenols (CSP), watermelon rind powder (WMRP), pineapple peel powder (PAPP), watermelon rind powder (WMRP) + *L. plantarum*, pineapple peel powder (PAPP) + *L. plantarum*, amla (*Phyllanthus emblica*) fruit extract, coffee silverskin (CSS), spent coffee grounds (SCG) (Van Doan et al. 2019; 2020a, b; 2021a, b, c, d; 2022a, b). Significant improvement in growth performance and FCR may be attributable to the bioactive compounds of these supplementations, which act not only as a nutrient source for fish but also are carbon sources for microbial protein production in biofloc systems. In studies using host-associated probiotics (*Bacillus altitudinis* B61-34b and *Lactobacillus paracasei* l61-27b), the authors indicated that supplementation of host-associated probiotics in indoor biofloc system resulted in better growth performance and feed utilization compared to the control group. This may be due to the complementary roles of biofloc and *B. altitudinis*. Numerous investigations have demonstrated that biofloc offers an essential nutrient source for tilapia (Ekasari et al. 2014a; Green et al. 2019). The addition of *Bacillus* spp. in cultured water or diets lowers ammonium levels in fish culture systems (Dash et al. 2018; Elsabagh et al. 2018). Furthermore, the presence of favorable microbial flocs and external probiotics will likely boost the number of valuable microbiota in the tilapia's digestive system (Rohani et al. 2022; de Sousa et al. 2019). Increased secretion of digestive enzymes through the colonization of bacteria facilitates the absorption of nutrients by the intestinal epithelial cells (Liu et al. 2017). *Bacillus* produces many biological substances, including cellulase, phytase, tannase, chitinase, xylanase, protease, amylase, and lipase (Ringø 2020). Favorable bacteria also release several nutrients, in particular vitamins, amino acids, and fatty acids, and diminish lethal feedstuffs and infectious bacteria (Zaineldin et al. 2021). Recently, the dietary inclusion of rambutan and long seed powder or rambutan peel in the Nile tilapia diet led to an increase in growth rate and feed utilization (Xuan et al. 2022; Wannavijit et al. 2022). It has been reported that these seeds are known as carbon sources (Yang et al. 2015; Lawtae and Tangsathitkulchai 2021) and has hence been used in biofloc system (Liu et al. 2019). It has been observed that adding carbon to a biofloc system causes heterotrophic bacteria to utilize the inorganic nitrogen by changing the water C: N ratio, resulting in a higher microbic protein source for the host and improved water quality (Guo et al. 2020). In addition, incorporating a carbon source leads to the formation of biofloc, a new protein source for fish (Krummenauer et al. 2020; Tinh et al. 2021). Additionally, these products also act as potential prebiotics or carbohydrate (Estrada-Gil et al. 2022; Jahurul et al. 2020). Similarly, supplementation of pizzeria by-product de Sousa et al. (2019); dietary digestible protein and digestible energy (Durigon et al. 2019); phosphatidylcholine (Sousa et al. 2020); *Tenebrio molitor* meal (TM) (Tubin et al. 2020); symbiotics (Laice et al. 2021); beneficial bacteria (Mohammadi et al. 2021); dietary phytase (Green et al. 2021), and chitosan (Chutia et al. 2022) led to higher growth rate and feed utilization.

Another common study aspect using biofloc technology is the application of different carbon sources and their effects on tilapia growth and feed conversion ratio. (Mirzakhani et al. (2019) reported that fish in a biofloc system with 100% of wheat flour at a C:N ratio of 15:1 showed the highest growth performance with improved intestine histoarchitecture. Wheat flour as a major source of starch and energy can also provide substantial amounts of other nutrients such as protein, vitamins, and phytochemicals and especially high fiber content (ca. 12%) compared to the lower fiber content in molasses (ca. 0.5%) (Shewry and Hey 2015). These nutrients might enhance the biochemical composition and bioactive compounds of biofloc. In addition, dispersed particles of wheat flour in water may provide a good substrate for the development and growth of microorganisms and bacteria because of which the nutrition value of the produced biofloc increases, ultimately influencing the fish growth and immune response (Mirzakhani et al. (2019). García-Ríos et al. (2019) indicated that the fingerlings obtained in BFT, with corn and sugar as C sources, had a similar growth rate to the control. However, the BFT promotes significant savings in feed (41.1 to 58.9%) and water (67.4 to 75.5%) compared to the traditional method. Similar results were in the study of Zaki et al. (2020), where the authors indicated that Increased growth and feed utilization were recorded in 40 fish per m^3 fed with broken rice flour. El-Hawarry et al. (2021) also found that the growth rate was improved in the groups of fish under low stocking density with molasses and glycerol as carbon sources. Carbon source affects the cultured species' growth depending on the formatted biofloc characteristics, such as its "volume, chemical composition, and ability to store bioactive compounds (Wang et al. 2015; Zhao et al. 2016). Additionally, microbial flocs which are formed from different carbon sources act as a supplemental food source that constantly provides additional essential amino acids profile (microbial protein), polyunsaturated fatty acids, minerals, vitamins, and an external source of digestive enzymes (Avnimelech 2007; Azim and Little 2008; Bakhshi et al. 2018; De Schryver and Verstraete 2009). In contrast, the use of biochar as an alternative carbon source for biofloc technology did not affect the growth rate of Nile tilapia (Abakari et al. 2020). Although biochar is regarded as a recalcitrant carbon source, the utilization of biochar-derived carbon by heterotrophic bacteria has been described (Farrell et al. 2013).

Effects of stocking density on Nile tilapia growth raised in biofloc systems have been conducted by several researchers. Liu et al. (2018) showed that low stocking density (166 fish/m^{-3}) improved growth performance and FCR of Nile tilapia raised in the biofloc system. Similarly, Manduca et al. (2021) reported that tilapia stocking density in BFT around 33 fish m^{-3} had higher profitability since it produces a large proportion of harvested fish that reach high body weights, and possibly high selling prices, combined with desirable biomass. Recently, Sarsangi Aliabad et al. (2022) also suggested that the stocking density of 1000/m^3 for larviculture of tilapia in BFT uses water and equipment more efficiently. Biofloc acts as the natural food that contributes significantly to the nutrition of tilapia fingerlings, allowing the reduction of the feeding rations. Biofloc consumption corresponds to 50% of the daily food of tilapia (Avnimelech 2007). Another study revealed that 25% of the protein requirement of tilapia could be provided by floc consumption (Avnimelech 2015). Besides, significantly improved growth performance, FCR, and digestive enzymes were observed in Nile tilapia raised in biofloc combined with different conditions, such as in-situ and ex-situ biofloc (Menaga et al. 2019), jaggery-based BFT (Elayaraja et al. 2020); light levels (Khanjani and Sharifinia 2021); salinities (Kumari et al. 2021); grade feeding rates (Oliveira et al. 2021), and 17-α-methyltestosterone (Costa e Silva et al. 2022).

9.2.3.2 Biofloc as Immunostimulants

Bioflocs also offer a lot of MAMPs (microbial-associated molecular patterns), which may be recognized as immunostimulants, resulting in higher resistance to diseases (Ekasari et al. 2014b, 2015a). Additionally, it consists of a wide range of organic compounds, such as carotenoids, chlorophylls, bromophenols, phytosterols, and antibacterials that have a positive effect on immune factors of cultivated aquatic species (Crab et al. 2010; Najdegerami et al. 2016; Bakhshi et al. 2018; Mirzakhani et al. 2019).

The effects of biofloc in combination with different functional feed additives on the immune response of Nile tilapia have been reported in previous studies (Doan et al. 2018; Van Doan et al. 2019, 2020a, b; 2021a, b, c, d; 2022a, b; Xuan et al. 2022; Le Xuan et al. 2022; Wannavijit et al. 2022; Table 9.2). Similar findings were observed in Nile tilapia fed in-situ and ex-situ biofloc (Menaga et al. 2019); phosphatidylcholine (Sousa et al. 2020); symbiotics (Laice et al. 2021); beneficial bacteria (Mohammadi et al. 2021), and probiotics (Bañuelos-Vargas et al. 2021). These substances act as immunostimulants and/or carbon sources for the proliferation of microbial proteins in the biofloc system.

Carbon source applications in the biofloc system could result in better immune response in Nile tilapia. Mirzakhani et al. (2019) indicated that fish reared in a biofloc system based on 100% wheat flour and a C/N ratio of 15 demonstrated the humoral immune response. Similarly, a significant increase in innate and specific immune responses was observed in Nile tilapia raised in biofloc with biochar (Abakari et al. 2020) and jaggery-based BFT (Elayaraja et al. 2020) as carbon sources. Carbon sources play a vital role in the proliferation of microbial protein in biofloc systems, which in turn act as immunostimulants for culture species (Panigrahi et al. 2019). Carbon sources and stocking densities also have a great impact on Nile tilapia's immune response raised under the biofloc system. El-Hawarry et al. (2021) indicated that the growth rate and growth-related genes were improved in the groups of fish under low stocking density (LSD) with molasses and glycerol as carbon sources. Recently, (Sarsangi Aliabad et al. 2022) also reported that the BFT system improved water quality, growth performance, and immune function of Nile tilapia fry.

9.2.3.3 Biofloc as Disease Prevention Techniques

In biofloc systems, aquaculture animals may also benefit from reduced pathogen pressure (Bossier and Ekasari 2017). Some studies demonstrated that the presence of potentially pathogenic bacteria might be reduced in biofloc systems (Gustilatov et al. 2022; de Lima Vieira et al. 2021; Khanjani et al. 2022; Table 9.3). Increase disease

Table 9.2 Immune responses of tilapia cultured biofloc system

Studied systems	Studies parameters	Fish species	Doses and supplementations duration	Results	References
BFT	Stocking density	Tilapias (0.51 ± 0.05 g)	166 orgs m^{-3} (LD, low density), 333 orgs m^{-3} (MD, middle density) and 600 orgs m^{-3} (HD, high density) 120 days	Lysozyme activity ↓ Complement 3 activity ↓ Glutathione level ↓	Liu et al. (2018)
BFT	Orange peels derived pectin	Nile tilapia (9.09 ± 0.05 g)	0 (control in clear water), 0 (control in biofloc system), 5, 10, and 20 g kg^{-1} OPDP for 8 weeks	Skin mucus immunity ↑ Serum immunity ↑	Doan et al. (2018)
BFT	*in-situ* and *ex-situ* biofloc	Nile tilapia (*O. niloticus*) (5.1 ± 0.05 g/fish)	T1-biofloc developed within the culture systems (*insitu*), T2-biofloc supplementation in fish feed (*exsitu*) and C- Control without biofloc	Immune gene expressions ↑	Menaga et al. (2019)
BFT	Carbon sources	Nile tilapia (*O. niloticus*) 2.7 ± 0.4 g	100% of MO, 75% of MO+ 25% of WF, 50% of MO+ 50% of WF, 25% of MO+ 75% of WF,and 100% of WF	Humoral immune responses ↑	Mirzakhani et al. (2019)
BFT	OPDP and *L. plantarum*	Nile tilapia (*O. niloticus*) (1.25 ± 0.15 g) 5.92 ± 0.08 g	Diet 1 (0 g kg^{-1} OPDP and 0 CFU g^{-1} *L. plantarum*), diet 2 (10 g kg^{-1} OPDP), diet 3 (108 CFU g^{-1} *L. plantarum*), and diet 4 (10 g kg^{-1} OPDP +108 CFU g^{-1} *L. plantarum*)	Skin mucus immunity ↑ Serum immunity ↑	Van Doan et al., (2019)
BFT	Biochar	Nile tilapia (*O. niloticus*) (36.2 g/fish)	B; only biochar, GB; biochar + glucose, while for the control (G) only glucose	Immune parameters ↑	Abakari et al. (2020)
BFT	Jaggery-based BFT	Nile tilapia (*O. niloticus*) (32.2 ± 10.1 g/fish)	22.5%, 27.7%, and 32.3% digestible protein (DP) and 6% lipid	Innate immunity ↑ Immune gene expressions ↑	Elayaraja et al. (2020)

(continued)

Table 9.2 (continued)

Studied systems	Studies parameters	Fish species	Doses and supplementations duration	Results	References
BFT	Chestnut polyphenols (CSP)	Nile tilapia (*O. niloticus*) (12.77 ± 0.17 g)	0, 1, 2, 4, and 8 g kg^{-1} of CSP	Skin mucus immunity ↑ Serum immunity ↑	Van Doan et al. (2020a)
BFT	Watermelon rind powder (WMRP)	Nile tilapia (*O. niloticus*) (17.14 ± 0.12 g)	0 (Diet 1- control), 20 g kg^{-1} WMRP (Diet 2), 40 g kg^{-1} WMRP (Diet 3), 80 g kg^{-1} WMRP (Diet 4), and 160 g kg^{-1} WMRP (Diet 5)	Immune responses ↑	Van Doan et al. (2020b)
BFT	Phosphatidylcholine	Nile tilapia (*O. niloticus*)	0, 400, 800 and 1200 mg/kg of feed	Antioxidant enzymes Lipoperoxidation in liver ↓	Sousa et al. (2020)
BFT	Pineapple peel powder	Nile tilapia (*O. niloticus*) (20.91 ± 0.11 g)	0, 10, 20, 30 and 40 g kg^{-1} PAPP 8 weeks	Skin mucus immunity ↑ Serum immunity ↑	Van Doan et al. (2021a)
BFT	Watermelon rind powder (WMRP) + *L. plantarum*	Nile tilapia (*O. niloticus*) (16.57 ± 0.14 g)	Diet 1 (0 g kg^{-1} WMRP and 0 CFU g^{-1} L. *plantarum*) (control), Diet 2 (40 g kg^{-1} WMRP), Diet 3 (10^8 CFU g^{-1} LP), and Diet 4 (40 g kg^{-1} WMRP + 10^8 CFU g^{-1} LP) 8 weeks	Skin mucus immunity ↑ Serum immunity ↑	Van Doan, Seyed Hossein Hoseinifar, et al. (2021b)
BFT	Pineapple peel powder (PAPP) + *L. plantarum*	Nile tilapia (*O. niloticus*) (20.91 ± 0.11 g)	Diet 1 (0 g kg^{-1} PAPP and 0 CFU g^{-1} L. *plantarum*) (control), Diet 2 (10 g kg^{-1} PAPP), Diet 3 (10^8 CFU g^{-1} LP), and Diet 4 (10 g kg^{-1} PAPP + 10^8 CFU g^{-1} LP) 8 weeks	Skin mucus immunity ↑ Serum immunity ↑	Van Doan et al. (2021d)
BFT	Amla (*Phyllanthus emblica*) fruit extract	Nile tilapia (*O. niloticus*) (10.48 ± 0.56 g)	0, 5, 10, 20, and 40 mg kg^{-1} AFE 8 weeks	Skin mucus immunity ↑ Serum immunity ↑	Van Doan et al. (2022b)

BFT	Symbiotics	Nile tilapia (*O. niloticus*) (30–35 g)	BFT with and without symbiotics 40 days	Hematological parameters ↑	Laice et al. (2021)
BFT	Beneficial bacteria	Nile tilapia (*O. niloticus*) (8.63 ± 3.35g)	T1, SR control; T2, SR + SSP; T3, SR + MSP; T4, BF + MSP; T5, BF + SSP; T6, BF control 112 days	Innate immune response ↑	Mohammadi et al. (2021)
BFT	Coffee silverskin (CSS)	Nile tilapia (*O. niloticus*) (15.54 ± 0.21 g)	CSS1 (Control), CSS2 (10 g kg−1), CSS3 (20 g kg−1), CSS4 (40 g kg−1), and CSS5 (80 g kg−1) 8 weeks	Skin mucus immunity ↑ Serum immunity ↑	Hien Van Doan et al. (2021c)
BFT	Carbon source and stocking density	Nile tilapia (*O. niloticus*) (5.15 ± 1.12 g)	Low stocking density (LSD), 140 fish /m^3 and high stocking density, (HSD), 280/m^3 98 days	Growth-related genes ↑	El-Hawarry et al. (2021)
BFT	Probiotics	Nile tilapia (*O. niloticus*) (6.7 ± 0.2 g)	Stocking densities (D1, 120 fish/m^3; D2, 240 fish/m^3) with biofloc plus probiotics	Immune response ↑	Bañuelos-Vargas et al. (2021)
BFT	Spent coffee grounds (SCG)	Nile tilapia (*O. niloticus*) (15.25 ± 0.07 g)	SCG1 (control), SCG2 (10 g kg^{-1}), SCG3 (20 g kg^{-1}), SCG4 (40 g kg^{-1}), and SCG5 (80 g kg^{-1}) 8 weeks	Skin mucus immunity ↑ Serum immunity ↑	Van Doan et al. (2022a)
BFT	Feeding levels and stocking densities	Nile tilapia (*O. niloticus*) (3.2 ± 0.05 g)	Feeding levels (0, 15, 30, 45 and 100) and 2 stocking densities (500 fish/m3 and 1000fish/m3)	Innate immune response ↑	Sarsangi Aliabad et al. (2022)
BFT	Rambutan seed (RS)	Nile tilapia (*O. niloticus*) (14.77 ± 0.80 g)	0, 5, 10, 20, and 40 g kg^{-1} of RS 8 weeks	Skin mucus immunity ↑ Serum immunity ↑ Gene expressions ↑	Xuan et al. (2022)

(continued)

Table 9.2 (continued)

Studied systems	Studies parameters	Fish species	Doses and supplementations duration	Results	References
BFT	Longan seed powder (LS)	Nile tilapia (*O. niloticus*) (13.82 ± 0.06 g)	Control (LS0), 10 (LS10), 20 (LS20), 40 (LS40), and 80 (LS80) g kg^{-1} LS 8 weeks	Skin mucus immunity ↑ Serum immunity ↑ Gene expressions ↑	Wannavijit et al. (2022)
BFT	Rambutan peel (RP)	Nile tilapia (*O. niloticus*) (17.14 ± 0.12 g)	0 g kg^{-1} (control – RP0); 10 g kg^{-1} (RP10); 20 g kg^{-1} (RP20); 40 g kg^{-1} (RP40), and 80 g kg^{-1} (RP80) 8 weeks	Skin mucus immunity ↑ Serum immunity ↑ Gene expressions ↑	Le Xuan et al. (2022)

Table 9.3 Increase disease resistance of tilapia cultured under biofloc system

Studied systems	Studies parameters	Fish species	Doses and supplementations duration	Results	References
BFT	Orange peels derived pectin	Nile tilapia (9.09 ± 0.05 g)	0 (control in clear water), 0 (control in biofloc system), 5, 10, and 20 g kg^{-1} OPDP for 8 weeks	Resistance to *S. agalactiae* ↑	Doan et al. (2018)
BFT	Dietary digestible protein and digestible energy	Nile tilapia (*Oreochromis niloticus*) (1.25 ± 0.15 g)	Digestible protein (22, 26 and 30% DP) and digestible energy (3000, 3150 and 3300 kcal kg^{-1}) 42 days	Ectoparasite spread ↓	Durigon et al. (2019)
BFT	*in-situ* and *ex-situ* biofloc	Nile tilapia (*O. niloticus*) (5.1 ± 0.05 g/fish)	T1-biofloc developed within the culture systems (*insitu*), T2-biofloc supplementation in fish feed (*exsitu*) and C- Control without biofloc	Resistance to *A. hydrophila* ↑	Menaga et al. (2019)
BFT	OPDP and *L. plantarum*	Nile tilapia (*O. niloticus*) (1.25 ± 0.15 g) 5.92 ± 0.08 g	Diet 1 (0 g kg^{-1} OPDP and 0 CFU g^{-1} *L. plantarum*), diet 2 (10 g kg^{-1} OPDP), diet 3 (108 CFU g^{-1} *L. plantarum*), and diet 4 (10 g kg^{-1} OPDP +108 CFU g^{-1} *L. plantarum*)	Resistance to *S. agalactiae* ↑	Van Doan et al. (2019)
BFT	Jaggery-based BFT	Nile tilapia (*O. niloticus*) (32.2 ± 10.1 g/fish)	22.5%, 27.7%, and 32.3% digestible protein (DP) and 6% lipid	Resistance to *A. hydrophila* ↑	Elayaraja et al. (2020)
BFT	Chestnut polyphenols (CSP)	Nile tilapia (*O. niloticus*) (12.77 ± 0.17 g)	0, 1, 2, 4, and 8 g kg^{-1} of CSP	Resistance to *S. agalactiae* ↑	Van Doan et al. (2020a)
BFT	Watermelon rind powder (WMRP)	Nile tilapia (*Oreochromis niloticus*) (17.14 ± 0.12 g)	0 (Diet 1- control), 20 g kg^{-1} WMRP (Diet 2), 40 g kg^{-1} WMRP (Diet 3), 80 g kg^{-1} WMRP (Diet 4), and 160 g kg^{-1} WMRP (Diet 5)	Resistance to *S. agalactiae* ↑	Van Doan et al. (2020b)
BFT	Pineapple peel powder (PAPP)	Nile tilapia (*O. niloticus*) (20.91 ± 0.11 g)	0, 10, 20, 30 and 40 g kg^{-1} PAPP 8 weeks	Resistance to *S. agalactiae* ↑	Van Doan et al. (2021a)

(continued)

Table 9.3 (continued)

Studied systems	Studies parameters	Fish species	Doses and supplementations duration	Results	References
BFT	Pineapple peel powder (PAPP) + *L. plantarum*	Nile tilapia (*O. niloticus*) (20.91 ± 0.11 g)	Diet 1 (0 g kg^{-1} PAPP and 0 CFU g^{-1} L. *plantarum*) (control), Diet 2 (10 g kg^{-1} PAPP), Diet 3 (10^8 CFU g^{-1} LP), and Diet 4 (10 g kg^{-1} PAPP + 10^8 CFU g^{-1} LP) 8 weeks	Resistance to *S. agalactiae* ↑	Van Doan et al. (2021d)
BFT	Coffee silverskin (CSS)	Nile tilapia (*O. niloticus*) (15.54 ± 0.21 g)	CSS1 (Control), CSS2 (10 g kg−1), CSS3 (20 g kg−1), CSS4 (40 g kg−1), and CSS5 (80 g kg−1) 8 weeks	Resistance to *S. agalactiae* ↑	Van Doan et al. (2021c)
BFT	Amla (*Phyllanthus emblica*) fruit extract	Nile tilapia (*O. niloticus*) (10.48 ± 0.56 g)	0, 5, 10, 20, and 40 mg kg^{-1} AFE 8 weeks	Resistance to *S. agalactiae* ↑	Van Doan et al. (2022b)
BFT	Beneficial bacteria	Nile tilapia (*O. niloticus*) (8.63 ± 3.35g)	T1, SR control; T2, SR + SSP; T3, SR + MSP; T4, BF + MSP; T5, BF + SSP; T6, BF control 112 days	Resistance to *A. hydrophila* ↑	Mohammadi et al. (2021)
BFT	Spent coffee grounds (SCG)	Nile tilapia (*O. niloticus*) (15.25 ± 0.07 g)	SCG1 (control), SCG2 (10 g kg^{-1}), SCG3 (20 g kg^{-1}), SCG4 (40 g kg^{-1}), and SCG5 (80 g kg^{-1}) 8 weeks	Resistance to *S. agalactiae* ↑	Van Doan et al. (2022a)

resistance against *Streptococcus agalactiae* and *Aeromonas hydrophila* have been reported in Nile tilapia-fed orange peels derived pectin (OPDP) (Doan et al. 2018); OPDP and *L. plantarum* (Van Doan et al. 2019); in-situ and ex-situ biofloc (Menaga et al. 2019); jaggery-based BFT (Elayaraja et al. 2020); chestnut polyphenols (CSP) (Van Doan et al. 2020a); watermelon rind powder (WMRP) (Van Doan et al. 2020b); pineapple peel powder (PAPP) (Van Doan et al. 2021a); pineapple peel powder (PAPP) + *L. plantarum* (Van Doan et al. 2021d); coffee silverskin (CSS) (Van Doan et al. 2021c); amla (*Phyllanthus emblica*) fruit extract (Van Doan et al. 2022b); beneficial bacteria (Mohammadi et al. 2021); spent coffee grounds (SCG) (Van Doan et al. 2022a). It has been also reported that a biofloc system could reduce ectoparasite spread Durigon et al. (2019). A significant increase in disease resistance may be attributable to the presence of MAMPs in the biofloc system, which may be recognized as immunostimulants, resulting in higher resistance to diseases (Ekasari et al. 2014b). In addition, it may be due to the prebiotic and probiotic properties of feed additives, which are known to enhance the immune response and disease resistance of Nile tilapia (Cavalcante et al. 2020; Cano-Lozano et al. 2022).

9.3 Recirculating Aquaculture Systems (RAS)

9.3.1 Brief History of Development

Though in its infancy and still considered to be a recent innovation, the basic technology of RAS has existed for over 65 years, with the first, pioneering RAS research activity being conducted in Japan in the 1950s (Murray et al. 2014; Saeki 1958). According to Espinal and Matulić (2019), the technology of RAS including aquaponics has been developed over the past 40 years. In the 1970s, a German program demonstrated the feasibility of intensive carp production in RAS, and subsequently, the Danish Aquaculture Institute undertook an innovative effort to develop further technical aspects of RAS (Goldman 2016). The idea for commercial fish production in RAS was first fostered in Denmark in the mid-1970s, and the first commercial RAS was then built in 1980 (Warrer-Hansen 2015). The Danish efforts supported the development of one of the initial commercial RAS industries, specifically for the production of European eel (Anguilla Anguilla) (Goldman 2016). This work inspired the subsequent further development and uptake of RAS in other European countries in the late 1980s and 1990s (Martins et al. 2010). Over the last 25–35 years, a significant and growing experience in designing, building, and operating RAS, particularly in Nordic countries, has been reported (Dalsgaard et al. 2013). The initial success of the RAS-based European eel industry also inspired to development of RAS in North America (Goldman 2016). In China, marine RAS was initiated in the 1980s, and since then China has made considerable progress in RAS (Ying et al. 2015). Since the 2000s, further development of RAS has occurred in Europe, North America, Australia, and other aquaculture-producing countries (Espinal and Matulić 2019; Martins et al. 2010).

A significant acceleration in the development of RAS technology has been observed over the last two decades (Espinal and Matulić 2019), and RAS have become popular in recent years. RAS has been developed to grow fish where inadequate biophysical conditions, water scarcity, poor water quality, and unfavorable environment exist (Murray et al. 2014). According to Malone (2013), RAS provides an alternative production method when environmental regulations, disease, land availability, salinity, temperature, and water supply prevent more cost-effective alternatives. However, other factors stimulated the development and implementation of RAS. For example, RAS are increasingly being used for Mediterranean marine fish and salmonid production cycle, particularly for juvenile stages, before being transferred into outdoor grow-out systems, such as cages or flow-through raceways (Bostock et al. 2016; Clarke and Bostock 2017; Terjesen et al. 2013). In fact, RAS can be used for broodstock and seedstock production, which can support cage and net-pen aquaculture (Malone 2013). In Europe and North America, RAS was developed as an alternative to the cage culture of salmon (Murray et al. 2014). RAS has also been developed to culture exotic fish species, to avoid adverse effects on native species and biodiversity (Malone 2013; Murray et al. 2014).

9.3.2 Basic Principles of RAS Operation and Production

RAS are land-based, indoor fish-rearing facilities, where fish are stocked in tanks within a controlled environment, and where filtration is applied to purify water by removing metabolic wastes of stock, before being recirculated into the system itself. Water purification is achieved through mechanical and/or biological filtration, sterilization, and oxygenation. Different levels of sophistication and efficiency can be achieved, but generally, all RAS have a high degree (>90%) of water reuse (Murray et al. 2014; Badiola et al. 2012). RAS provides opportunities to enhance waste management, reduce water usage, and nutrient recycling (Martins et al. 2010; Murray et al. 2014; van Rijn 2013).

Although RAS have been initially developed and are ideally suited to produce freshwater as well as warm water fish species, RAS is flexible and can be modified and adapted to be operated with brackish and marine water as well as cold water species (Helfrich and Libey 1991). Therefore, by decoupling fish production from the marine environment, RAS may offer an alternative to traditional and net pen aquaculture (O'Shea et al. 2019). RAS can also provide suitable environmental conditions for fish species that are sensitive to water quality (Zhang et al. 2011). Despite RAS the potential to produce diverse seafood products, RAS is generally utilized to culture high-value fish, with high stocking densities and year-round production to offset high operational costs (Dalsgaard et al. 2013; Martins et al. 2010; Murray et al. 2014). Nevertheless, other fish species, including arctic char, clarias, halibut, pangasius, tilapia, and turbot are also commonly produced in RAS (Badiola et al. 2018; Ngoc et al. 2016a; Ngoc et al. 2016b; Summerfelt and Vinci 2008). The selection of fish species can be "market-driven" due to a high return on investment to keep the RAS profitable (Badiola et al. 2017). The choice of fish also depends on the fast-growing and hardy fish in RAS (Badiola et al. 2018).

RAS can be categorized into five types: (1) hatchery and grow-out, (2) breeding, (3) long-term holding, (4) short-term holding, and (5) display (Yanong 2012). Moreover, RAS can be incorporated into an "integrated agriculture-aquaculture" system, which is known as aquaponics (Martins et al. 2010). Aquaponics is considered a particular type of RAS, where vegetable plants are included with fish to provide water filtration and crop diversification (Goddek et al. 2019). RAS have greater control over production outcomes, and the productivity of RAS depends on culture species, stocking densities, feeding rate, duration of the production cycle, and other management aspects. According to available scientific literature, the stocking densities of RAS range from 70 to 120 kg/m^3 with feed conversion ratio (FCR) values from 0.8 to 1.1. RAS can be of various sizes including small, medium, and large (Helfrich and Libey 1991), with a large-scale RAS typically being able to produce 400e500 tons of fish per annum (Murray et al. 2014). However, even higher stocking densities and total production values are currently reported by some commercial producers. According to (Bregnballe 2015), RAS are highly productive intensive farming, which generates vast quantities of fish (500 tons/ha/year) in a comparatively small volume of water. Because of higher production, RAS is often referred to as "hyper" or "super" intensive farming (O'Shea et al. 2019).

9.3.3 RAS in Tilapia Culture

There is limited information regarding the application of RAS in Nile tilapia farming. It has been reported that RAS is a costly engineering approach, with a high initial investment in installation and operation (Murray et al. 2014). The reported annual production cost of RAS (US$2250e8800 per ton) is considerably higher than conventional pond aquaculture (US$2000 per ton) (Waite et al. 2014). Additionally, the Economic viability of RAS requires a long payback period, on average 8 years (Badiola et al. 2012). Wang et al. (2020) indicated that light intensity and photoperiod manipulation did not cause a significant chronic stress response in tilapia. This study demonstrated that light intensity, especially at 2000 lx, and photoperiod manipulation could stimulate the growth of tilapia in the RAS and significantly affect economic profitability. Another study using a different magnetic field (Hassan et al. 2018) showed that based on the growth, water properties, and serum biochemistry, it was concluded that magnetized water at 0.15 T intensity may improve tilapia growth in recirculating aquaculture systems. Recently, Deng et al. (2022) indicated that rearing fish larvae in RAS supports better survival compared to the flow-through system, while dietary probiotic supplementation further modulates the gut bacterial composition and stimulates the presence of beneficial bacteria during early life. It has been reported that RAS has a more stable and diverse microbial community composition, which could result in better growth performance compared to other culture systems (Deng et al. 2022). In addition, RAS could reduce the ammonium and toxic gas in the culture system (Villar-Navarro et al. 2021; Nguyen et al. 2021).

A comparison between RAS, clear water, and biofloc system has been conducted in Nile tilapia. (Fleckenstein et al. 2018) indicated that clear water or hybrid systems may be a better choice for tilapia nurseries than chemoautotrophic biofloc systems due to the short-term periods in which nurseries operate and the volatility of nitrification in biofloc systems. In another study, Cao et al. (2020) indicated that There was no significant difference between the RASs and BFT aquaculture systems in terms of P recovery rate. The regular backwashing of the drum filter and biological filter in RAS accounted for 41 ± 2% of input N and 39 ± 2% of input P. Approximately 54% of unassimilated nitrogen N was removed by nitrification in the BFT aquaculture systems. The results from the present study suggest that nitrification may be the dominant pathway for ammonia removal in a BFT aquaculture system rather than by heterotrophic bacterial assimilation. RAS is characterized as a closed aquaculture and water reuse system; however, without an efficient and effective system for the treatment of discharged water and solids, this characterization only seems to indicate potential. Significantly, the treatment of the solids and water discharged from RASs has been suggested by (Luo et al. 2013). In BFT systems, most unused N and P are retained in biofloc and nitrate in tanks. The biofloc can be used for shrimp feed (Ray et al. 2017) or for feeding Artemia (Luo et al. 2017). Nitrate may be reduced by denitrification and dissimilatory nitrate reduction to ammonium (DNRA) activities in the BFT systems (Chutivisut et al. 2014). In these respects, the production activity in BFT systems may be more closed than that of RAS (Cao et al. 2020).

References

Abakari G, Luo G, Meng H, Yang Z, Owusu-Afriyie G, Kombat EO, Alhassan EH (2020) The use of biochar in the production of tilapia (Oreochromis niloticus) in a biofloc technology system–BFT. Aquac Eng 91:102123. https://doi.org/10.1016/j.aquaeng.2020.102123

Adegbeye MJ, Elghandour MMMY, Monroy JC, Abegunde TO, Salem AZM, Barbabosa-Pliego A, Faniyi TO (2019) Potential influence of yucca extract as feed additive on greenhouse gases emission for a cleaner livestock and aquaculture farming–a review. J Clean Prod 239:118074. https://doi.org/10.1016/j.jclepro.2019.118074

Ahmad AL, Chin JY, Mohd Harun MHZ, Low SC (2022) Environmental impacts and imperative technologies towards sustainable treatment of aquaculture wastewater: a review. J Water Process Eng 46:102553. https://doi.org/10.1016/j.jwpe.2021.102553

Ahmad I, Babitha Rani AM, Verma AK, Maqsood M (2017) Biofloc technology: an emerging avenue in aquatic animal healthcare and nutrition. Aquac Int 25(3):1215–1226. https://doi.org/10.1007/s10499-016-0108-8

Ahmed N, Thompson S, Glaser M (2019) Global aquaculture productivity, environmental sustainability, and climate change adaptability. Environ Manag 63(2):159–172. https://doi.org/10.1007/s00267-018-1117-3

Ahmed N, Turchini GM (2021) Recirculating aquaculture systems (RAS): environmental solution and climate change adaptation. J Clean Prod 297:126604. https://doi.org/10.1016/j.jclepro.2021.126604

Anand PSS, Kohli MPS, Kumar S, Sundaray JK, Roy SD, Venkateshwarlu G, Sinha A, Pailan GH (2014) Effect of dietary supplementation of biofloc on growth performance and digestive enzyme activities in *Penaeus monodon*. Aquaculture 418-419:108–115. https://doi.org/10.1016/j.aquaculture.2013.09.051

Avnimelech Y (2007) Feeding with microbial flocs by tilapia in minimal discharge bio-flocs technology ponds. Aquaculture 264(1–4):140–147. https://doi.org/10.1016/j.aquaculture.2006.11.025

Avnimelech Y (2015) Biofloc technology: a practical guide book. World Aquaculture Society

Azim ME, Little DC (2008) The biofloc technology (BFT) in indoor tanks: water quality, biofloc composition, and growth and welfare of Nile tilapia (Oreochromis niloticus). Aquaculture 283(1–4):29–35. https://doi.org/10.1016/j.aquaculture.2008.06.036

Baag S, Mandal S (2022) Combined effects of ocean warming and acidification on marine fish and shellfish: a molecule to ecosystem perspective. Sci Total Environ 802:149807. https://doi.org/10.1016/j.scitotenv.2021.149807

Badiola M, Basurko OC, Gabiña G, Mendiola D (2017) Integration of energy audits in the life cycle assessment methodology to improve the environmental performance assessment of recirculating aquaculture systems. J Clean Prod 157:155–166. https://doi.org/10.1016/j.jclepro.2017.04.139

Badiola M, Basurko OC, Piedrahita R, Hundley P, Mendiola D (2018) Energy use in recirculating aquaculture systems (RAS): a review. Aquac Eng 81:57–70. https://doi.org/10.1016/j.aquaeng.2018.03.003

Badiola M, Mendiola D, Bostock J (2012) Recirculating aquaculture systems (RAS) analysis: main issues on management and future challenges. Aquac Eng 51:26–35. https://doi.org/10.1016/j.aquaeng.2012.07.004

Bahar NHA, Lo M, Sanjaya M, Van Vianen J, Alexander P, Ickowitz A, Sunderland T (2020) Meeting the food security challenge for nine billion people in 2050: what impact on forests? Glob Environ Chang 62:102056. https://doi.org/10.1016/j.gloenvcha.2020.102056

Bakhshi FH, Najdegerami E, Manaffar R, Tokmechi A, Rahmani Farah K, Shalizar Jalali A (2018) Growth performance, haematology, antioxidant status, immune response and histology of common carp (Cyprinus carpio L.) fed biofloc grown on different carbon sources. Aquac Res 49(1):393–403

Banha F, Diniz AM, Olivo del Amo R, Oliva-Paterna FJ, Anastácio PM (2022) Perceptions and risk behaviors regarding biological invasions in inland aquatic ecosystems. J Environ Manag 308: 114632. https://doi.org/10.1016/j.jenvman.2022.114632

Bañuelos-Vargas I, Rodríguez-Montes de Oca GA, Martínez-Montaño E, Pérez-Jiménez A, Mendoza-Gamboa OA, Estrada-Godínez JA, Hernández C (2021) Antioxidant and immune response of juvenile red tilapia (Oreochromis sp) cultured at different densities in sea water with biofloc plus probiotics. Aquaculture 544:737112. https://doi.org/10.1016/j.aquaculture.2021.737112

Bashir I, Lone FA, Bhat RA, Mir SA, Dar ZA, Dar SA (2020) Concerns and threats of contamination on aquatic ecosystems. Bioremed Biotechnol:1–26. https://doi.org/10.1007/978-3-030-35691-0_1

Berners-Lee M, Kennelly C, Watson R, Hewitt CN (2018) Current global food production is sufficient to meet human nutritional needs in 2050 provided there is radical societal adaptation. Elementa 6. https://doi.org/10.1525/elementa.310

Bossier P, Ekasari J (2017) Biofloc technology application in aquaculture to support sustainable development goals. Microb Biotechnol 10(5):1012–1016. https://doi.org/10.1111/1751-7915.12836

Bostock J, Lane A, Hough C, Yamamoto K (2016) An assessment of the economic contribution of EU aquaculture production and the influence of policies for its sustainable development. Aquacult Int 24(3):699–733. https://doi.org/10.1007/s10499-016-9992-1

Boyd CE, D'Abramo LR, Glencross BD, Huyben DC, Juarez LM, Lockwood GS, McNevin AA, Tacon AGJ, Teletchea F, Tomasso JR Jr, Tucker CS, Valenti WC (2020) Achieving sustainable

aquaculture: historical and current perspectives and future needs and challenges. J World Aquacult Soc 51(3):578–633. https://doi.org/10.1111/jwas.12714
Bregnballe J (2015) A guide to recirculation aquaculture. FAO Eurofish Rep 100
Cano-Lozano JA, Villamil Diaz LM, Melo Bolivar JF, Hume ME, Ruiz Pardo RY (2022) Probiotics in tilapia (Oreochromis niloticus) culture: Potential probiotic Lactococcus lactis culture conditions. J Biosci Bioeng 133(3):187–194. https://doi.org/10.1016/j.jbiosc.2021.11.004
Cao B, Abakari G, Luo G, Tan H, Xia W (2020) Comparative analysis of nitrogen and phosphorus budgets in a bioflocs aquaculture system and recirculation aquaculture system during overwintering of tilapia (GIFT, Oreochromis niloticus). Aquac Eng 89:102026. https://doi.org/10.1016/j.aquaeng.2019.102026
Carballeira Braña CB, Cerbule K, Senff P, Stolz IK (2021) Towards environmental sustainability in marine finfish aquaculture. Front Mar Sci 8(343). https://doi.org/10.3389/fmars.2021.666662
Cavalcante RB, Telli GS, Tachibana L, Dias DC, Oshiro E, Natori MM, Silva WF, Ranzani-Paiva MJ (2020) Probiotics, prebiotics and Synbiotics for Nile tilapia: growth performance and protection against Aeromonas hydrophila infection. Aquacult Rep 17:100343. https://doi.org/10.1016/j.aqrep.2020.100343
Chutia A, Xavier KAM, Shamna N, Rani AMB (2022) Application of bioflocculating agent in inoculum enhances quality of biofloc and influences growth, feed utilization and stress responses of GIFT tilapia reared in-situ. Aquaculture 553:738050. https://doi.org/10.1016/j.aquaculture.2022.738050
Chutivisut P, Pungrasmi W, Powtongsook S (2014) Denitrification and dissimilatory nitrate reduction to ammonium (DNRA) activities in freshwater sludge and biofloc from Nile tilapia aquaculture systems. J Water Environ Technol 12(4):347–356
Clarke R, Bostock J (2017) Regional review on status and trends in aquaculture development in Europe-2015. Food and Agriculture Organisation of the United Nations
Costa e Silva RZ, Alvarenga ÉR, Matta SV, Alves GFO, Manduca LG, Silva MA, Yoshinaga TT, Fernandes AFA, Turra EM (2022) Masculinization protocol for Nile tilapia (O. niloticus) in biofloc technology using 17-α-methyltestosterone in the diet. Aquaculture 547:737470. https://doi.org/10.1016/j.aquaculture.2021.737470
Crab R, Chielens B, Wille M, Bossier P, Verstraete W (2010) The effect of different carbon sources on the nutritional value of bioflocs, a feed for *Macrobrachium rosenbergii* postlarvae. Aquac Res 41(4):559–567. https://doi.org/10.1111/j.1365-2109.2009.02353.x
Dalsgaard J, Lund I, Thorarinsdottir R, Drengstig A, Arvonen K, Pedersen PB (2013) Farming different species in RAS in Nordic countries: current status and future perspectives. Aquac Eng 53:2–13. https://doi.org/10.1016/j.aquaeng.2012.11.008
Dash P, Tandel RS, Bhat RAH, Mallik S, Pandey NN, Singh AK, Sarma D (2018) The addition of probiotic bacteria to microbial floc: water quality, growth, non-specific immune response and disease resistance of *Cyprinus carpio* in mid-Himalayan altitude. Aquaculture 495:961–969. https://doi.org/10.1016/j.aquaculture.2018.06.056
Dauda AB, Ajadi A, Tola-Fabunmi AS, Akinwole AO (2019) Waste production in aquaculture: sources, components and managements in different culture systems. Aquac Fish 4(3):81–88. https://doi.org/10.1016/j.aaf.2018.10.002
de Lima Vieira J, dos Santos NL, Rodrigues de Menezes FG, Vieira de Mendonça K, Viana de Sousa O (2021) An integrated approach to analyzing the effect of biofloc and probiotic technologies on sustainability and food safety in shrimp farming systems. J Clean Prod 318: 128618. https://doi.org/10.1016/j.jclepro.2021.128618
De Schryver P, Verstraete W (2009) Nitrogen removal from aquaculture pond water by heterotrophic nitrogen assimilation in lab-scale sequencing batch reactors. Bioresour Technol 100(3): 1162–1167. https://doi.org/10.1016/j.biortech.2008.08.043
de Sousa AA, Pinho SM, Rombenso AN, de Mello GL, Emerenciano MGC (2019) Pizzeria by-product: a complementary feed source for Nile tilapia (*Oreochromis niloticus*) raised in biofloc technology? Aquaculture 501:359–367. https://doi.org/10.1016/j.aquaculture.2018.11.055

Deng Y, Verdegem MCJ, Eding E, Kokou F (2022) Effect of rearing systems and dietary probiotic supplementation on the growth and gut microbiota of Nile tilapia (Oreochromis niloticus) larvae. Aquaculture 546:737297. https://doi.org/10.1016/j.aquaculture.2021.737297

Devi A, Kurup BM (2015) Biofloc technology: an overview and its application in animal food industry. Int J Fish Aquac Sci 5(1):1–20

Doan HV, Hoseinifar SH, Elumalai P, Tongsiri S, Chitmanat C, Jaturasitha S, Doolgindachbaporn S (2018) Effects of orange peels derived pectin on innate immune response, disease resistance and growth performance of Nile tilapia (*Oreochromis niloticus*) cultured under indoor biofloc system. Fish Shellfish Immunol 80:56–62. https://doi.org/10.1016/j.fsi.2018.05.049

Durigon EG, Almeida APG, Jerônimo GT, Baldisserotto B, Emerenciano MGC (2019) Digestive enzymes and parasitology of Nile tilapia juveniles raised in brackish biofloc water and fed with different digestible protein and digestible energy levels. Aquaculture 506:35–41. https://doi.org/10.1016/j.aquaculture.2019.03.022

Durigon EG, Lazzari R, Uczay J, Lopes DLA, Jerônimo GT, Sgnaulin T, Emerenciano MGC (2020) Biofloc technology (BFT): adjusting the levels of digestible protein and digestible energy in diets of Nile tilapia juveniles raised in brackish water. Aquac Fish 5(1):42–51. https://doi.org/10.1016/j.aaf.2019.07.001

Eding EH, Kamstra A, Verreth JAJ, Huisman EA, Klapwijk A (2006) Design and operation of nitrifying trickling filters in recirculating aquaculture: a review. Aquac Eng 34(3):234–260. https://doi.org/10.1016/j.aquaeng.2005.09.007

Ekasari J, Angela D, Waluyo SH, Bachtiar T, Surawidjaja EH, Bossier P, De Schryver P (2014a) The size of biofloc determines the nutritional composition and the nitrogen recovery by aquaculture animals. Aquaculture 426:105–111

Ekasari J, Hanif Azhar M, Surawidjaja EH, Nuryati S, De Schryver P, Bossier P (2014b) Immune response and disease resistance of shrimp fed biofloc grown on different carbon sources. Fish Shellfish Immunol 41(2):332–339. https://doi.org/10.1016/j.fsi.2014.09.004

Ekasari J, Rivandi DR, Firdausi AP, Surawidjaja EH, Zairin M Jr, Bossier P, De Schryver P (2015a) Biofloc technology positively affects Nile tilapia (Oreochromis niloticus) larvae performance. Aquaculture 441(0):72–77. https://doi.org/10.1016/j.aquaculture.2015.02.019

Ekasari J, Zairin M Jr, Putri DU, Sari NP, Surawidjaja EH, Bossier P (2015b) Biofloc-based reproductive performance of Nile tilapia Oreochromis niloticus L. broodstock. Aquac Res 46(2):509–512

El-Hawarry WN, Shourbela RM, Haraz YG, Khatab SA, Dawood MAO (2021) The influence of carbon source on growth, feed efficiency, and growth-related genes in Nile tilapia (Oreochromis niloticus) reared under biofloc conditions and high stocking density. Aquaculture 542:736919. https://doi.org/10.1016/j.aquaculture.2021.736919

El-Sayed AFM (2021) Use of biofloc technology in shrimp aquaculture: a comprehensive review, with emphasis on the last decade. Rev Aquac 13(1):676–705

Elayaraja S, Mabrok M, Algammal A, Sabitha E, Rajeswari MV, Zágoršek K, Ye Z, Zhu S, Rodkhum C (2020) Potential influence of jaggery-based biofloc technology at different C:N ratios on water quality, growth performance, innate immunity, immune-related genes expression profiles, and disease resistance against Aeromonas hydrophila in Nile tilapia (Oreochromis niloticus). Fish Shellfish Immunol 107:118–128. https://doi.org/10.1016/j.fsi.2020.09.023

Elsabagh M, Mohamed R, Moustafa EM, Hamza A, Farrag F, Decamp O, Dawood MA, Eltholth M (2018) Assessing the impact of *bacillus strains* mixture probiotic on water quality, growth performance, blood profile and intestinal morphology of Nile tilapia, *Oreochromis niloticus*. Aquac Nutri 24(6):1613–1622

Emerenciano M, Cuzon G, Goguenheim J, Gaxiola G, Aquacop (2012) Floc contribution on spawning performance of blue shrimp Litopenaeus stylirostris. Aquac Res 44(1):75–85. https://doi.org/10.1111/j.1365-2109.2011.03012.x

Emerenciano M, Gaxiola G, Cuzon G (2013) Biofloc technology (BFT): a review for aquaculture application and animal food industry. In: Biomass now-cultivation and utilization, pp 301–328

Emerenciano MGC, Miranda-Baeza A, Martínez-Porchas M, Poli MA, Vieira FN (2021) Biofloc technology (BFT) in shrimp farming: past and present shaping the future. Front Mar Sci 8. https://doi.org/10.3389/fmars.2021.813091

Espinal CA, Matulić D (2019) Recirculating aquaculture technologies. In: Aquaponics food production systems, pp 35–76

Estrada-Gil L, Contreras-Esquivel JC, Flores-Gallegos C, Zugasti-Cruz A, Govea-Salas M, Mata-Gómez MA, Rodríguez-Herrera R, Ascacio-Valdés JA (2022) Recovery of bioactive Ellagitannins by ultrasound/microwave-assisted extraction from Mexican Rambutan Peel (Nephelium lappaceum L.). Molecules 27(5):1592

FAO (2020a) Cultured Aquatic Species Information Programme *Labeo rohita* (Hamilton, 1822)

FAO (2020b) The state of World fisheries and aquaculture (SOFIA). FAO, Rome, Italy

Farrell M, Kuhn TK, Macdonald LM, Maddern TM, Murphy DV, Hall PA, Singh BP, Baumann K, Krull ES, Baldock JA (2013) Microbial utilisation of biochar-derived carbon. Sci Total Environ 465:288–297. https://doi.org/10.1016/j.scitotenv.2013.03.090

Felix S, Menaga M (2021) Applied aquaculture biofloc technology. CRC Press

Fleckenstein LJ, Tierney TW, Ray AJ (2018) Comparing biofloc, clear-water, and hybrid recirculating nursery systems (part II): tilapia (Oreochromis niloticus) production and water quality dynamics. Aquac Eng 82:80–85. https://doi.org/10.1016/j.aquaeng.2018.06.006

García-Ríos L, Miranda-Baeza A, Coelho-Emerenciano MG, Huerta-Rábago JA, Osuna-Amarillas P (2019) Biofloc technology (BFT) applied to tilapia fingerlings production using different carbon sources: emphasis on commercial applications. Aquaculture 502:26–31. https://doi.org/10.1016/j.aquaculture.2018.11.057

Goddek S, Joyce A, Kotzen B, Burnell GM (2019) Aquaponics food production systems: combined aquaculture and hydroponic production technologies for the future. Springer Nature

Goldman J (2016) So you want to be a fish farmer. World Aquaculture 25

Green BW, Rawles SD, Schrader KK, Gaylord TG, McEntire ME (2019) Effects of dietary protein content on hybrid tilapia (Oreochromis aureus× O. niloticus) performance, common microbial off-flavor compounds, and water quality dynamics in an outdoor biofloc technology production system. Aquaculture 503:571–582

Green BW, Rawles SD, Schrader KK, McEntire ME, Abernathy J, Ray CL, Gaylord TG, Lange MD, Webster CD (2021) Impact of dietary phytase on tilapia performance and biofloc water quality. Aquaculture 541:736845. https://doi.org/10.1016/j.aquaculture.2021.736845

Guo H, Huang L, Hu S, Chen C, Huang X, Liu W, Wang S, Zhu Y, Zhao Y, Zhang D (2020) Effects of carbon/nitrogen ratio on growth, intestinal microbiota and metabolome of shrimp (Litopenaeus vannamei). Front Microbiol 11:652–652. https://doi.org/10.3389/fmicb.2020.00652

Gustilatov M, Widanarni EJ, Pande GSJ (2022) Protective effects of the biofloc system in Pacific white shrimp (Penaeus vannamei) culture against pathogenic Vibrio parahaemolyticus infection. Fish Shellfish Immunol 124:66–73. https://doi.org/10.1016/j.fsi.2022.03.037

Hassan SM, Sulaiman MA, Rahman RA, Kamaruddin R (2018) Effects of long term and continuous magnetic field exposure on the water properties, growth performance, plasma biochemistry and body composition of tilapia in a recirculating aquaculture system. Aquac Eng 83:76–84. https://doi.org/10.1016/j.aquaeng.2018.09.003

Helfrich LA, Libey GS (1991) Fish farming in recirculating aquaculture systems (RAS). Virginia Cooperative Extension,

Jahurul MHA, Azzatul FS, Sharifudin MS, Norliza MJ, Hasmadi M, Lee JS, Patricia M, Jinap S, Ramlah George MR, Firoz Khan M, Zaidul ISM (2020) Functional and nutritional properties of rambutan (*Nephelium lappaceum* L.) seed and its industrial application: a review. Trends Food Sci Technol 99:367–374. https://doi.org/10.1016/j.tifs.2020.03.016

Jamal MT, Broom M, Al-Mur BA, Al Harbi M, Ghandourah M, Al Otaibi A, Haque MF (2020) Biofloc technology: emerging microbial biotechnology for the improvement of aquaculture productivity. Pol J Microbiol 69(4):401–409. https://doi.org/10.33073/pjm-2020-049

Kaleem O, Bio Singou Sabi A-F (2021) Overview of aquaculture systems in Egypt and Nigeria, prospects, potentials, and constraints. Aquac Fish 6(6):535–547. https://doi.org/10.1016/j.aaf.2020.07.017

Khanjani MH, Alizadeh M, Mohammadi M, Aliabad HS (2021) The effect of adding molasses in different times on performance of Nile tilapia (Oreochromis niloticus) raised in a Low-salinity biofloc system. Ann Anim Sci 21(4):1435–1454. https://doi.org/10.2478/aoas-2021-0011

Khanjani MH, Sharifinia M (2020) Biofloc technology as a promising tool to improve aquaculture production. Rev Aquac n/a (n/a) https://doi.org/10.1111/raq.12412

Khanjani MH, Sharifinia M (2021) Production of Nile tilapia Oreochromis niloticus reared in a limited water exchange system: the effect of different light levels. Aquaculture 542:736912. https://doi.org/10.1016/j.aquaculture.2021.736912

Khanjani MH, Sharifinia M, Hajirezaee S (2022) Recent progress towards the application of biofloc technology for tilapia farming. Aquaculture 552:738021. https://doi.org/10.1016/j.aquaculture.2022.738021

Kiruba-Sankar R, Praveen Raj J, Saravanan K, Lohith Kumar K, Raymond Jani Angel J, Velmurugan A, Dam Roy S (2018) Chapter 9–invasive species in freshwater ecosystems–threats to ecosystem services. In: Sivaperuman C, Velmurugan A, Singh AK, Jaisankar I (eds) Biodiversity and climate change adaptation in tropical Islands. Academic Press, pp 257–296. https://doi.org/10.1016/B978-0-12-813064-3.00009-0

Kosten S, Almeida RM, Barbosa I, Mendonça R, Santos Muzitano I, Sobreira Oliveira-Junior E, Vroom RJE, Wang H-J, Barros N (2020) Better assessments of greenhouse gas emissions from global fish ponds needed to adequately evaluate aquaculture footprint. Sci Total Environ 748:141247. https://doi.org/10.1016/j.scitotenv.2020.141247

Krummenauer D, Abreu PC, Poersch L, Reis PACP, Suita SM, dos Reis WG, Wasielesky W (2020) The relationship between shrimp (Litopenaeus vannamei) size and biofloc consumption determined by the stable isotope technique. Aquaculture 529:735635. https://doi.org/10.1016/j.aquaculture.2020.735635

Kuhn DD, Boardman GD, Lawrence AL, Marsh L, Flick GJ Jr (2009) Microbial floc meal as a replacement ingredient for fish meal and soybean protein in shrimp feed. Aquaculture 296(1–2):51–57

Kuhn DD, Lawrence AL, Boardman GD, Patnaik S, Marsh L, Flick GJ (2010) Evaluation of two types of bioflocs derived from biological treatment of fish effluent as feed ingredients for Pacific white shrimp, Litopenaeus vannamei. Aquaculture 303(1):28–33. https://doi.org/10.1016/j.aquaculture.2010.03.001

Kumari S, Harikrishna V, Surasani VKR, Balange AK, Babitha Rani AM (2021) Growth, biochemical indices and carcass quality of red tilapia reared in zero water discharge based biofloc system in various salinities using inland saline ground water. Aquaculture 540:736730. https://doi.org/10.1016/j.aquaculture.2021.736730

Laice LM, Corrêa Filho RAC, Ventura AS, Farias KNN, Silva ALN, Fernandes CE, Silva ACF, Barbosa PTL, de Souza AI, Emerenciano MGC, Povh JA (2021) Use of symbiotics in biofloc (BFT)-based Nile tilapia culture: production performance, intestinal morphometry and hematological parameters. Aquaculture 530:735715. https://doi.org/10.1016/j.aquaculture.2020.735715

Lawtae P, Tangsathitkulchai C (2021) The use of high surface area mesoporous-activated carbon from Longan seed biomass for increasing capacity and kinetics of methylene blue adsorption from aqueous solution. Molecules 26(21):6521

Le Xuan C, Wannavijit S, Outama P, Montha N, Lumsangkul C, Tongsiri S, Chitmanat C, Hoseinifar SH, Van Doan H (2022) Effects of dietary rambutan (Nephelium lappaceum L.) peel powder on growth performance, immune response and immune-related gene expressions of striped catfish (Pangasianodon hypophthalmus) raised in biofloc system. Fish Shellfish Immunol 124:134. https://doi.org/10.1016/j.fsi.2022.03.039

Liu G, Deng Y, Verdegem M, Ye Z, Zhu S (2019) Using poly(-β-hydroxybutyrate-β-hydroxyvalerate) as carbon source in biofloc-systems: nitrogen dynamics and shift of Oreochromis niloticus gut microbiota. Sci Total Environ 694:133664. https://doi.org/10.1016/j.scitotenv.2019.133664

Liu G, Ye Z, Liu D, Zhao J, Sivaramasamy E, Deng Y, Zhu S (2018) Influence of stocking density on growth, digestive enzyme activities, immune responses, antioxidant of Oreochromis niloticus fingerlings in biofloc systems. Fish Shellfish Immunol 81:416–422. https://doi.org/10.1016/j.fsi.2018.07.047

Liu H, Wang S, Cai Y, Guo X, Cao Z, Zhang Y, Liu S, Yuan W, Zhu W, Zheng Y, Xie Z, Guo W, Zhou Y (2017) Dietary administration of *Bacillus subtilis* HAINUP40 enhances growth, digestive enzyme activities, innate immune responses and disease resistance of tilapia, *Oreochromis niloticus*. Fish Shellfish Immunol 60:326–333. https://doi.org/10.1016/j.fsi.2016.12.003

Luo G, Liu G, Hx T (2013) Effects of stocking density and food deprivation-related stress on the physiology and growth in adult Scortum barcoo (McCulloch & Waite). Aquac Res 44(6): 885–894

Luo G, Zhang N, Cai S, Tan H, Liu Z (2017) Nitrogen dynamics, bacterial community composition and biofloc quality in biofloc-based systems cultured Oreochromis niloticus with poly-β-hydroxybutyric and polycaprolactone as external carbohydrates. Aquaculture 479:732–741. https://doi.org/10.1016/j.aquaculture.2017.07.017

Malone R (2013) Recirculating aquaculture tank production systems: a review of current design practice (453)

Manduca LG, Silva MA, Alvarenga ÉR, Alves GFO, Ferreira NH, Teixeira EA, Fernandes AFA, MdAe S, Turra EM (2021) Effects of different stocking densities on Nile tilapia performance and profitability of a biofloc system with a minimum water exchange. Aquaculture 530:735814. https://doi.org/10.1016/j.aquaculture.2020.735814

Martins CIM, Eding EH, Verdegem MCJ, Heinsbroek LTN, Schneider O, Blancheton JP, d'Orbcastel ER, Verreth JAJ (2010) New developments in recirculating aquaculture systems in Europe: a perspective on environmental sustainability. Aquac Eng 43(3):83–93. https://doi.org/10.1016/j.aquaeng.2010.09.002

Maulu S, Hasimuna OJ, Haambiya LH, Monde C, Musuka CG, Makorwa TH, Munganga BP, Phiri KJ, Nsekanabo JD (2021) Climate change effects on aquaculture production: sustainability implications, mitigation, and adaptations. Front Sustain Food Sys 5(70). https://doi.org/10.3389/fsufs.2021.609097

Menaga M, Felix S, Charulatha M, Gopalakannan A, Panigrahi A (2019) Effect of in-situ and ex-situ biofloc on immune response of genetically improved farmed tilapia. Fish Shellfish Immunol 92:698–705. https://doi.org/10.1016/j.fsi.2019.06.031

Mirzakhani N, Ebrahimi E, Jalali SAH, Ekasari J (2019) Growth performance, intestinal morphology and nonspecific immunity response of Nile tilapia (Oreochromis niloticus) fry cultured in biofloc systems with different carbon sources and input C:N ratios. Aquaculture 512:734235. https://doi.org/10.1016/j.aquaculture.2019.734235

Mohammadi G, Rafiee G, Tavabe KR, Abdel-Latif HMR, Dawood MAO (2021) The enrichment of diet with beneficial bacteria (single- or multi- strain) in biofloc system enhanced the water quality, growth performance, immune responses, and disease resistance of Nile tilapia (Oreochromis niloticus). Aquaculture 539:736640. https://doi.org/10.1016/j.aquaculture.2021.736640

Moyo NAG, Rapatsa MM (2021) A review of the factors affecting tilapia aquaculture production in southern Africa. Aquaculture 535:736386. https://doi.org/10.1016/j.aquaculture.2021.736386

Murray F, Bostock J, Fletcher D (2014) Review of recirculation aquaculture system technologies and their commercial application

Najdegerami EH, Bakhshi F, Lakani FB (2016) Effects of biofloc on growth performance, digestive enzyme activities and liver histology of common carp (*Cyprinus carpio* L.) fingerlings in zero-water exchange system. Fish Physiol Biochem 42(2):457–465

Ngoc PTA, Meuwissen MPM, Cong Tru L, Bosma RH, Verreth J, Lansink AO (2016a) Economic feasibility of recirculating aquaculture systems in pangasius farming. Aquac Econ Manag 20(2): 185–200. https://doi.org/10.1080/13657305.2016.1156190

Ngoc PTA, Meuwissen MPM, Le TC, Bosma RH, Verreth J, Lansink AO (2016b) Adoption of recirculating aquaculture systems in large pangasius farms: a choice experiment. Aquaculture 460:90–97. https://doi.org/10.1016/j.aquaculture.2016.03.055

Nguyen HYN, Trinh TL, Baruah K, Lundh T, Kiessling A (2021) Growth and feed utilisation of Nile tilapia (Oreochromis niloticus) fed different protein levels in a clear-water or biofloc-RAS system. Aquaculture 536:736404. https://doi.org/10.1016/j.aquaculture.2021.736404

Nisar U, Peng D, Mu Y, Sun Y (2022) A solution for sustainable utilization of aquaculture waste: a comprehensive review of biofloc technology and Aquamimicry. Front Nutr 8. https://doi.org/10.3389/fnut.2021.791738

O'Shea T, Jones R, Markham A, Norell E, Scott J, Theuerkauf S, Waters T (2019) Towards a blue revolution: catalyzing private investment in sustainable aquaculture production systems. The Nature Conservancy and Encourage Capital, Arlington, VA

Ogello EO, Outa NO, Obiero KO, Kyule DN, Munguti JM (2021) The prospects of biofloc technology (BFT) for sustainable aquaculture development. Scientific African 14:e01053. https://doi.org/10.1016/j.sciaf.2021.e01053

Oliveira LK, Pilz L, Furtado PS, Ballester ELC, Bicudo ÁJA (2021) Growth, nutritional efficiency, and profitability of juvenile GIFT strain of Nile tilapia (Oreochromis niloticus) reared in biofloc system on graded feeding rates. Aquaculture 541:736830. https://doi.org/10.1016/j.aquaculture.2021.736830

Panigrahi A, Sundaram M, Saranya C, Swain S, Dash RR, Dayal JS (2019) Carbohydrate sources deferentially influence growth performances, microbial dynamics and immunomodulation in Pacific white shrimp (*Litopenaeus vannamei*) under biofloc system. Fish Shellfish Immunol 86: 1207–1216. https://doi.org/10.1016/j.fsi.2018.12.040

Prabu E, Rajagopalsamy C, Ahilan B, Jeevagan IJMA, Renuhadevi M (2019) Tilapia–an excellent candidate species for World aquaculture: a review. Ann Res Rev Biol:1–14

Pradeepkiran JA (2019) Aquaculture role in global food security with nutritional value: a review. Transl Anim Sci 3(2):903–910. https://doi.org/10.1093/tas/txz012

Ray AJ, Drury TH, Cecil A (2017) Comparing clear-water RAS and biofloc systems: shrimp (Litopenaeus vannamei) production, water quality, and biofloc nutritional contributions estimated using stable isotopes. Aquac Eng 77:9–14

Ringø E (2020) Probiotics in shellfish aquaculture. Aquac Fish 5(1):1–27. https://doi.org/10.1016/j.aaf.2019.12.001

Rohani MF, Islam SMM, Hossain MK, Ferdous Z, Siddik MAB, Nuruzzaman M, Padeniya U, Brown C, Shahjahan M (2022) Probiotics, prebiotics and synbiotics improved the functionality of aquafeed: upgrading growth, reproduction, immunity and disease resistance in fish. Fish Shellfish Immunol 120:569–589. https://doi.org/10.1016/j.fsi.2021.12.037

Saeki A (1958) Studies on fish culture in aquarium of closed-circulating system. Its fundamental theory and standard plan. Nippon Suisan Gakkaishi 23(11):684–695

Samocha TM, Prangnell DI, Castro LF (2019) Chapter 3–Biofloc. In: Samocha TM (ed) Sustainable biofloc systems for marine shrimp. Academic Press, pp 29–36. https://doi.org/10.1016/B978-0-12-818040-2.00003-4

Sarsangi Aliabad H, Naji A, Mortezaei SRS, Sourinejad I, Akbarzadeh A (2022) Effects of restricted feeding levels and stocking densities on water quality, growth performance, body composition and mucosal innate immunity of Nile tilapia (Oreochromis niloticus) fry in a biofloc system. Aquaculture 546:737320. https://doi.org/10.1016/j.aquaculture.2021.737320

Shaaban NA, Tawfik S, El-Tarras W, El-Sayed Ali T (2021) Potential health risk assessment of some bioaccumulated metals in Nile tilapia (Oreochromis niloticus) cultured in Kafr El-Shaikh farms, Egypt. Environ Res 200:111358. https://doi.org/10.1016/j.envres.2021.111358

Shewry PR, Hey SJ (2015) The contribution of wheat to human diet and health. Food Energy Security 4(3):178–202

Sousa AA, Lopes DLA, Emerenciano MGC, Nora L, Souza CF, Baldissera MD, Baldisserotto B, Alba DF, Da Silva AS (2020) Phosphatidylcholine in diets of juvenile Nile tilapia in a biofloc technology system: effects on performance, energy metabolism and the antioxidant system. Aquaculture 515:734574. https://doi.org/10.1016/j.aquaculture.2019.734574

Summerfelt ST, Vinci BJ (2008) Better management practices for recirculating aquaculture systems. In: Environmental best management practices for aquaculture, pp 389–426. https://doi.org/10.1002/9780813818672.ch10

Terjesen BF, Summerfelt ST, Nerland S, Ulgenes Y, Fjæra SO, Megård Reiten BK, Selset R, Kolarevic J, Brunsvik P, Bæverfjord G, Takle H, Kittelsen AH, Åsgård T (2013) Design, dimensioning, and performance of a research facility for studies on the requirements of fish in RAS environments. Aquac Eng 54:49–63. https://doi.org/10.1016/j.aquaeng.2012.11.002

Thong PY (2014) Biofloc technology in shrimp farming: success and failure. Aquacult Asia Pac 10: 13–16

Tinh TH, Momoh TA, Kokou F, Hai TN, Schrama JW, Verreth JAJ, Verdegem MCJ (2021) Effects of carbohydrate addition methods on Pacific white shrimp (Litopenaeus vannamei). Aquaculture 543:736890. https://doi.org/10.1016/j.aquaculture.2021.736890

Tom AP, Jayakumar JS, Biju M, Somarajan J, Ibrahim MA (2021) Aquaculture wastewater treatment technologies and their sustainability: a review. Energy Nexus 4:100022. https://doi.org/10.1016/j.nexus.2021.100022

Tubin JSB, Paiano D, Hashimoto GSO, Furtado WE, Martins ML, Durigon E, Emerenciano MGC (2020) Tenebrio molitor meal in diets for Nile tilapia juveniles reared in biofloc system. Aquaculture 519:734763. https://doi.org/10.1016/j.aquaculture.2019.734763

Ulloa Walker DA, Morales Suazo MC, Emerenciano MGC (2020) Biofloc technology: principles focused on potential species and the case study of Chilean river shrimp Cryphiops caementarius. Rev Aquac 12(3):1759–1782

Van Doan H, Hoseinifar SH, Harikrishnan R, Khamlor T, Punyatong M, Tapingkae W, Yousefi M, Palma J, El-Haroun E (2021a) Impacts of pineapple peel powder on growth performance, innate immunity, disease resistance, and relative immune gene expression of Nile tilapia, Oreochromis niloticus. Fish Shellfish Immunol 114:311–319. https://doi.org/10.1016/j.fsi.2021.04.002

Van Doan H, Hoseinifar SH, Hung TQ, Lumsangkul C, Jaturasitha S, Ehab E-H, Paolucci M (2020a) Dietary inclusion of chestnut (Castanea sativa) polyphenols to Nile tilapia reared in biofloc technology: impacts on growth, immunity, and disease resistance against Streptococcus agalactiae. Fish Shellfish Immunol 105:319–326. https://doi.org/10.1016/j.fsi.2020.07.010

Van Doan H, Hoseinifar SH, Naraballobh W, Jaturasitha S, Tongsiri S, Chitmanat C, Ringø E (2019) Dietary inclusion of Orange peels derived pectin and lactobacillus plantarum for Nile tilapia (Oreochromis niloticus) cultured under indoor biofloc systems. Aquaculture 508:98–105

Van Doan H, Hoseinifar SH, Naraballobh W, Paolucci M, Wongmaneeprateep S, Charoenwattanasak S, Dawood MAO, Abdel-Tawwab M (2021b) Dietary inclusion of watermelon rind powder and lactobacillus plantarum: effects on Nile tilapia's growth, skin mucus and serum immunities, and disease resistance. Fish Shellfish Immunol 116:107–114. https://doi.org/10.1016/j.fsi.2021.07.003

Van Doan H, Lumsangkul C, Hoseinifar SH, Harikrishnan R, Balasundaram C, Jaturasitha S (2021c) Effects of coffee silverskin on growth performance, immune response, and disease resistance of Nile tilapia culture under biofloc system. Aquaculture 543:736995

Van Doan H, Lumsangkul C, Hoseinifar SH, Hung TQ, Stejskal V, Ringø E, Dawood MAO, Esteban MÁ (2020b) Administration of watermelon rind powder to Nile tilapia (*Oreochromis niloticus*) culture under biofloc system: effect on growth performance, innate immune response, and disease resistance. Aquaculture 528:735574. https://doi.org/10.1016/j.aquaculture.2020.735574

Van Doan H, Lumsangkul C, Hoseinifar SH, Jaturasitha S, Tran HQ, Chanbang Y, Ringø E, Stejskal V (2022a) Influences of spent coffee grounds on skin mucosal and serum immunities, disease resistance, and growth rate of Nile tilapia (Oreochromis niloticus) reared under biofloc system. Fish Shellfish Immunol 120:67–74

Van Doan H, Lumsangkul C, Hoseinifar SH, Tongsiri S, Chitmanat C, Musthafa MS, El-Haroun E, Ringo E (2021d) Modulation of growth, innate immunity, and disease resistance of Nile tilapia (Oreochromis niloticus) culture under biofloc system by supplementing pineapple peel powder and lactobacillus plantarum. Fish Shellfish Immunol 115:212–220

Van Doan H, Lumsangkul C, Ruangwong OU, Meidong R, Hoseinifar SH, Dawood MA, Carnevali O (2021e) Effects of host-associated probiotic Bacillus altitudinis B61-34b on growth performance, immune response and disease resistance of Nile tilapia (Oreochromis niloticus) raised under biofloc system. Aquaculture Nutrition 27:61–72

Van Doan H, Lumsangkul C, Sringarm K, Hoseinifar SH, Dawood MAO, El-Haroun E, Harikrishnan R, Jaturasitha S, Paolucci M (2022b) Impacts of Amla (Phyllanthus emblica) fruit extract on growth, skin mucosal and serum immunities, and disease resistance of Nile tilapia (Oreochromis niloticus) raised under biofloc system. Aquac Rep 22:100953. https://doi.org/10.1016/j.aqrep.2021.100953

van Rijn J (2013) Waste treatment in recirculating aquaculture systems. Aquac Eng 53:49–56. https://doi.org/10.1016/j.aquaeng.2012.11.010

Villar-Navarro E, Garrido-Pérez C, Perales JA (2021) Recycling "waste" nutrients back into RAS and FTS marine aquaculture facilities from the perspective of the circular economy. Sci Total Environ 762:143057. https://doi.org/10.1016/j.scitotenv.2020.143057

Vyas A (2020) Chapter 3–Biofloc systems in aquaculture: global status and trends. In: Rastegari AA, Yadav AN, Yadav N (eds) New and future developments in microbial biotechnology and bioengineering. Elsevier, pp 31–42. https://doi.org/10.1016/B978-0-12-820528-0.00004-1

Waite R, Beveridge M, Brummett R, Castine S, Chaiyawannakarn N, Kaushik S, Mungkung R, Nawapakpilai S, Phillips M (2014) Improving productivity and environmental performance of aquaculture. World Fish

Wang G, Yu E, Xie J, Yu D, Li Z, Luo W, Qiu L, Zheng Z (2015) Effect of C/N ratio on water quality in zero-water exchange tanks and the biofloc supplementation in feed on the growth performance of crucian carp, Carassius auratus. Aquaculture 443:98–104. https://doi.org/10.1016/j.aquaculture.2015.03.015

Wang K, Li K, Liu L, Tanase C, Mols R, van der Meer M (2020) Effects of light intensity and photoperiod on the growth and stress response of juvenile Nile tilapia (*Oreochromis niloticus*) in a recirculating aquaculture system. Aquac Fish 8:85. https://doi.org/10.1016/j.aaf.2020.03.001

Wannavijit S, Outama P, Le Xuan C, Lumsangkul C, Lengkidworraphiphat P, Tongsiri S, Chitmanat C, Doan HV (2022) Modulatory effects of longan seed powder on growth performance, immune response, and immune-antioxidant related gene expression in Nile tilapia (Oreochromis niloticus) raised under biofloc system. Fish Shellfish Immunol 123:460–468. https://doi.org/10.1016/j.fsi.2022.03.029

Warrer-Hansen I (2015) Potential for land based salmon grow-out in recirculating aquaculture systems (RAS) in Ireland. IFA Aquaculture, Netherlands

Xuan CL, Wannavijit S, Outama P, Lumsangkul C, Tongsiri S, Chitmanat C, Doan HV (2022) Dietary inclusion of rambutan (Nephelium lappaceum L.) seed to Nile tilapia (Oreochromis niloticus) reared in biofloc system: impacts on growth, immunity, and immune-antioxidant gene expression. Fish Shellfish Immunol 122:215–224. https://doi.org/10.1016/j.fsi.2022.01.020

Yang J, Yu M, Chen W (2015) Adsorption of hexavalent chromium from aqueous solution by activated carbon prepared from longan seed: kinetics, equilibrium and thermodynamics. J Ind Eng Chem 21:414–422. https://doi.org/10.1016/j.jiec.2014.02.054

Yanong RP (2012) Biosecurity in aquaculture, Part 2: Recirculating aquaculture systems. Retrieved November 29:2018

Ying L, Baoliang L, Ce S, Guoxiang S (2015) Recirculating aquaculture systems in China-current application and prospects. Fish Aquac J 6(3):1

Zaineldin AI, Hegazi S, Koshio S, Ishikawa M, Dawood MA, Dossou S, Yukun Z, Mzengereza K (2021) Singular effects of Bacillus subtilis C-3102 or Saccharomyces cerevisiae type 1 on the growth, gut morphology, immunity, and stress resistance of red sea bream (Pagrus major). Ann Anim Sci 21(2):589–608

Zaki MAA, Alabssawy AN, Nour AE-AM, El Basuini MF, Dawood MAO, Alkahtani S, Abdel-Daim MM (2020) The impact of stocking density and dietary carbon sources on the growth, oxidative status and stress markers of Nile tilapia (*Oreochromis niloticus*) reared under biofloc conditions. Aquac Rep 16:100282. https://doi.org/10.1016/j.aqrep.2020.100282

Zhang S-Y, Li G, Wu H-B, Liu X-G, Yao Y-H, Tao L, Liu H (2011) An integrated recirculating aquaculture system (RAS) for land-based fish farming: the effects on water quality and fish production. Aquac Eng 45(3):93–102. https://doi.org/10.1016/j.aquaeng.2011.08.001

Zhao D, Pan L, Huang F, Wang C, Xu W (2016) Effects of different carbon sources on bioactive compound production of biofloc, immune response, antioxidant level, and growth performance of Litopenaeus vannamei in zero-water exchange culture tanks. J World Aquacult Soc 47(4): 566–576